MW00582104

Kit Airplane Construction

Kit Airplane Construction

Third Edition

Ronald J. Wanttaja

McGraw-Hill

New York   Chicago   San Francisco   Lisbon   London   Madrid
Mexico City   Milan   New Delhi   San Juan   Seoul
Singapore   Sydney   Toronto

The McGraw·Hill Companies

Cataloging-in-Publication Data is on file with the Library of Congress.

Copyright © 2006 by The McGraw-Hill Companies, Inc. All rights reserved. Except as permitted under the United States Copyright Act of 1976, no part of this publication may be reproduced or distributed in any form or by any means, or stored in a data base or retrieval system, without the prior written permission of the publisher.

1 2 3 4 5 6 7 8 9 0   DOC/DOC   0 10 9 8 7 6 5

ISBN 0-07-145973-1

The sponsoring editor for this book was Steve Chapman, the editing supervisor was Caroline Levine, and the production supervisor was Richard C. Ruzycka. The art director for the cover was Anthony Landi. It was set in ITC Slimbach Std. by Wayne A. Palmer of McGraw-Hill Professional's Hightstown, N.J., composition unit.

McGraw-Hill books are available at special quantity discounts to use as premiums and sales promotions, or for use in corporate training programs. For more information, please write to the Director of Special Sales, McGraw-Hill Professional, Two Penn Plaza, New York, NY 10121-2298. Or contact your local bookstore.

Information contained in this work has been obtained by The McGraw-Hill Companies, Inc. ("McGraw-Hill") from sources believed to be reliable. However, neither McGraw-Hill nor its authors guarantee the accuracy or completeness of any information published herein and neither McGraw-Hill nor its authors shall be responsible for any errors, omissions, or damages arising out of use of this information. This work is published with the understanding that McGraw-Hill and its authors are supplying information but are not attempting to render engineering or other professional services. If such services are required, the assistance of an appropriate professional should be sought.

Contents

Preface

THIS BOOK IS A GUIDE TO CHOOSING AND BUILDING a kit homebuilt airplane, or kitplane. The implied simplicity of a kit appeals especially to those who have no experience in aircraft construction or repair. But the first-time builder generally has no idea of the skills or techniques necessary, nor the actual time or money required.

Kit Airplane Construction, Third Edition, isn't intended to replace the kit's instructions, nor does it provide information on building a cockpit latch or landing gear assemblies. The EAA offers excellent books on specifics such as these.

Rather, this book concentrates on describing the basic skills required to build composite, metal, tube-and-fabric, and wooden airplanes.

This book helps the builder in three ways: First, it provides sufficient background to help choose a kit that matches the prospective builder' preferences and existing skills; second, the information provides a solid background on construction techniques, which helps the builder understand kitplane plans and reduces uncertainty and error; and third, the book gives the builder enough knowledge to recognize problems and ask intelligent questions.

Ronald J. Wanttaja

Acknowledgments

A BOOK IS ONLY AS GOOD AS its author's knowledge and its author's sources of information. Over the course of this and the previous editions, several men helped plug the all-too-frequent gaps in my own education. I gratefully acknowledge the assistance of Cecil Hendricks, EAA technical counselor and flight advisor; Ed Ullrich, engineer and A&P; and Terry Dazey, award-winning builder of Long-EZ N86TD.

Special thanks, too, to the participants in case studies past and present: Mike Sabourin, Chuck Bailey, Neil and Marty Bryant, Mike and Arlene Doherty, Jerry Parrish, Kirk McCarty, Dave Nason, and Bruce Bateman.

A final "thank you" to the members of EAA Chapters 26 and 441 and especially to the brave souls who let me prowl their workshops with notebook and camera.

1

Beginnings

BREATHES THERE A PILOT WHO NEVER MADE A MODEL AIRPLANE?

I doubt it.

Who could resist the models shelves at the local toy or hobby store? The multi-colored boxes featured pictures of bombers soaring, fighters scrapping, and evil Huns going down in flames. Sure, there were more model cars than airplanes, and the boats took up a lot of room, too. But the section with the airplane models always had that special look (Fig. 1-1).

Deciding was agony. Get an ME-109? A P-51 to protect your B-17? Or an all-black Stealth Fighter? Finally, the allowance savings were doled out and the precious package tied to the bike for the trip home.

Fig. 1-1. *The model shelves of the local hobby store launched many a homebuilder.*

The magic moment came when the last decal was slipped into place. You'd hold the plane steady and stare into the cockpit. It was like you could feel the stick in your hand and see the wings stretching to either side. The roar of the engine. The whooooooosh of the slipstream. You could almost taste the pure, delirious joy of flight.

Then it was time for a strafing run across the desktop or a low-level mission against the family dog.

When you were a kid, all you had to worry about was getting glue on your fingertips. Mistakes could be painted over. If all else failed, a botched kit always could fall victim to that great allowance-waster, Fourth of July fireworks.

It's one thing for a 10-year-old to blow his allowance on the newest product by Revell or Tamiya. But now as an adult, you're considering dropping a year's salary on a kit for a homebuilt airplane. Page through the homebuilders' magazines. Every advertisement extols the virtues of a kitplane, from a fabric-covered fun plane to a fiberglass speedster. The phrases "Easy-to-build," "Requires only average skills," and "Average construction time 400 hours" leap out at you.

The 10-year-old in you says, "Go for it!" But decades of life's lessons have made you more cautious. How much do you really *know* about aircraft construction? Can you rivet? What does "prepreg" mean? Where are you going to get an engine? How much is everything going to cost? Do you know enough to even ask the right questions?

It's tough to get straight answers in the kitplane field. Kit advertisements can stretch the truth just like any other ad. Most articles in homebuilders' magazines are aimed at projects already well along: problem solving rather than information for beginners. They don't answer the basic question of the first-time builder: "Can I successfully build a homebuilt airplane?"

I can't answer that. Neither can anyone else.

But what I *can* do is provide enough information to allow you to make that decision. I'll help you to select the kitplane that is appropriate to your needs and skills. I'll provide an honest and realistic appraisal of the amount of time, money, and tools required. I'll show you the requirements of aircraft-quality workmanship. And finally, I'll demonstrate basic building operations of the various types of construction.

This last point requires some explanation. Don't look in the Contents and expect to find a chapter entitled, "How to Build a Lancair" or "How to Build an RV-9." To exceed the detail given in the manufacturer's plans, this entire book would have to be dedicated to a single kitplane.

Instead, detail is given on the typical operations required by various construction types. By reading the composite construction chapter (Chap. 8), for example, you'll see how to prepare the surfaces for bonding, how to make the layups, and how to prevent or correct typical errors. Similar chapters explain metal, tube, and wood construction.

Maybe you've already decided on composite construction or a simple tube-and-fabric airplane. Why bother with chapters on metal or wood construction?

Few kits use only one construction material. Metal airplanes, for example, usually have composite cowlings and fairings. Composite airplanes make extensive use of steel and aluminum fittings, panels, and brackets. Wooden kitplanes and many tube-and-fabric types use all materials: wood, metal, and composites.

By the end of this book, you'll know how to tell a good kit from a bad one. You'll have a fairly accurate estimate on the total cost of the whole project. Engine selection will be easier. You'll handle tools with new confidence. During construction, fewer surprises will sap your budget and enthusiasm.

Best of all, your chance of successfully completing the airplane will be far higher.

DEFINITIONS

Did you know that, legally, there's no such thing as a "homebuilt" aircraft?

Before going any further, then, let's define what we're talking about.

All aircraft licensed in the United States must hold an *airworthiness certificate*. There are two kinds: *standard* and *special* (Fig. 1-2).

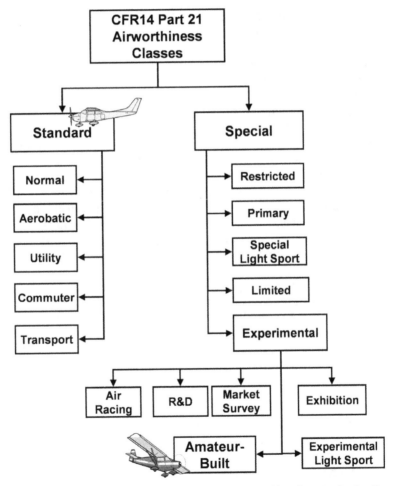

Fig. 1-2. *All aircraft in the United States fall under either the standard or the special airworthiness category. Amateur-built aircraft fall under special.*

Standard airworthiness

Aircraft with *standard* airworthiness certificates are your garden-variety factory-built aircraft. These must be designed to particular standards, and complying with those standards is neither easy nor cheap.

The manufacturer must prove that the aircraft design is adequately strong, both by analysis and by ground testing. An extensive flight-test program is necessary to demonstrate that the aircraft meets certain stability standards, is controllable through all normal phases of flight, and can withstand the designed airspeed range, G-loading, and the like.

When the Federal Aviation Administration (FAA) is satisfied, it awards the aircraft a type certificate in the appropriate category, such as normal, utility, aerobatic, and so on. This type certificate is the buyer's assurance that the plane meets FAA standards.

However, the type certificate also places requirements on the owner. First and foremost, most maintenance must be performed by a licensed airframe and power-plant mechanic (A&P). Annual inspections require an A&P with inspection authoriza-tion (IA). Any parts installed must be approved under a technical service order (TSO).

Not all problems will be discovered during the test phase. The existence of a type certificate grants the government certain powers. When safety-related problems arise, the FAA can issue an airworthiness directive (AD). Compliance with the provisions of the AD is mandatory.

Prior to award of the type certificate, an aircraft is tested in a variety of configura-tions. After certification, any changes that affect the flight characteristics or operation must be fully documented and proven safe, and aircraft operation with the changes is restricted until the FAA awards a *supplemental type certificate* (STC).

Common STCs include upgrading to a larger engine, installing speed modifica-tions such as gap seals and special fairings, and tailwheel conversions of trigear air-craft. Changes with only minor effect (such as a radio installation) can be approved as a minor or field alteration, but any change to the engine, airframe, or control system requires an STC.

To the uninitiated observer, this process can be drawn to ridiculous lengths. A number of years ago, an inventor developed an improved aircraft oil filter. The filter element was an ordinary roll of toilet tissue! The element didn't last very long, but one would expect it would be dirt cheap to replace.

Alas, 'twas not to be. Any component installed in a type-certificated aircraft's engine must be TSO'd, and the average supermarket doesn't sell TSO'd toilet paper.

Special airworthiness

Aircraft design is a study in compromise. The designer must take the performance objectives, customer preferences, and marketing inputs and generate a successful aircraft design. He cannot hope to please everyone; rather, the designer's goal is to create an aircraft that each can accept with minimal grumbling.

But sometimes the requirements for a standard airworthiness certificate conflict with the job the owner needs the airplane to perform. A plane designed to break speed records doesn't need the drag produced by the large tail surfaces necessary to meet the stability standards for standard airworthiness. An aerial applicator can tolerate the higher stall speeds that will result if more than a given amount of chemicals are loaded into her crop-spraying airplane.

Or . . . well, sometimes, it's just logic. When a brand-new airliner rolls out the Boeing factory door, the FAA can't just grant it a standard airworthiness certificate. After all, since it hasn't flown yet, it hasn't yet demonstrated that it meets the requirements!

To cover these and other situations, the FAA will grant a *special* airworthiness certificate. Just as with the standard certificate, a number of aircraft categories fall under "special." These include *restricted, limited, primary, special light sport,* and *experimental.*

The experimental category

In most of the special categories, the aircraft receive type certificates similar to those in the standard categories. They usually require that some analysis and testing be performed prior to the first flight, and they are subject to mandatory maintenance items and limitations on owner modifications.

But the *experimental* category offers an escape from the restrictions of type certification: There are no STCs and no ADs, and airworthiness is proven aloft, not on the ground.

In every other category, the FAA's main aim is to maximize the safety of the pilot and passengers of the aircraft. In the experimental category, the FAA concerns itself mostly with ensuring that the aircraft in question doesn't endanger anyone else.

In the experimental category, designers can escape from the limitations imposed on "normal" airplanes. Design compromises can be taken on the side of performance. Individual aircraft can be modified to the owner's requirements. The qualifications for maintaining the planes sometimes are relaxed.

There are a number of subcategories under experimental, including:

- *Racing.* Aircraft intended for speed competition.
- *Research and development.* Aircraft undergoing flight test for either pure research or for certification purposes.
- *Exhibition.* Aircraft used for special purposes, such as movies or air shows. An example might be a highly modified Stearman used in an aerobatic routine.
- *Amateur-built.* Aircraft built for recreational and educational purposes.

All groups come with restrictions on their operation. Aircraft certified as *experimental/exhibition,* for example, usually can only be flown at shows, to and from shows, or for necessary pilot proficiency flights. You can't legally take anyone for a joyride. Similarly, under the *research and development* classification, you must present the FAA with your proposed test schedule. Flights outside that schedule are prohibited.

The category with the least restrictions? *Experimental/amateur-built*—the traditional "homebuilt" category.

While homebuilt aircraft have been around since the days of the Wright brothers, gaining the freedoms we hold today has been a long, difficult process.

HISTORY

In the years prior to World War I, all aircraft were essentially "homebuilt." They might have been made by a factory, but the techniques used were the same as those of the shade-tree aircraft builder.

And nearly all of them were delivered as kits. When it took 84 days for Cal Rogers to fly the "Vin Fizz" from New York to California, it's easily understood why the aircraft factories didn't bother to deliver newly purchased planes to their owners by air. Especially when it is realized that, by the time it reached California, the only original components left of Rogers' airplane were the rudder and one strut!

So when a sportsman ordered an aircraft, it arrived by train or wagon. If the sportsman were especially wealthy, factory mechanics would assemble his plane for him. In any case, though, pilot training at the factory would have included how to assemble and set up the aircraft. The new owner would consult his notes and grab his tools.

But not everyone could afford factory airplanes. They cost around $5,000 to $10,000, the equivalent of $100,000 to $200,000 today.

So hundreds of aviation enthusiasts began building their planes from scratch. Plans for "production" aircraft were readily available, and Demoiselles and Bleriots popped up in places the factories had never heard of. The only things needed were mechanical aptitude and piano wire from the neighborhood hardware store. Little separated the production aircraft from the homebuilt.

This situation didn't change during the Great War; the greatest advance was the mass production of aircraft engines, which benefited the homebuilder as well. Design innovations arose, but airplane manufacturing took just shade-tree mechanics and a well-stocked general store.

Proof can be found at just about any fly-in. None of the World War I biplanes you see (Fig. 1-3) are original. Most are subscale or updated replicas, but some are built from original plans by typical homebuilders.

Emerging from the shade

Aviation left these low-tech roots as a byproduct of the striving for faster and bigger machines. By substituting cast or machined fittings for those made out of sheet stock or angle iron, the designers could greatly reduce weight and increase strength. Construction began to require presses, milling machines, and custom forgings and castings. Financially, the processes only made sense for large production runs. Shade-tree amateur builders couldn't afford it.

But homebuilders benefited. For example, manufacturers needed something stronger and more reliable than piano wire, and thus came the development of strong, flexible aircraft cable for bracing and control wires. Some detractors might say that aircraft homebuilders have been riding the coattails of the "legitimate" aircraft industry. But homebuilders still lead the way, as witnessed by the composite revolution of the 1970s and 1980s. Aircraft such as the VariEze and Glasair have led the way for acceptance of composite parts in aircraft such as the B-2 stealth bomber and the Boeing 787.

Fig. 1-3. *All flying World War I fighters are modern replicas, such as this Sopwith Camel. The skills and techniques used to build these aircraft originally are the same used today by homebuilders.*

Fig. 1-4. *The Heath Parasol was the first mass-marketed kitplane. This is a Super Parasol, slightly larger and more powerful than the original.*

The biggest mass-marketed kitplane of the early years was the brainchild of a man named Ed Heath. In the late 1920s and early 1930s, he sold a number of his Heath Parasol kits (Fig. 1-4), which initially used a converted motorcycle engine as a powerplant. The Parasol had a ready-made market because most other production aircraft were aimed at the wealthy sportsman rather than the everyday fun flyer. You could buy the completed aircraft for $975 or a kit for $199 (Fig. 1-5). After several ownership and name changes, Heath Aircraft eventually became the nucleus of

General Price of Material Needed in the Construction of the "Parasol"

Fuselage	$32.87	Instrument and cowl support	$ 5.12
Controls	9.15	Tail unit	9.50
Pilot seat	.90	Landing gear	41.32
Tail skid	1.27	Center wing supports	3.11
Cowling	17.55	Outer wing supports	10.52
Engine mounting	3.44	Wings, gas tank, ailerons	76.00

TOTAL COST IF PURCHASED SEPARATELY (varies slightly with market) $247.00

SPECIAL PRICE ON COMPLETE BILL OF MATERIAL, $199.00; BOXING, $5.00

FOR GROUP PRICES SEE BLUE PRICE LIST

All Prices Subject to Change Without Notice

General Specifications of the "Parasol"

Span	25 ft.	Stabilizer area	5.5 sq. ft.	Useful load	300 lbs.
Chord	4 ft. 6 in.	Rudder area	3.8 sq. ft.	Gas capacity	5 gals.
Angle of incidence	4 degrees	Length over all	17 ft.	Oil capacity	6 qts.
Wing area	110 sq. ft.	Height over all	6 ft.	High speed	85 m.p.h.
Aileron area	10 sq. ft.	Weight, empty	285 lbs.	Landing speed	28 m.p.h.
Elevator area	5.2 sq. ft.	Rate of climb (first minute)	600 ft.	Cruising radius	200 miles

Skiis

The use of skies on your airplane doubles the usefulness of the craft. These handy appliances which you can attach in ten minutes can be bought ready built for $25.00 or you can buy the material and blueprints for $10.75 and make them yourself. You are never "snowed in" with skiis.

Complete "Parasol" Prices

COMPLETELY ASSEMBLED, READY TO FLY AWAY AT CHICAGO.

Equipped with Heath B-4 motor and Walnut prop	$975.00
Equipped with wheel brakes	add 35.00
Equipped with motor starter	add 25.00
Equipped with metal propeller	add 35.00
Heath Parasol without motor	640.00
Crating complete plane	18.00
Boxing, motor only	3.00

Notice

ALL ORDERS MUST BE ACCOMPANIED BY AT LEAST ONE-THIRD IN CASH; BALANCE C. O. D. OR SIGHT DRAFT.

Export orders must be accompanied by Cash in Full, and a reasonable amount for crating and shipping charges must be included. Personal checks must be certified.

The duty on aircraft material to most foreign countries, such as Canada, is in the neighborhood of 30 per cent.

Anyone that places an order for $100.00 worth of merchandise from our price list will be supplied with a set of blueprints free or will get credit for the blueprints already purchased.

Any parts that are found defective may be returned to us upon receiving shipping instructions from us.

In case of damage in transit, notify your express or freight agent at once and have him verify damage, as our responsibility ceases with delivery of merchandise in good order to the common carrier.

Price, Complete Bill of Material, $199.00

HEATH AIRCRAFT CORPORATION

LINCOLN 6196-6197   1721-29 Sedgwick Street   Chicago, Illinois

Fig. 1-5. *The Parasol was the first modern kitplane, at least from the packaging point of view.* EAA.

the Heathkit home electronics company. The company didn't keep selling airplanes. Pity.

For its day, the Parasol wasn't bad. It had a cruise speed of 70 mph and a stall speed of about 25 mph, and it got about 45 miles per gallon. The Parasol kit resembled several of today's kits: It could be built from plans, from material packages, or from a complete kit including prewelded fuselage and tail feathers.

With Lindberg's flight in 1927, interest in aviation exploded across America. Many enthusiasts couldn't afford the factory airplanes then being produced, and interest in homebuilt aircraft rose even higher.

Squeezed out of existence—almost

In 1926, the newly formed Aeronautics Branch of the Department of Commerce began to define regulations governing the licensing of pilots and certificating of aircraft designs. But the Aeronautics Branch [renamed the Bureau of Air Commerce in 1934 and split from the Department of Commerce in 1938 as the Civil Aeronautics Authority (CAA)] didn't have any direct effect on homebuilding. It set strict engineering and test standards for aircraft.

But its jurisdiction was limited to, essentially, commercial operations—pilots flying for hire or aircraft offered for sale. Individuals could fly private aircraft without a license, and homebuilt aircraft weren't restricted.

Unfortunately, as often happens, homebuilding became a victim of its own popularity. The accident rates surged as marginal designs hit the market and people who didn't know any better used substandard materials in construction.

Appalled by the carnage, individual states stepped in. They passed laws requiring that all aircraft meet CAA design requirements. Then, as now, proving that an aircraft design met the federal standards was an expensive, time-consuming process. Most homebuilders couldn't afford it.

The result was inevitable. By World War II, homebuilt aircraft had been banned in every state of the Union.

Except Oregon. The state's laws still permitted people like Les Long to design, build, and fly their own airplanes. Just prior to World War II, Tom Story built one of Long's designs, a low-wing, 65-hp single-seat airplane called "Wimpy." After the war, a man named George Bogardus bought the plane, modified it to some extent, and enjoyed many hours of happy flying the plane he'd named, "Little Gee Bee" (Fig. 1-6).

But George still couldn't fly "Little Gee Bee" outside Oregon owing to the laws in all the other states. As the president of the American Airmen's Association, he lobbied long and hard to get the federal government to institute a certification category for homebuilts. The Civil Aeronautics Board (CAB), the rule-making arm of the CAA,

Fig. 1-6. *George Bogardus' "Little Gee Bee" demonstrated that homebuilts could be reliable and safe. Bogardus' work led to establishment of the amateur-built category.*

had a certification category for experimental aircraft, but it was aimed at exhibition and racing machines. Bogardus and the American Airmen's Association pushed for an *amateur-built* category under experimental.

In 1947, the CAB was engaged in a massive rework of the Federal Air Regulations, and Bogardus was invited to come to Washington and discuss the matter. Bogardus flew "Little Gee Bee" all the way across the country, from Portland, Oregon, to Washington, D.C., to prove that homebuilt aircraft could be safe and reliable.

He made his point. The CAB's updating of the regulations eventually included commonsense rules to allow amateurs to build and fly aircraft for educational and recreational purposes. As federal law, it trumped all the antihomebuilt state regulations—and homebuilding was back on track. Paul Poberezny then founded the Experimental Aircraft Association, and scratch-built planes such as the Pober Pixie started going together in various garages.

In the late 1950s, Ray Stits produced a kit of his SA-6B Flut-R-Bug, a two-seat tandem shoulder-winged monoplane. This kit included everything except the engine and propeller. The steel-tube fuselage came prewelded. Twenty-seven were sold at $1,100 each, as well as more than 1,200 sets of plans.

The small number of kits sold doesn't reflect on the quality of the aircraft or the lack of a market. Then, as now, it's often cheaper to buy a used production aircraft than a homebuilt kit. Back then, a Flut-R-Bug kit cost the same as a good used Aeronca Champ and required years of building time.

The surge of the 1970s

To be successful, the kitplane designer obviously had to offer pizzazz and superior performance at a low price. Jim Bede promised that, and more, in 1969. He announced the development of the BD-5 Micro (Fig. 1-7), with a design goal of 215 mph on a 35-hp two-stroke snowmobile engine. Barely 13 feet long, with a wingspan of just 17 feet, this single-seat bullet seemed to match every pilot's dream. The promotional material listed only 300 hours to build the kit!

Thousands of homebuilders plopped $500 deposits toward the BD-5 kit. The project consisted of a number of subkits, and the contract called for payment of the entire balance ($2,000 total) on delivery of the first subkit.

Controversy soon erupted. No one really doubted Jim Bede's ability to design fast airplanes. After all, his BD-1 homebuilt had just entered production as the American Yankee, and the Yankee went like scat. But Bede's claims for BD-5 performance seemed too good to be true. Skepticism grew to the point where *Flying* magazine offered a reward for the first BD-5 to exceed 200 mph in level flight.

Sales continued. Many initial subkits were delivered, obligating the buyers to remit the balance. But deliveries slowed. Bede held off delivering the powerplant subkits.

The BD-5 had engine problems. Pusher aircraft require special attention to engine cooling because the prop blast isn't available. The Micro's first engine suffered from overheating and seizing. In addition, the BD-5 used a variable-speed reduction drive

Fig. 1-7. *The BD-5 (foreground) took the aviation world by storm, promising 200 mph on 35 hp. Engine problems prevented its success. A number of BD-5s have been finished by builders in the years since the company's collapse. The aircraft behind the BD-5 is a later Bede design, the BD-17.*

and shaft to transmit power from the centrally mounted engine to the tail propeller. Problems with such a drive shaft are a recurring theme throughout the history of homebuilding. Several other engines were tried, but other problems cropped up.

Yet the hype continued. Bede demonstrated a jet version, the BD-5J. The company announced a sailplane version, the BD-5S, with a wingspan 10 feet longer. Plans were made to sell the ready-to-fly BD-5B for $4,000. But the company still couldn't field a reliable propeller-driven model.

Buyers became restless waiting for the rest of the kit. If a builder insisted, Bede would supply the remainder of the subkits, based on whatever engine was currently in vogue. A few owner-built BD-5s were completed with other engines, only to fall victim to the same cooling and drive-shaft problems the factory suffered. An alarming percentage of crashes ended with fatalities (a BD-5 pilot sits in the very front of the airplane with little crash protection).

The company danced on the verge of failure for a few years and then went bankrupt in 1979. Many hopeful builders lost their deposits, and some were stuck with partial kits. All were disillusioned. If it hadn't been for four men, the kit aircraft concept could have suffered a fatal blow.

Modern pioneers

Burt Rutan worked for Bede for a few years during the early 1970s and then broke off to build his own designs. After success with his unconventionally configured yet conventionally constructed VariViggen, he designed the VariEze—"very easy" to build and fly. The Eze was a pusher, like the BD-5, but its tail-first *canard* configuration allowed the engine to be mounted all the way aft. By avoiding a midships-mounted engine, Rutan sidestepped the BD-5's cooling and drive-shaft troubles.

Rutan introduced another innovation as well. Rather than the traditional built-up wing of wood or metal, the builder carved the airfoil shape from closed-cell foam using a heated wire and then applied multiple layers of fiberglass.

Ken Rand took a slightly different approach. His KR-1 and KR-2 were of conventional configuration, with wood fuselage and wing spars, but used foam and fiberglass to achieve a sleek, fast, and tiny airframe (a KR-1 stands only 3½ feet high, with 17-foot wingspan). Powered by a converted Volkswagen engine, a well-built KR cruises at a blistering 180 mph.

Even as the Bede empire tottered, the VariEze and the KR series stepped in to feed the growing market for a fast-building sport aircraft. Kits appeared, consisting of the material packages similar to those offered by Heath 40 years earlier. Other composite designs and kits appeared, such as the Polliwagen and the Quickie.

It took Frank Christensen to introduce the modern kitplane. He was a competition aerobatic pilot whose expertise in semiconductor manufacturing technology allowed him to retire at the age of 32. His Christen Eagle biplane was developed for one simple reason: Curtis Pitts wouldn't sell him the rights to the Pitts Special. Christensen decided to market a kitplane that matched the Pitts' aerobatic prowess yet included the creature comforts a customer with $25,000 to burn would expect. Note that this is $25,000 in 1970s dollars, when a loaded Chevy Camaro went for one-quarter of the Eagle's price; the price soon shot upward.

The Christen Eagle kit was *Complete*—with a capital C. Building time is estimated at 2,500 hours, which is much more than the VariEze, for example, but the kit contained *everything* required to build. The parts came bubble-wrapped to cardboard backing, and even the razor blades to remove the parts were included. Christensen single-handedly set the tone for the modern kitplane industry. Eagle sales continue today, and Christensen finally obtained the rights to the Pitts Special.

But long before the BD-5 fiasco, another revolution was brewing—in Oregon, fittingly enough. A young engineer named Richard VanGrunsven had purchased a flying example of another of Leroy Stits' designs, the Playboy. Dissatisfied with the aircraft's performance, he installed a larger engine and worked on reducing drag. He didn't get what he wanted, though, until he'd replaced the Stits' strut-braced wood-and-fabric wing with an all-aluminum cantilever design. He dubbed it the "RV-1."

VanGrunsven flew the RV-1 for over 500 hours but still wanted more. He went to the drawing board and then the workshop, and the new all-metal RV-3 made its first flight in 1971. Sales of plans and kits for the RV-3 (Fig. 1-8) kept him busy, but he was inundated with requests for a two-seat version. The tandem-seating RV-4 flew in 1979, and Van's Aircraft took off.

The appearance of the RV-4 coincided with a surge in homebuilt aviation. Tom Hamilton, working from a gravel strip nicknamed, "The Pig Farm," from its former use, developed the Glasair (Fig. 1-9). Instead of the moldless technique used by Rutan, the

Fig. 1-8. *The RV-3 was the first homebuilt design that Dick VanGrunsven put on the market. It led to a long series of all-metal homebuilts offering what Vans' Aircraft calls, "Total Performance."*

Fig 1-9. *The Glasair was one of the first kits that could be assembled like a plastic model. In the background sits the prototype Bowers Fly Baby, an all-wood homebuilt dating from the 1960s.*

Glasair kit delivered molded composite fuselage, wing, and tail surface components very similar to those of plastic scale models. The Avid Flyer arrived for those pilots wanting a fast-building kit with Piper Cub–like performance. The ultralight movement produced a market for *air recreational vehicles* (ARVs), aircraft with ultralight simplicity but closer in appearance and appointments to a regular airplane.

Homebuilts expanded in the other direction as well, with the 1990s seeing great strides in homebuilt speeds and capacity. The turn of the century saw a surge in four-seat bush planes such as the Zenith CH801 (Fig. 1-10) and the Murphy Moose and

Fig. 1-10. *Unlike the two-seaters of 20 years ago, modern homebuilts such as the Zenith CH801 can be used to transport the entire family.* Photo from Zenith Aircraft Company.

the appearance of the first four-seat VanGrunsven aircraft, the RV-10. Leaders, variations, and competitors arose, as did a support industry of suppliers and dealers.

More than a century since the Wright brothers began tinkering in their bicycle shop, homebuilding still goes on. The tinkering might be with epoxy or aluminum rather than bamboo and muslin, but the spirit is the same.

2

Fundamentals

HOMEBUILT AIRCRAFT ARE LICENSED by the Federal Aviation Administration (FAA) as experimental/amateur-built aircraft, as discussed in Chapter 1. To qualify for certification in this category, the "major portion" of the aircraft must have been ". . . fabricated and assembled by persons who undertook the construction project for their own education or recreation" (14CFR Part 21.191).

The builder is legally the manufacturer of the aircraft and receives a repairman certificate that allows him to perform all inspections and maintenance on that aircraft.

The FAA requirement that the "major portion" of the aircraft be fabricated by amateurs is the basis for the famous "51 percent rule." An aircraft cannot be certified in this category unless an amateur builder performed the majority of the construction.

The rule stems from problems that occurred during the early days of the modern homebuilt era. When the rules were first instituted in the early 1950s, some people took production aircraft such as J-3 Cubs, modified them to some extent, and received certification as amateur-built aircraft. The FAA's predecessor, the Civil Aeronautics Authority (CAA), instituted the "major portion" clause to close this loophole.

The FAA's interpretation of the 51 percent rule has evolved over the years. Frank Christensen made hurried changes to the kit for the Christen Eagle (Fig. 2-1) when an FAA representative decided that the Eagle violated the rule. He received approval by including a rib-building kit and components rather than completed ribs.

Since then, the FAA's attitude has gradually relaxed. After all, molded composite kits don't contain ribs in the traditional sense, just cutout pieces of foam. These days, the FAA routinely approves "quick build" kits that not only include prebuilt ribs but also have them already riveted or bonded in place!

What happened? The FAA has transitioned to a process where the builder is expected to perform at least 51 percent of the *tasks,* not the individual pieces of labor. There is no additional education benefit in building 40 ribs versus a single rib. Hence quick-build kits include a lot of nearly finished components, but each usually includes some key task that the builder must complete.

Fig. 2-1. *The original Christen Eagle kit had to be changed to comply with the then-interpretation of the 51 percent rule. Today, the standards are different.*

ANSWERING COMMON QUESTIONS

Pilot licensing

To fly an aircraft licensed in the experimental/amateur-built category, the operator must possess at least a properly endorsed student pilot certificate. If the airplane qualifies as a light sport aircraft, a student sport pilot license is sufficient. Aircraft that conform to FAA Part 103 are ultralight air vehicles, and no license is required. Part 103 is fairly restrictive, though.

Light sport aircraft and ultralights are discussed in Chap. 3.

FAA inspections during construction

Older references to constructing homebuilts refer to sequential inspections by FAA inspectors, especially a "precover inspection." Most of these FAA inspections were eliminated in the early 1980s. Now the FAA (or its designated representative) only inspects the aircraft when it is ready for its first flight.

Aircraft equipment requirements

Homebuilts must comply with the same aircraft equipment requirements as production aircraft. They must include the basic instruments specified by 14CFR Part 91.205, a transponder if operated in the appropriate airspace, and so on.

Annual inspections

Homebuilt aircraft (and their pilots) must comply with applicable FAA regulations. However, there are several regulations that exclude experimental aircraft. For instance 14CFR Part 91.409 requires annual inspections of all aircraft—*except* experimentals!

Unfortunately, this doesn't mean that the homebuilt aircraft owner is off the hook. When the FAA awards an airworthiness certificate in the experimental/amateur-built category, the owner of the aircraft is assigned a set of operating limitations. These limitations invariably include the requirement for ". . . an annual condition inspection to the scope of 14CFR Part 43." This must be performed either by the builder of the aircraft (under the builder's repairman certificate) or a licensed airframe and power-plant mechanic (A&P).

Maintenance

14CFR Part 43, which governs aircraft maintenance, does not apply to homebuilts at all ("This part does not apply to any aircraft for which an experimental certificate has been issued." 14CFR Part 43.1).

Therefore, *anyone* can perform maintenance and modifications on a homebuilt aircraft. They do not have to be the original builder, nor must they possess any sort of license.

However, recall that the aircraft still requires an annual condition inspection. At this event, the inspector must ensure that the aircraft is airworthy. If a modification is unsafe or previous maintenance work is sloppy, the inspector can refuse to sign off on the inspection unless the condition is corrected.

Modifying production aircraft

As mentioned earlier, the 51 percent rule came into play to prevent licensing modified production planes as experimental/amateur-built aircraft. Fifty years later, such modifications are *still* the subject of very common question regarding homebuilts.

Certified aircraft engines are expensive. If the owner of a Cessna 152 finds that she must replace her engine, the bill might well run in excess of $10,000.

But one can buy some very nice conversions of Subaru automobile engines that produce more power than the Cessna's Lycoming and weigh less besides—all for less cost than a rebuilt Lycoming O-235 engine and with more fuel efficiency to boot.

And so the questions gets asked: "Can I replace my run-out Lycoming with a converted Subaru and license the Cessna as a homebuilt?"

The simple answer: No. The 51 percent rule comes into play. Unless you can convince the FAA that you've built more than 51 percent of the Cessna (lotsa' luck!), the FAA will not allow the airplane to be licensed as experimental/amateur-built. It could be licensed in one of the other experimental subcategories (such as research and development), but none of these other areas allow recreational use of such a modified aircraft.

Experimental light sport aircraft

When the sport pilot and light sport aircraft (LSA) regulation changes were implemented in 2004, a new type of experimental aircraft also was added: the *experimental light sport aircraft* (ELSA). The primary difference from experimental/amateur-built aircraft is that there is no 51 percent rule for ELSAs. Kits may be produced at any level of completion. However, designs for new ELSA aircraft must go through the same certification process as production-type LSAs.

Both production LSAs and ELSAs are discussed further in Chap. 3.

WHY BUILD?

Why do people build their own aircraft? There are a number of reasons.

Many people build because they enjoy the complexity and challenge of building their own plane. They delight in working with their hands and seeing the project come together in the shop. Often, these people fly their new planes only for a year or so before missing the building experience and starting a new one.

Similarly, some folks like to be able to maintain their own aircraft rather than having to hire an A&P mechanic. Any homebuilt owner is allowed to maintain his aircraft, and the builder can receive a repairman certificate that allows him to perform the annual inspection.

Some folks build homebuilt aircraft because it provides cost-effective style. If you want an open-cockpit biplane for less than $40,000, a homebuilt is your only option. A full-size P-51 Mustang might cost a million dollars; Titan's all-metal replica (Fig. 2-2) sells for less than a twentieth of that amount.

Others build to be able to customize a plane to their liking. Automobile owners can modify their cars to match their personalities, but it's difficult to go through the same process on a factory-built airplane. As mentioned in Chap. 1, any modifications have to be approved by the FAA, and the changes cannot affect the basic characteristics of the aircraft. Homebuilt aircraft builders aren't restricted by the FAA's certification rules.

Fig. 2-2. *Real P-51 Mustangs cost millions of dollars, but a homebuilder can construct one of these Titan T-51s for far, far less.*

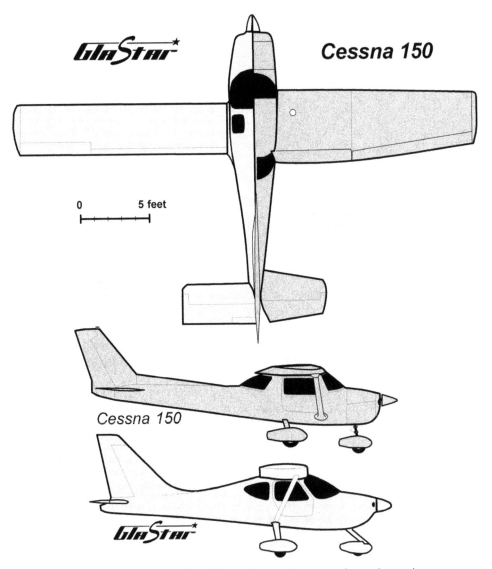

Fig. 2-3. *While the Stoddard-Hamilton GlaStar is about the same size and same horsepower as the Cessna 150, improved aerodynamics of the homebuilt results in a definite performance increase.*

Homebuilts usually offer an increase in performance (especially speed) versus a production aircraft mounting the same engine. Compare the Cessna 150 and the GlaStar (Fig. 2-3). They're about the same size, except the GlaStar has a slightly longer wing with a narrower chord. The homebuilt has a cabin that is 8 inches wider than the production plane, with a much larger baggage compartment. Let's compare the numbers:

	GlaStar	1977 Cessna 150
Empty weight	1,100 lb	1,100 lb
Gross weight	1,900 lb	1,600 lb
Top speed	170 mph	160 mph
Cruise speed	140 mph	120 mph
Rate of climb	1,500 fpm	670 fpm
Takeoff distance	300 feet	735 feet
Engine	125 hp	100 hp

The GlaStar has a slightly larger engine, but the performance increase is much more than you'd expect from an additional 25 hp. Double the rate of climb, 20 mph faster at cruise, half the takeoff distance—all in an airplane with a larger, more comfortable cabin.

This higher performance usually comes at a lower cost than a factory aircraft. To get the performance of some of the kitplane speedsters on the market, one would have to buy a Bonanza or other complex aircraft. The fixed-gear Glasair FT, for instance, cruises at more than 200 mph.

In addition, homebuilts typically are less expensive to own and operate. As mentioned earlier, a homebuilt owner can perform her own maintenance. This can save thousands of dollars per year.

Parts and accessories installed on homebuilts do not need FAA approval, as on production airplanes. This can save a considerable amount of money. Twenty years ago, the pull-cable that activated the starter on my old Cessna 150 broke. Cessna wanted $85 for a replacement. Recently, a similar cable broke on the homebuilt Fly Baby I now own. The replacement cost $7 at the local auto-parts store.

Finally, there's the ego gratification of homebuilt ownership. Arrive at the gas pumps in a Cessna 150, and nobody notices. Roll up in a Velocity, though. . . .

DRAWBACKS

Now, let's point out what's wrong with homebuilts.

Don't take it personally. Homebuilts are great; most of my flying time is in one homebuilt or another. But my eyes are open wide to the dangers of the hobby. Don't step into kit building with the attitude that the airplane will be the same as a factory model. By understanding the problems of homebuilt aircraft, you can take steps to avoid the dangers.

Stability and handling

On the surface, it appears that the experimental/amateur-built category has all the advantages. It's cheaper to build an airplane than to buy a new one. By performing all maintenance, you save the cost of an A&P.

There's the performance advantages, too. The typical kitplane is faster and climbs better than a comparable factory-built airplane. And with few exceptions, factory planes are stodgy and boring.

But there's a dark side to experimental aircraft, one that must be understood before construction is started.

Federal requirements that a type-certificated airplane must meet have been mentioned. Many of those requirements have to do with stability and handling.

Sporty handling is one reason homebuilts are popular. Most feature sensitive controls and a "fighter-like" feel. Some have control forces so low that they don't even need cockpit-adjustable elevator trim. But such sensitivity is often gained at the cost of *stability*.

Remember, aircraft design is the art of compromise. Designers of production aircraft usually sacrifice performance in the interest of meeting the FAA's handling and stability requirements. Kitplane designers can decide not to compromise. The gentle stall characteristics of the Cessna series aren't an accident; the airfoils and wing shapes selected might not be the fastest, but they ensure good stall behavior. A Cessna 172 might not demonstrate the control response of a Pitts Special, but it won't start rolling if you glance down to unfold a chart.

There are various types of stability. *Pitch stability*, for instance, is defined as the willingness to return to the trim airspeed when disturbed. Assume that an aircraft is trimmed out at 100 mph; pull back the stick momentarily and release. An aircraft with positive stability would soon reassume 100 mph. One with neutral stability will maintain whatever speed was reached before the stick was released. Aircraft with negative pitch stability keep slowing until the stall occurs.

Production aircraft *must* demonstrate positive pitch stability. But these regulations don't apply to homebuilts. *Aviation Consumer* magazine tested one of the most popular kitplanes:

> [It] clearly has no desire to maintain its trim speed—a fault that would immediately flunk it for FAA certification.
>
> The poor little autopilot trimmed its brains out trying to hold altitude, but never did manage anything better than an endless 500-fpm roller-coaster ride.
>
> The stall itself—at least the nibble-at-the-edge variety we tried—was docile enough. . . . [The kit manufacturer's test pilot] said he preferred not to try full stick-to-the-stop-and-hold-it stalls, as required by FAA certification testing, without a parachute.

Several popular homebuilts have run afoul of Australia's regulations regarding aircraft stability. Until recently, all aircraft designs (including homebuilts) had to undergo government testing to be licensed in Australia.

One popular American homebuilt flunked the stability test. During stall testing, the aircraft flipped inverted and dropped a thousand feet—*with a professional test pilot at the controls.*

Scared? Don't be. Just don't expect Cessna-like handling or stability in a homebuilt. The aircraft aren't unflyable; in fact, the preceding aircraft is a real joy to fly, a minifighter. The article just quoted also says: ". . . we think [it] is a fine airplane overall. If by magic we could suddenly conjure up the spare time. . . ."

A few last points on the stability and handling issues:

- The time to find out about handling quirks is before you buy the kit. Get a test flight. Read pilot reports. One magazine had these comments from a kitplane's designer: "I won't permit deep stalls. . . . I don't think you can get it out of a spin. . . . As far as unusual-attitude flight testing, we're not under an obligation to keep some crazy from killing himself." I don't know about you, but this doesn't give me a warm, fuzzy feeling. Note that it is a different airplane from the one mentioned in the earlier set of quotes.

- A proper flying checkout is a must. Don't assume that you can test-fly your new homebuilt with just a few Warrior hours under your belt. Find someone with the same model, and get some copilot time. Most kit manufacturers offer demonstration rides at the factory; the makers of high-performance kits usually can arrange a formal checkout. In fact, you may not be able to get insurance for some of the hottest kitplanes without undergoing such a program.

- If you plan on equipping your aircraft for Instrument Flight Rules (IFR) flight, select a kitplane with adequate stability.

Stalls and stall warning

The maximum allowable stall speed for type-certificated small aircraft is 66 mph. There is, of course, no such rule for homebuilts. One way to achieve a high cruise speed is to install a tiny wing, which, of course, increases stall speed. Some versions of the BD-5 stall at over 100 mph.

In addition, your favorite kitplane might not stall as gently as the trainer you learned to fly in. It might break more sharply, tend to drop a wing, and require more altitude to recover.

It also won't have a stall horn. Type-certificated planes must have some sort of warning that occurs at least 5 mph above stall. The requirement can be satisfied with either an aerodynamic buffet or a mechanical system such as a horn. Conscientious kitplane designers try to ensure some aerodynamic buffeting prior to stall, but it isn't required by regulations.

Maintenance drawbacks

Once the builder of a kitplane is awarded a repairman certificate, he is fully authorized to perform all maintenance and inspection of the aircraft built.

The cost savings are obvious: At least $400 for the annual inspection (even more for retractable-geared aircraft) and probably at least that much more in labor costs for routine maintenance.

The drawback is sweat—annual inspections are hard work! The annual of a simple aircraft such as a Cessna 150 takes an A&P at least 6 hours; a simple homebuilt should take at least that long. Complex planes take longer. For instance, retractable-geared aircraft must be set on jacks for gear tests.

The repairman certificate makes it legal. But being legal isn't the same as being safe: Are you competent to maintain and inspect the aircraft? Sure, you know this specific airplane inside and out, far better than any A&P you might hire.

But that A&P knows aircraft; she knows how they wear and can recognize the subtle signs of corrosion or impending failure. I went over our club Fly Baby before the annual one-year inspection and didn't find much amiss. But the A&P found a crack in the control stick. Not a major one, and not one that caused anything beyond a small amount of play. But with years of experience inspecting hundreds of airplanes, he was suspicious enough to chase down the problem.

Also, most homebuilders don't really know engines. We buy them and install them, but we don't really understand maintaining them. A pilot can be a shade-tree mechanic with the family car and risk nothing more than a little shoe leather. Aircraft engine maintenance requires a little finer attention to detail.

For safety's sake, the Experimental Aircraft Association (EAA) recommends that every other annual inspection be performed by an A&P. Take advantage of the experience contained in your local EAA chapter. EAA technical counselors are A&Ps or experienced homebuilders. They can give advice on the annuals or perhaps direct you to an A&P who'll help for little or no cost.

Configuration quirks

One feature that has made homebuilts popular is their unusual configurations. These features sometimes cause some unexpected operational "gotchas."

For instance, homebuilts generally are smaller than their factory counterparts. The result? They require even more care in the weight and balance department. For a worst-case example, examine the center of gravity (CG) envelope of the Cessna 150 versus that of the BD-5. The Cessna 150 has an allowable CG range of 6.5 inches, but the BD-5's is *2.5 inches*—about four fingers wide.

The BD-5 is an unusual case. Still, most homebuilts are "short coupled"; that is, they don't have much tail moment arm. Because their handling habits generally are worse to start with, they can get absolutely appalling when loaded out of limits.

Other aspects of homebuilt designs sometimes work against them. The Quickie, for example, was a Rutan design featuring the main landing gear mounted on the tips of the canard. This was a good design solution, perhaps, and definitely an eye-catcher. More than 1,500 one- and two-seat Quickies have flown. However, more than a thousand have been involved in accidents, mostly ground loops. The survivors are rapidly being converted to a more stable gear configuration.

Homebuilt safety

After all this discussion on how homebuilts are trickier to fly, how do the numbers really stack up?

Based on a comparative analyses of aircraft accidents during the 1998–2000 time period, the homebuilt accident rate is about 50 percent higher than the overall rate. About 7 of every 1,000 U.S.-registered aircraft crash in a given year versus 10 of every 1,000 homebuilts (0.72 versus 1.05 percent). Figure 2-4 illustrates the homebuilt accident rate versus several other types of general aviation aircraft.

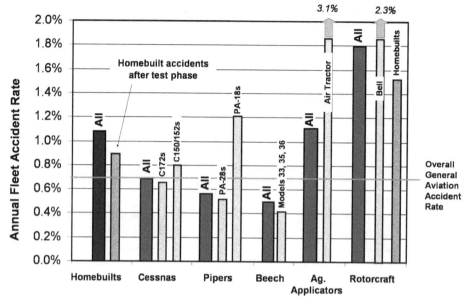

Fig. 2-4. *Even after the initial test period is complete, homebuilt aircraft have a higher accident rate than the rest of the general aviation fleet.*

Of course, a number of the homebuilts that crashed were in their test phases. About 6 percent of the homebuilt crashes occurred on the first flight; about 21 percent happened during the typical 40-hour test period.

Once their teething pains have been taken care of, the homebuilt accident rate drops significantly. Even so, the rate is still almost 20 percent higher than the overall U.S. rate.

What are the differences? Mechanical failures, either due to builder mistakes or mistakes made during maintenance. Almost one in every six homebuilt accidents involve mistakes by the builder or maintainer; in more than 10 percent of homebuilt accidents, such errors were the direct cause of the accident.

The systems affected by builder errors and the types of mistakes made are shown in Fig. 2-5. Two-thirds of the errors involved either the engine or the fuel system. Twenty percent of the errors affected the control system, and about 12 percent involved the airframe.

What kinds of mistakes did the builders make? Almost 40 percent of the cases featured improper installation of an off-the-shelf component. These include such items as fuel valves and propellers. About a quarter of the time the builder made some sort of change to the design that didn't pan out, such as omitting installation of a fuel boost pump. Almost 20 percent of the cases involved poor workmanship, such as a bad composite layup or faulty nicopress fitting or failure to follow the construction instructions.

Builders improperly performed initial setups or adjustments in 15 percent of the builder-error cases. This includes such items as control rigging and rotor tracking.

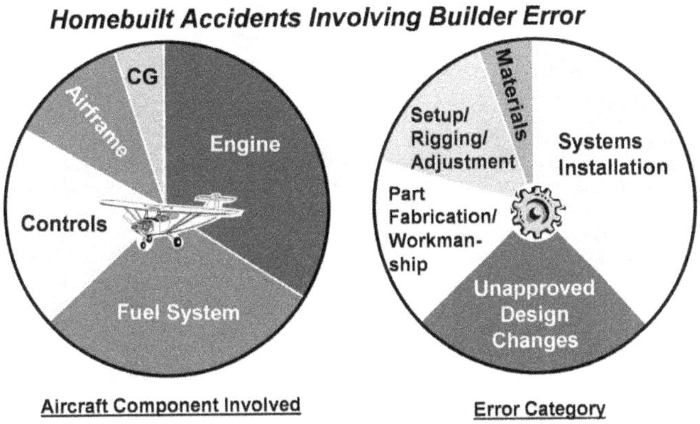

Homebuilt Accidents Involving Builder Error

Aircraft Component Involved | Error Category

Fig. 2-5. *Typical builder mistakes leading to accidents.*

Finally, about 5 percent resulted from the use of materials or parts that were inadequate to the task.

Insurance woes

During the late 1990s, the insurance industry became alarmed at the claim trends for homebuilt aircraft. Not only were accidents occurring more often, but also far more expensive aircraft were involved. As mentioned earlier, homebuilts sometimes can be more challenging to fly than certified aircraft. The drive to wring the last ounce of speed out of a design can result in a plane that requires a higher level of piloting skill. And usually, these ultra-high-performance aircraft are very expensive, too.

As more of these high-performance aircraft were completed, the accident rate started to climb. At one point, it became almost impossible to insure some of the more expensive homebuilts. It's a fairly daunting situation to have over $100,000 in a homebuilt aircraft and be unable to obtain insurance.

The kit companies were forced to react. Training programs were initiated, as were recurrent safety training and aircraft inspection programs. By jumping carefully through the right hoops, builders could obtain insurance from major underwriters through the kit manufacturer. These programs generally prohibit alternate engines (e.g., no auto-engine conversions) and may require that the final inspection and the flight testing be performed by specific individuals.

Insurance for the simpler aircraft is still available. But many policies won't cover the aircraft during the test period. Some companies will provide coverage if the builder takes advantage of training offered by the kit manufacturer or the EAA Technical Counselor and Flight Advisor programs.

THE KITBUILDING PROCESS

In case you haven't figured it out by now, your life is going to depend on your ability to build your aircraft safely. Let's take an in-depth look into kit aircraft construction.

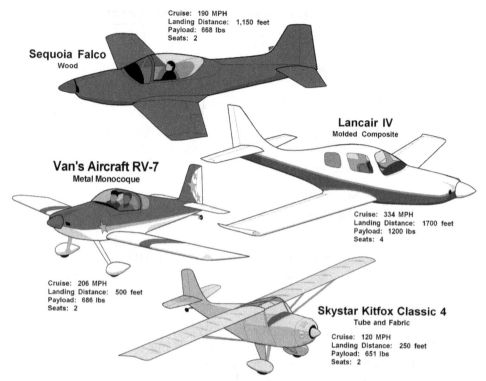

Cruise: 190 MPH
Landing Distance: 1,150 feet
Payload: 668 lbs
Seats: 2

Sequoia Falco
Wood

Lancair IV
Molded Composite

Van's Aircraft RV-7
Metal Monocoque

Cruise: 334 MPH
Landing Distance: 1700 feet
Payload: 1200 lbs
Seats: 4

Cruise: 206 MPH
Landing Distance: 500 feet
Payload: 686 lbs
Seats: 2

Skystar Kitfox Classic 4
Tube and Fabric

Cruise: 120 MPH
Landing Distance: 250 feet
Payload: 651 lbs
Seats: 2

Fig. 2-6. *The four major types of homebuilt aircraft construction.*

Types of construction

To some people, the term "kitplane" is synonymous with "composite airplane." As Fig. 2-6 illustrates, there are several types of construction common to kit aircraft. Let's look at composite first.

A composite material is nonhomogeneous; it doesn't consist of solely one component. Wood is a composite—not only does it contain air bubbles, which allow it to float, but the differences in grain provide the same effect. As far as aircraft construction is concerned, a composite material is made artificially by bonding materials together. It doesn't have to be high-tech. One early example was an aluminum and balsa sandwich. Plywood is a composite material because it consists of several thin layers of wood held together by glue.

Composite construction works by combining two (or more) materials whose advantages complement each other and whose deficiencies cancel out. In the simplest form, strong but flexible fiberglass cloth is soaked in stiff but brittle epoxy (itself a mixture of resin and hardener) to form strong-and-stiff components such as cowlings and wheel pants.

If such a composite isn't strong enough, layers of cloth and epoxy can be separated by a sheet of foam. Composite sandwiches like these form the basic structure of a number of kitplanes.

How does it work? Imagine an ordinary piece of paper. It's easy to tear and fold, but it resists stretching. Imagine, now, a sheet of Styrofoam. The foam resists bending and crushing but crumbles when stretched.

Glue a sheet of paper on either side of the foam. Apply a bending load, and the tension strength of the paper resists. Inserting the foam has changed the bending moment (to which the paper has little resistance) to a stretching moment.

The advantage of composite construction is the easy workability of the component materials. The fiberglass cloth can be cut by a scissors, the foam can be cut or shaped by ordinary hand tools or a hot-wire "cheese slicer," and the epoxies are easily mixed and applied.

The two basic types of aircraft composite construction are *molded* and *moldless*. The first uses a mold to define the shape of the structure; fiberglass layups are made directly on the mold, allowed to harden, and then removed; the mold is then ready to make another piece. These are your fundamental composite kitplanes, where fuselage halves and wing panels are supplied with the kit. Because the molds are expensive and time-consuming to make, molded construction is best suited for mass-produced kitplanes. Examples include the Lancair IV, the Glasair, and several other aircraft.

Moldless composite aircraft can be built in a number of ways. Most common is to build the basic structure out of wood, glue foam to the structure, then carve out the desired shape, and apply the fiberglass and epoxy. Or one can eliminate the wood and carve the shape from foam alone. The moldless method is used for the Velocity, the Co-Z, and the Vision.

There is actually a third type of composite, called *Taylor paper/glass* (TPG) after its inventor, the legendary Molt Taylor. TPG, used on Taylor designs such as the Mini-Imp (Fig. 2-7), uses a type of cardboard for its core material.

Metal-monocoque construction is used by aircraft manufacturers from Piper to Boeing. A thin sheet of flat metal is pretty flimsy. But roll the metal into a wide tube, and it becomes stiff. Attach some bulkheads inside to prevent the tube from collapsing under heavy loads, and you've got a light, strong structure ideal for aircraft.

The metal of choice is aluminum that is alloyed with other metals to optimize its characteristics. Typical aluminum homebuilts include the Van's Aircraft RV series, the Murphy Rebel, and the Zenith line.

Tube-and-fabric construction is another of the traditional ways to make light aircraft. The main structural shape of the fuselage is defined by a metal truss. The wing can be built in one of several methods. The spars, for example, can be solid wood, built-up wooden boxes, or metal extrusions. The ribs can be stamped metal, cut plywood, built-up shapes, or even foam. Fabric is applied to the structure, hence the term "ragwings," and then sealed with dope to produce an enclosed, streamlined shape. Commercial tube-and-fabric light planes include the Piper Cub series and the Aviat Husky.

Tube-and-fabric kitplanes use prewelded steel tubing, such as the Kitfox, or aluminum tube with gussets or extrusions pop-riveted in place, such as the Murphy Renegade.

Wooden aircraft are built just like scale balsa models. The fuselage structure consists of longerons and bulkheads glued into the proper shape. As with tube-and-fabric

Fig. 2-7. *The Mini-Imp uses a third method of composite construction, Taylor paper/glass.*
Mini-Imp Aircraft Company.

aircraft, the wings can be built in several ways and are either sheathed in plywood or covered with fabric. Wooden kitplanes include the Fisher R-80 Tiger Moth replica and the Sequoia Falco.

The preceding is only a general guide—all kitplanes incorporate welded-steel components, require builders to fabricate fittings from aluminum, and usually need some fiberglass work (Fig. 2-8). Sometimes major components even use different construction modes. The fuselage of the GlaStar consists of a steel-tube framework covered by a composite shell; its wings and tail are of conventional metal-monocoque construction. Even composite airplanes such as the Velocity embed wood hardpoints in the fiberglass for bolt attachments.

Generally, designers keep to one major mode of construction throughout. Not only does it make the parts list simpler, but also it reduces the construction time because the builder won't have to learn two unrelated skills.

Advantages/disadvantages

Each type of construction method has its own advantages and disadvantages both during construction and afterwards.

Composite construction is the most controversial. There's no question that the most streamlined shapes are produced by composites and that it's far easier to bond two fuselage halves than to jig up bulkheads and drive 10,000 rivets. Composites

Fig 2-8. *Just because you pick one type of construction, don't expect to escape working with other materials. This wing assembly features aluminum-tube spars, wooden ribs, and a fiberglass wingtip.*

don't rot like wood; they don't corrode like metal. And no one doubts the strength of composite aircraft.

Curiously, its very strength works against it. Controversy rages regarding the crashworthiness of composite airframes. Composites have no "give." A metal aircraft slightly deforms on impact and absorbs some of the crash forces before they can affect the occupants. Composite structures maintain their shape against high forces and then shatter, allowing those forces to be transmitted to the passengers. Yet this doesn't always seem to be true. In one well-publicized case, a composite aircraft prototype crashed into a housing development after an engine failure. Two houses, a van, and the plane were wrecked, but the aircraft occupants walked away.

However, another important point is *repairability.* Major damage to a structural component usually will require replacement of the entire component. One then must hope that the kitplane manufacturer is still in business and still retains the molds for one's particular aircraft model. This doesn't mean that minor damage isn't repairable, though. With new production composite aircraft such as the Cirrus and the Diamond, the repair techniques are fairly well established.

Yet another drawback is temperature sensitivity. Some composite formulations lose strength when warmed excessively, such as might happen if the plane sits outdoors in the sun for long periods. The FAA requires that certificated composite aircraft, mostly sailplanes, be painted white to reduce this problem.

Speaking of sensitivity, some composite kitplane builders develop allergies to the materials used. The introduction of safer epoxies has reduced this occurrence, but always follow handling and safety instructions.

One thing that can't be escaped is the odor. The epoxies have a strong chemical smell and require excellent ventilation of the workspace. In addition, these chemicals must be used within particular temperature ranges, and the workshop therefore might have to be heated during the winter months.

Metal-monocoque and tube-and-fabric construction methods have several advantages over composite. Crashworthiness is good, aluminum or steel allergies are almost unheard of, and aluminum or fabric-covered aircraft can be painted any desired color. However, metal is not as easily formed into the swoopy curves necessary for high-speed aircraft. Aluminum can be bent into complex shapes, but the necessary skills take time to learn.

On the plus side, aluminum doesn't care what the temperature is, so one doesn't have to heat the workshop as long as the builder doesn't mind bundling up. However, if the exterior skin (aluminum or fabric) is installed in the cold, wrinkles can appear during the summer. Unlike composites, there are no special restrictions on exposure to the elements. But outside storage in coastal areas can accelerate the corrosion process.

Approved procedures exist for repair of damaged monocoque or tube aircraft. However, if a metal tube kit comes with a prewelded fuselage, you won't acquire the skills necessary to repair the fuselage should it get damaged. For persons both accident-prone and pain-sensitive, aluminum construction uses many sharp tools and creates sharp edges on metal.

Wood has a combination of the advantages and disadvantages of the other construction modes. Like composites, wood construction requires a climate-controlled workshop, and the finished aircraft must be protected from the elements. Like metal construction, wooden aircraft have good crashworthiness, and approved damage repair methods assist reconstruction. Because of the nature of the material, wooden kitplanes include fewer precut parts and generally require more work on the part of the builder. Additional skills might be required because the builder must learn to make scarf joints in plywood and to gusset and nail other joints and the myriad other tasks of the woodworker. However, this problem is reduced in modern kit aircraft such as the Loehle line of World War II fighter replicas (Fig. 2-9).

Some advantages of wood are subjective in nature. Sawdust is a far more pleasant aroma than that of composite epoxies and is easier to vacuum than metal chips. Wooden airplanes seem more solid and quieter than other types. While wood rot is still somewhat of a problem, modern preservatives drastically reduce the danger.

The 51 percent rule and you

Recall the various subgroups under the experimental category; experimental/amateur-built aircraft are under the strict limitation of the 51 percent rule.

The rule exists to prevent the sale of nearly-ready-to-fly aircraft that do not meet the requirements of standard category. We've already explained the process involved in certificating a factory aircraft and how this process ensures stable and predictably

Fig 2-9. *Many modern wooden kits, such as this Loehle P-40, feature prefabricated components and reduced construction time.*

handling aircraft. Without the 51 percent rule, a manufacturer could bypass the regulations merely by requiring the installation of a few small parts. Such kits are allowed in the light sport aircraft category, but more stringent controls are applied to LSA kits (see Chap. 3).

As mentioned earlier, the 51 percent rule came about in response to some poorly engineered modifications of existing planes. However, this *doesn't* mean that you can't use components of existing aircraft. Few FAA inspectors will complain if the entire firewall-forward section of a factory airplane appears on your homebuilt. Several older homebuilt designs, such as the Breezy, even use complete wings from factory planes. However, these homebuilts are complex enough to make up for the existing parts.

One good fallout from the popularity of homebuilt aircraft has been a certain amount of standardization and centralization on the part of the FAA. The FAA now has a standard process for determining whether a given kit qualifies under the 51 percent rule, and the agency publishes this list online as the "Revised Listing of Amateur-Built Aircraft Kits." Using a search engine such as Google or Yahoo, search for "amateur-built aircraft kits." The FAA list typically comes up very early in the search window; look for a link to the file name "ama-kits.pdf."

Note that if a kit *isn't* on the list, that doesn't mean that it can't be certified as an experimental/amateur-built airplane. The list is just designed to make things easier for the individual FAA regions.

While it's up to the kit manufacturer to ensure that its product meets the rule, you're the one who'll suffer if the FAA refuses to license your airplane as amateur-built. All you could ever expect to recover is the cost of the kit, and kitplane manufacturers go out of business with sad regularity.

But there's another way the 51 percent rule can cause you problems, although not with the FAA. Let's take a look at the rule from another perspective. Let's say that a kit manufacturer proudly states, "Our kit meets the 51 percent rule." So you order the kit, and when you open the box you find

- Sketches of an aircraft design
- The deeds for a bauxite mine in Oregon and an iron mine in Minnesota
- Six spruce logs and a chainsaw
- A dead cow and instructions for how to make glue from the carcass

This meets the 51 percent rule—and how!

Truism Number One. "The 51 percent rule means that the builder must perform *between* 51 and 100 percent of the total work."

This makes it hard to compare kits—they all meet the 51 percent rule, but which kits require more work? Careful selection analysis is required. Hints on determining actual construction methods can be found in Chap. 4.

The kit manufacturer's dilemma

The cheapest (legal) way to acquire an airplane is to purchase a set of plans and then convert raw materials into aircraft components. The most expensive way is to buy a manufactured model such as a Piper or a Cessna.

By regulation, the kit company can do between 0 and 49 percent of the work. Obviously, the less work left for the builder, the better most buyers will like the kit. But the kit manufacturer has to charge more, which will reduce interest in the product. Figuring out the breakeven point is a major source of ulcers in the kit industry.

Truism Number Two. "If it costs a kitplane manufacturer x dollars to include a step that reduces building time, the additional cost to the builder is at least twice that amount."

In other words, if a manufacturer decides to predrill critical holes, and that costs him $100 per airplane, the price of the kit increases by at least $200.

When subcontractors are involved, the cost to the kit buyer increases geometrically. Super Subcontractor Incorporated supplies wing spars to the FlySoon Kit Company. FlySoon decides to reduce the building time by having the holes predrilled through the spars and inserting bushings into the landing-gear mounts. It costs Super Subcontractor $250 to perform the operation, and it bills FlySoon for $500. The price of FlySoon's kit then rises by $1,000 to cover the additional expense.

Sometimes it's worth it. Minor changes at the manufacturer's level can cause drastic reductions in building time. Stoddard-Hamilton Aircraft greatly reduced the building time between the Glasair I and Glasair II kits. The kit price also doubled, but much of that was due to inflation. However, some of the rise must be laid at the door of the more complete kit.

Some kit companies do little actual manufacturing at their "factory." Subcontractors produce the parts in accordance with the computer files generated by the kit company and then ship stacks of parts to the company facility. There, personnel inspect the components and then place them in inventory. When an order arrives, the appropriate components are crated and shipped to the customer.

The builder

Beyond examination of homebuilts in general and kits in particular, it's time to examine the poor slob who has to build the darn thing.

It's hard for a prospective builder to understand what kitbuilding is really like. The kit manufacturers make it look so easy: "Only 600 hours to construct using only average workshop skills." If that's the case, why aren't more kitplanes flying? Why do you see ads for partially completed kits?

Later chapters explain the physical requirements of building; for now, let's concentrate on the mental preparations required.

Two basic points must be kept foremost in your mind during the construction process:

1. You are building an aircraft in which you plan to fly at heights and speeds incompatible with survival if major failures occur.

2. A kitplane project requires a major financial investment on your part. Houses cost more, and so do many cars. But inattention and errors during the acquisition and construction phases could cause a major financial loss with little chance to recoup.

To reiterate something mentioned earlier: *I am* NOT *trying to talk you out of building a kitplane.* Rather, you must understand the physical, emotional, and financial costs involved before you are committed. Thousands of men and women have built experimental aircraft even before the advent of the modern kitplane. In all likelihood, you can too. I hope to prepare you for the effort.

The personal cost of kitbuilding

Too often, when a kitbuilder is asked how much his pride and joy cost, the response is, "Twenty thousand dollars and one marriage."

How free is your life? How much of it can you spend in the garage for the next couple of years?

Clearly, if you're the married sort, some type of compromise will be necessary. We'll assume that your spouse is at least not hostile to your building an airplane; otherwise, you have *big* problems.

Spousal resentment of a project stems from two points: money and neglect. It's been said that an airplane is a hole in the sky into which you throw money. A homebuilt project is even worse, and it's not yet even a hole in the sky. One's spouse can become justifiably testy if the money needed for a new couch or car tires somehow gets appropriated for a transponder. The solution, of course, is careful budgeting and the understanding that funds set aside for the kitplane might be raided for emergencies.

But that'll set back the completion date, you say?

One of the most absurd things you can do is set a deadline for completion. It seems to be standard practice to take the advertised construction time and divide by the number of weeks until the next Oshkosh fly-in to determine the hours you'll work per week.

But while you're toiling on the kitplane, the grass still grows, the kids still have to be taken to soccer practice, you haven't seen Auntie Grizzelda for 10 years, and your family would like to see more of you than just a mound of sawdust at mealtimes.

A deadline can mess you up in at least two ways. First, it can make you rush and make mistakes. Second, falling behind schedule is depressing. It becomes harder and harder to face the thing in the garage. Work grinds to a halt.

Set a realistic goal for the number of hours per week spent on the project. What's realistic? Count on no more than 11 hours on the weekend and maybe 3 hours on each of three weekday evenings. That leaves enough time for family activities on the weekend and keeps two weekday evenings open.

Of that 20 hours a week, count on only 15 or so of useful work. A lot of time is spent preparing, cleaning up, and so forth. Note that you don't have to be restricted to the schedule because any unexpected free time can be converted into workshop hours. If dinner will be delayed by half an hour, duck off to the workshop and buff those aluminum parts or deburr those holes.

Don't underestimate your spouse either. Many husbands and wives have spent hundreds of hours bucking their mate's rivets or mixing epoxy for the layup crew. Sometimes the kit just becomes another aspect of family life (Fig. 2-10). If your life-mate pitches in willingly, you've got a gem beyond compare.

Setting goals

Obviously, with only 15 effective hours per week, it'll take a while to complete all but the most basic kitplane. Common wisdom says that the kitplane manufacturer's estimate of construction time is half the actual time; if the manufacturer says 500 hours, it'll take closer to 1,000. If you can get only 10 effective hours of work in per week (a pretty realistic figure), that's two years to complete the kit.

How do you maintain enthusiasm? Simply forget that you are building an airplane.

Instead, you're building aircraft subassemblies. Set goals based on the completion of individual items. It's like flying a cross-country. If you expect to climb to altitude, spot your destination, and fly directly to it, you won't get far. Instead, you fly to intermediate locations on the way to the target.

If the plans say to build an aileron, don't look beyond its completion. You've got all the time in the world, after all; making an aileron is easy. When it's done, admire it for a few moments, and then carefully store it and look in the plans for the next project.

Glamour versus drudgery

There's a famous saying in the homebuilt world: "The last 10 percent takes 90 percent of the construction time." The airframe is the easiest part of 99 percent of

Fig 2-10. *Every year, artist Sarah Tracy draws a typical at-home scene for the family Christmas card. This one shows husband Dan's RV-7 parts stored on the wall of the living room.* Sarah Tracy.

homebuilt projects. For example, one tube-and-fabric design has a prewelded fuselage available. The plane takes 1,200 hours to build, but buying the prewelded fuselage only cuts 100 hours from the total time.

What gives?

Truism Number Three. "Eighty percent of the work and 95 percent of the sheer drudgery are involved with subsystems, not structure."

Structural construction is the glamour-child of homebuilding. Few people appreciate a finely crafted canopy latch, but everyone loves to run their hands over a just-completed fuselage. A wide-open garage door revealing a bunch of nondescript parts doesn't grab the neighborhood's attention like a sleek white fuselage on shiny aluminum landing gear.

Imagine that you're building the newest composite kit. Bond the fuselage halves together by applying epoxy and cloth to either side of the join line. Bond a few bulkheads inside as well. After a few hour's work, the casual observer might think the fuselage is done.

But let's follow the installation of one subsystem: wheel brakes. First, let's assume that the landing gear, wheels, and rudder pedals are in place. Follow the progress in Fig. 2-11.

The wheel incorporates a brake disk, but the brake assembly itself must be installed on the gear leg. Because the leg is just a blank slab, you have to drill the

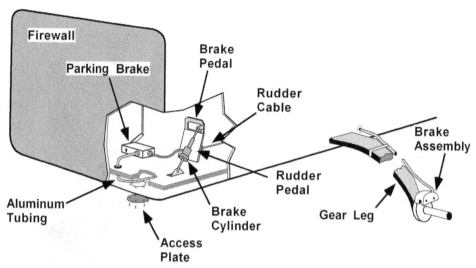

Fig. 2-11. *No matter what construction method a kitplane uses, the brake systems are nearly identical.*

holes to mount the brake assembly. This demands precision—if the brake assembly is askew, wheel shimmy, poor braking, and abnormal pad wear can result. To be on the safe side, temporarily clamp the assembly in place, and then jack the plane off the floor so that the wheel can be spun to check pad clearance. You'll need a hydraulic hand pump or a similar device to work the pads while the wheel is spinning.

Once the brake assembly is bolted in place, it's time to install the brake master cylinders. The kit manufacturer should have supplied the brackets and everything to mount the cylinders to the rudder pedals. Once the cylinder is mounted, you've got to run a brake line through the parking brake valve to the wheel. On this aircraft, the line must run down through the cockpit floorboards, outboard to the fuselage side, aft through two bulkheads, out the bottom of the fuselage, then along the gear leg to the wheel cylinder. All in all, this amounts to about 4 feet of brake line with at least four bends. Some kits might give the exact instructions for brake line routing; let's assume that this kit says, "Route the brake line from the pedal to the wheel."

You don't want to do it as one piece. If it ever broke in service, replacement would be difficult. Lay out a piece of rope to figure out the length of each segment, and then use a tubing cutter on the stock supply of brake line. Each segment end must have a pipe fitting installed, requiring flaring with a special tool. Brake lines are made of soft aluminum, but they can be ruptured or pinched if bent carelessly or if you push too hard trying to work them around tight quarters. You'll have to make a guess at the bend, try to install the line segment, and then withdraw it and adjust the bend angle. You might have to redo one or two pieces that develop crimps or come out too short.

Once the rest of the interior is installed, the brake lines will be tough to get at. So, whether the plans call for it or not, you should protect the lines from any hazards and make sure that you can get to them when necessary. For instance, add chafe protection where the line passes close to any metal fittings, and install grommets anywhere the line passes through a bulkhead or the side of the airplane. Add access/inspection panels where necessary because 5 years from now you might want to remove brake lines without disassembling the airplane.

When the line is connected, fill the brake system with hydraulic fluid, bleed the air, and test for leaks.

Done? Sigh, wipe the sweat off your brow, and then start all over on the *other wheel's* brake!

There's not a kit in the world delivered with the brake system installed. Similar complex processes must be performed on other components and subsystems: controls, radios, instruments, doors, cockpit heat, fuel systems, and the like. And it's a whole 'nother problem forward of the engine mount. Figure 2-12 gives a brief summary of necessary subsystem work.

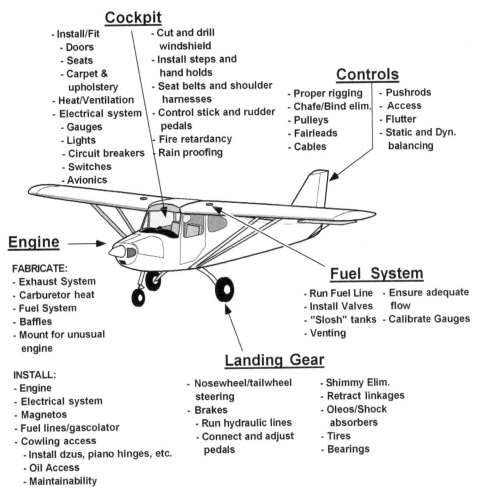

Cockpit
- Install/Fit
  - Doors
  - Seats
  - Carpet & upholstery
- Heat/Ventilation
- Electrical system
  - Gauges
  - Lights
  - Circuit breakers
  - Switches
  - Avionics
- Cut and drill windshield
- Install steps and hand holds
- Seat belts and shoulder harnesses
- Control stick and rudder pedals
- Fire retardancy
- Rain proofing

Controls
- Proper rigging
- Chafe/Bind elim.
- Pulleys
- Fairleads
- Cables
- Pushrods
- Access
- Flutter
- Static and Dyn. balancing

Engine
FABRICATE:
- Exhaust System
- Carburetor heat
- Fuel System
- Baffles
- Mount for unusual engine

INSTALL:
- Engine
- Electrical system
- Magnetos
- Fuel lines/gascolator
- Cowling access
  - Install dzus, piano hinges, etc.
  - Oil Access
  - Maintainability

Fuel System
- Run Fuel Line
- Install Valves
- "Slosh" tanks
- Venting
- Ensure adequate flow
- Calibrate Gauges

Landing Gear
- Nosewheel/tailwheel steering
- Brakes
  - Run hydraulic lines
  - Connect and adjust pedals
- Shimmy Elim.
- Retract linkages
- Oleos/Shock absorbers
- Tires
- Bearings

Fig. 2-12. *Building the structure of a kitplane is only the beginning of the job. No matter how fast the airframe goes together, the myriad little details of the systems will take most of the construction time.*

The worst by-product of subsystem work is the apparent lack of progress. A buddy will stop by, glance at the project, and say, "Haven't been working on it, huh?" Even if you point out the newly installed brakes, he'll examine the completed system and say, "So what?" To a nonbuilder, it looks easy.

It's tough to keep your enthusiasm going. Again, make all your goals short-term ones. If one particular goal is getting you down, leave it and build something else. Or contact another builder and have him help you through.

The builder as inspector

There's one major ability necessary to complete a safe homebuilt: the willingness to look at a newly completed part, spot an error, say, "This is garbage," and then chuck it in the trash and start over again.

Homebuilders generally build two airplanes: one in the hangar and one in the scrap bin. Everyone makes mistakes, especially when starting out. I cut and drilled three rudder spars for my project before I was satisfied.

Modern kitplanes are, for the most part, wonders of foolproof design. The designers work overtime to reduce the chance of the builder making a mistake.

But if you blow it, scrap it. It's your life on the line. If the part is made from raw materials, start over again. If it's a prefabbed kit item, call the manufacturer and order another one. All legitimate kit manufacturers sell individual parts to builders for precisely this reason. If many folks seem to stumble over the same step, a revised design quite often results.

Does this mean that only perfect workmanship is acceptable? No. Replacing a part for purely cosmetic reasons is up to you. Small errors might be fixable or might not affect strength or operation. Contact the kit manufacturer if you have any doubt and/or talk to an EAA technical counselor.

But remember, *you* are the final quality inspector. A friend of mine went out one day for the first flight of his homebuilt aircraft. He and a partner had spent 12 years building this very complex retractable-gear amphibian aircraft. Countless visits had been made by fellow builders. The plane had passed a final inspection not only by the FAA inspector but also by several knowledgeable friends.

During his pretakeoff check next to the runway, the man had a feeling that something was wrong. He carefully examined the entire aircraft and discovered the problem: The ailerons were working backwards.

He taxied back to the ramp. It turned out that the ailerons were built *according to the designer's instructions*. The builders had wanted a stick-type control instead of a yoke, and the designer had sent them modified plans that inadvertently reversed the aileron linkages. The plane was so large that the wings hadn't been installed until close to flight time. Everyone checked that the aileron controls were free of binding, but it didn't occur to anyone that the direction of motion might be wrong.

This was an unusual situation—you're not likely to be the first person to fly a particular type of aircraft. But the kit manufacturers don't always anticipate everything. Comply with the instructions, but if you have a question or concern, talk to the kit manufacturer or other builders.

Don't make changes without checking with the kit manufacturer. Ensure that every part installed in your aircraft is the best that can be made. Don't rely on your tech counselor or the FAA inspector to find mistakes. It's your keister in the pilot's seat.

Modifications

Builders occasionally modify their aircraft during construction. Nothing scares kitplane designers more. As mentioned earlier, aircraft design is a study in compromise, and an unknowing change made by the builder might reduce safety margins to dangerous levels.

For example, take a kit designed to be powered by a Rotax 503 engine. Some people don't like two-stroke engines. Because the Volkswagen four-stroke engine is approximately the same horsepower, why not use it instead of the Rotax?

In the first place, the VW engine weighs at least 40 pounds more than the Rotax. Rotax 503–powered aircraft generally weigh less than 400 pounds, so the empty weight has increased by 10 percent. The builder might be tempted to boost the gross weight to maintain the same empty weight; this is perfectly legitimate because legally, the builder is the manufacturer and can claim whatever gross weight he wishes.

Yet the extra weight decreases the G-load factor. If the aircraft at the design gross has a limit load factor of + 5.5 Gs, increasing the gross might drop the load factor to less than 5. The stall speed increases, and the forward shift in CG increases the elevator loads.

There are other problems as well, but you get the point. This does not mean that you cannot make *any* changes. But check with the kit manufacturer first, and don't be too let down if the manufacturer refuses to approve it. Manufacturers get inundated with suggestions, and their liability status is precarious enough without granting quick approval to some stranger's idea.

The dark side

Careful inspections during construction and no unauthorized modifications seem like reasonable precautions. Yet too often the wreckage of a homebuilt aircraft reveals evidence of violation of those basic precepts:

- A wing separates in flight. Investigators find that the builder *never installed the wing bolts;* the *holes* hadn't even been drilled. The aircraft flew 15 hours with only a layer of fiberglass holding the wing in place.
- A wing disintegrates on a high-speed pass. On investigation, it is found that the builder had skimped on fiberglass cloth.
- A small homebuilt spins in on final during its first flight. Examiners determine the CG to have been at least 2 inches behind the aft limit in part owing to unauthorized modifications. In fact, the builder expended considerable effort to *hide* the CG problem—he "fudged" the weight-and-balance paperwork presented to the FAA.

Why? What made these builders commit such fundamental errors as omitting wing bolts or modifying their airplanes past the limits and beyond? No one will ever know. Each answer died in the respective wreckage.

Your airplane won't be perfect; you will make mistakes. Expect them, and *correct* them. Don't let it get to you; just redo the part or ask for advice.

The skills required

The biggest question you're asking yourself right now is whether you have the skills to build the aircraft. That's not actually as serious a consideration; problems are more likely to occur as a result of deliberate actions rather than accidentally because of a lack of skill.

But what level of skill is required? Many kitplane ads use the term "average skills." What constitutes average? Twenty years ago, most high-school boys took shop classes and learned the basics of machine work and welding. In our school, we melted aluminum scrap, cast rods in sand molds, chucked them into lathes, and turned cufflinks from the castings.

Things have changed. Today's high-schooler spends more hours behind a computer keyboard than a lathe.

What does "average skills" mean to a kitplane manufacturer? Manufacturers spend much of their time immersed in homebuilt aircraft construction. Bandsaw operation is second nature. They know which size drill bits are used for rivet holes. (*Of course* the holes have to be deburred; everybody knows that!)

As far as I'm concerned, "average skills" means that you know that the pointy end of the drill makes the holes. It means that you know nothing about aluminum alloys, epoxies, AN hardware, or fiberglass cloth. That's what this book is about: presentation of the basic techniques and processes in a form that those with "average skills" can understand. With this help, and a little practice, you'll be ready to tackle just about any kit. One of the benefits of kitbuilding is that you'll end up with "above average" skills. Another benefit is that you will have the tools and experience to tackle those household projects you used to shy away from.

Can a person with "average skills" build a modern kitplane?

Certainly. The better kitplanes come with critical holes drilled, and complex parts are prefabricated. Successful completion requires careful study of the instructions and diagrams and cautious execution of each construction step.

But your ultimate goal isn't to build an airplane. That's far easier than you think. Your goal is to build a *safe* airplane. Read the instructions carefully, and follow the plans.

A note about the case studies

The best way to illustrate some of the points in this book is to look at the experiences of average everyday builders. Scattered throughout this book are a set of "case studies" where we do exactly that.

Here's the first:

CASE STUDY: THE FAMILY LONG-EZ

Taking three years of spare time to build your own airplane can have quite an impact on your family.

How, then, does a family stay together when the construction takes *four times* longer?

The Sabourin family Long-EZ shows how a homebuilt doesn't have to be a home-wrecker. Not only that, but looking at Long-EZ N747MS shows how making the right decisions right from the very start helps result in a successful conclusion.

Mike Sabourin always liked unusual airplanes. Like many homebuilders, he started out in radio-controlled (RC) models. He saw the canard revolution of the late 1970s and got an itch to build his own canard-type airplane.

He was familiar with the Long-EZ designs, of course. But at the time, there was a slick little canard ultralight that looked just right for him. He even drove from Seattle to California to visit the factory.

The plane was all right—but he wasn't too impressed with the two-stroke engine that powered it. When the company president admitted that he never flew the plane beyond gliding range of a landing place, Mike decided that it wasn't the plane for a married man with a 2-year-old daughter.

The drive home from the factory clinched it for him: "If I want a real airplane, I'll build a real airplane" is how he remembers his decision.

He ordered a set of Rutan Long-EZ plans in late 1983. He discussed the project with his wife, Maureen. "I made it clear that this was a hobby," he says. "Other things came first. It took longer, but I had a happy family."

The Sabourin family (Fig. 2-13) became quite involved in the process, as we'll see in a little bit.

Fig. 2-13. *Building Long-EZ N747MS was a family effort. Mike Sabourin, his wife Maureen, and daughters Annie (right) and Amy (front).*

CONSTRUCTION

Mike studied his plans for almost a year. He joined the local EAA chapter. He introduced himself as a potential Long-EZ builder and found several others within the chapter. These men would become his friends and advisors over the long construction period ahead.

By fall of 1984, he was ready to build. His EAA connections came through: A fellow member knew of a Long-EZ materials kit being sold by a man who'd suffered some financial reverses. Mike bought a complete kit of materials for $5,000, about a 25 percent discount over the new price.

He was able to carry the makings of his future airplane home with several round trips in his station wagon. The purchase didn't include an engine, but some of the fiberglass parts had been purchased from commercial vendors, such as the wing strakes (wing root area forward on either side of the cockpit forward of the wing) and main landing gear. The epoxy was a bit old; rather than trusting it, Mike bought fresh.

Then began 11 long years of construction. Mike had promised not to neglect his family, and he kept his promise. In addition, other things came up. He went back to school for his masters' degree in 1986. During the 4 years of school, he only got 200 hours or so of construction time in on the Long-EZ. Can't blame it all on school, though—daughter Amy was born in 1987.

Still, Mike never lost sight of the goal. Or his promise. "Maureen was pretty supportive from the first," recalls Mike. "And got more and more supportive as it neared completion." There's always something that needs to be turned or held while building an airplane, and Mike's wife was always ready. Even his daughters got into the act (Fig. 2-14). There was always some noncritical sanding that could be done. At one point, Mike needed to install a nut in a very tight location and just couldn't get to it. Somehow, he and Maureen managed to coax 2-year-old Amy to put it in place.

Finally, after almost 11 years, the Long-EZ was ready to move to the airport. The local FAA required Mike to fly his test period at a strip that was an hour's drive from home. The typical last-minute adjustments and tests promised to keep Mike from his family for a number of weekends.

Instead, Maureen and the kids joined Mike at the airport. They slept in tents inside the hangar.

September 1, 1995, saw the moment of truth. With his family watching, Mike flew Long-EZ N747MS on a flawless 80-minute first flight. We'll talk about his first-flight experience in Chap. 12.

Builder's background and facility

As mentioned, Mike had considerable experience building RC models. While most of his friends had done ordinary wood construction, he'd used some fiberglass on a couple of small boats he built in high school. He'd been happy enough with the results to decide on a composite kit.

In addition, he'd just finished remodeling his house. Not only did it give him experience with tools, but he also used the opportunity to put in a 14 × 26-foot workshop adjoining the recreation room. The workshop was a necessity because

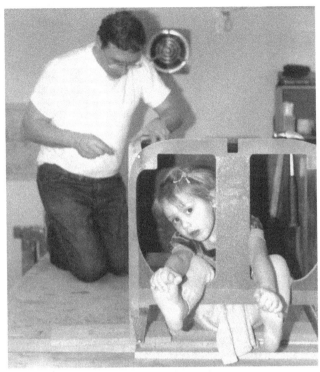

Fig. 2-14. *Mike's workshop adjoined the recreation room, and his daughters grew up with the aircraft. Annie Sabourin inspects her father's workmanship in the forward fuselage of the Long-EZ.*
Mike Sabourin.

his house didn't have a garage. Being actually within the house, the workshop was adequately heated—a must for a composite airplane. And it encouraged his family to stay in touch with the project.

One slight drawback: The shop was on the second floor! The window was the only way out for some kit components (Fig. 2-15).

He did erect a small shed in the back yard where he could store parts when the shop became too cluttered. The 13 × 19-foot shed gave him just enough room to fit the fuselage after he installed the 9-foot-wide center-section spar. The wings were hung from the rafters, and the canard hung from the wall. The shed kept the rain off Mike while installing the systems and during final finishing.

However, it certainly didn't have enough room for him to paint the entire aircraft. When painting time came, Mike erected a temporary 25 × 20-foot paint booth off the side of his shed. The booth was a simple framework of 2 × 4 lumber covered with sheet plastic. It worked well enough, but Mike says that he's not sure he'd go through the hassle again on another airplane. He'd probably just have the plane painted by a professional.

Even so, it turned out pretty well. Long-EZ N747MS took the composite "Champion Custom-Built, Plans-Built" trophy at its first airshow.

Fig. 2-15. *Since Mike's workshop was on the second floor, some aerobatics were requried to remove parts via the window.* Mike Sabourin.

Builder's experience

Mike's biggest surprise, he says, is "... how long everything took to put together. The structure went together faster than I expected, but the finishing work (installing systems, wiring, painting, and so on) took much longer than I expected."

He didn't really have to pick up many new skills. "I'd done most of the stuff before," he says, "but only on a smaller scale."

Tools included a drill press and an upright belt sander. The belt sander was perfect for finishing aluminum parts—he could cut out the piece with his jigsaw and then trim to final shape with the sander.

His most vital tool was a Dremel Moto-Tool, a small hand-held motor than can take a variety of bits for a number of jobs. Mike went through three of them in the course of building his aircraft.

He received plenty of local advice, and he subscribed to two newsletters aimed at EZ builders. He never had to call the Rutan factory for advice, depending instead on his local contacts.

He bought his engine five years before the first flight. The Lycoming O-235-L2C had been purchased new by the then-owner, who had installed it on his own Long-EZ and had flown 700 trouble-free hours. The owner decided to upgrade to a larger engine and advertised his old one in *Trade-A-Plane*. It cost Mike $5,000 plus $300 shipping.

Flight experience

Details of Mike's first flight are given in Chap. 12. In the years since, he has put over 500 hours on his airplane, with only minor troubles. Even better, his plane has turned out to be an award winner—he's won a number of awards at regional fly-ins and a "Lindy" workmanship award at EAA Airventure.

Advice

While some builders are distracted by visitors to his shop, Mike welcomed them. "Share the experience," he says. "It's been a lot of fun having people over to help."

When it comes to family relationships while the plane is being built, Mike's advice for prospective builders is clear: "Let it be an enjoyable experience, not something that dominates you and hurts your family."

Finally: "Don't compromise safety."

Did Mike's experience get you pumped to get started—or are you starting to worry a bit? Let's take a look at some alternative ways of getting into the air.

3

Alternatives

FOR ONE REASON OR THE OTHER, BUILDING A KITPLANE MAY NOT BE THE BEST
SOLUTION TO YOUR NEEDS. Let's take a moment to examine some alternatives.

PRODUCTION AIRCRAFT

Looking for a kitplane for fast cross-country travel? How does this sound: four seats,
160 mph cruise speed, retractable gear, and fully Instrument Flight Rules (IFR)–
equipped. The cost is $50,000, including a midtime engine, constant-speed propeller,
instruments, radios, paint, wheels, and upholstery.

Building time? Zero hours—the plane is a used Mooney, of mid-1960s vintage
(Fig. 3-1).

Note that $50,000 also will buy a Lancair ES kit, but without engine, propeller,
avionics, and so on. You may end up paying at least another $40,000. The Lancair
will cruise faster—but $40,000 faster?

Homebuilts became popular for two reasons. First, they offered features un-
available in production aircraft, such as aerobatic capability or open-cockpit flying.
Second, their performance was out of proportion to their cost. Back when a new
Cessna 150 was priced at $10,000, a Thorp T-18 could be built for half that cost and
cruised 50 percent faster. The Thorp was designed for a Lycoming O-290G taken
from a military-surplus ground power unit.

Things are a bit different now. Kit aircraft aren't cheap, and inexpensive surplus engines are long gone. Kitplanes still offer greater performance for the money,
but the performance increase over production aircraft isn't as great as it used to be.
Modified versions of some kitplane designs, such as the GlaStar and the Lancair IV,
have even been certified and are available as production aircraft.

If you're looking *solely* for an aircraft with particular features, you're better off
buying a good used factory aircraft. Here are the relative advantages of used production aircraft over kitplanes:

1. *Better overall flying characteristics.* Even older production planes are less
   demanding to fly than some kitplanes. As mentioned in Chap. 2, sometimes
   kitplanes get great performance at the expense of handling.

Fig. 3-1. *If you'd rather fly than build, why not spend the same amount of money on a used factory aircraft rather than a kitplane? Performance might be slightly less, but you'll be flying immediately instead of next year. And the doors probably leak no less than that of an average kitplane.*

2. *Instant availability.* Drop the money on the barrelhead, and you can fly away in your new airplane rather than toiling in the garage for the next five years.

3. *Wide selection and a wide range of prices.* Find the exact airplane you need. The selection of kitplanes is still rather limited, and discounts are rare.

4. *Known resale.* If you need to sell the airplane, market value is easily determined. Kitplane resale varies; sometimes the owner can't get more than the cost of the kit itself.

5. *Loan availability.* Assuming that you can qualify financially, you can easily get a loan to buy a production airplane, using the plane itself as collateral. Kitplane loans have become available, but there are only a few sources.

6. *Off-the-shelf maintenance.* If a production airplane needs repair, you can buy ready-to-install parts. While parts for older aircraft might be harder to obtain, it's still easier than trying to remember how a certain kitplane part is made. You can save money and do some of the work because the Federal Aviation Administration (FAA) allows owners of production aircraft to perform certain preventative maintenance operations (Fig. 3-2). In case of major damage, a wing or tail section can be bought from salvage and bolted into place. Carry insurance, and the company will pay someone to do the work, too.

7. *No emotional resistance to outside tiedowns.* After spending several years building a kitplane, most builders want inside hangarage for their pride and

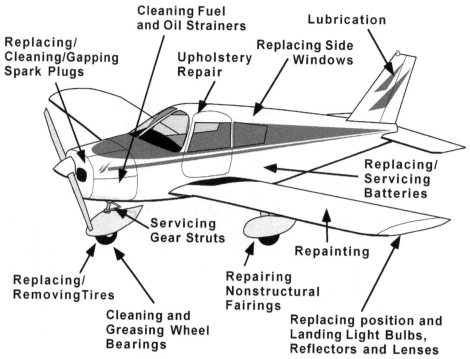

Fig. 3-2. *Preventative maintenance tasks can be performed by the owner of a production aircraft.*

joy. This is expensive at most airports. But few pilots have compunctions about leaving a used production airplane outside. This can reduce the fixed costs of ownership significantly. For example, a closed hangar at a local airport rents for $180 a month; tiedowns are $120 less.

Because this is a book on building kitplanes, recall the advantages of the home-built aircraft:

1. You can get a new aircraft for the price of a used production model.

2. You can save maintenance costs by legally performing your own inspection and repairs. Also, parts prices for production aircraft generally are exorbitant.

3. The plane will have exactly the equipment you want in exactly the right paint scheme.

4. Planes that come with individually available subkits have the equivalent of interest-free financing because you can buy a subkit when you have the money for it.

5. If finances get tight after completion, the plane can be disassembled and stored at home to save tiedown and insurance costs.

6. Performance generally is better than that of an equivalently priced used aircraft.
7. Styling and appearance are far better than stodgy production models.
8. There is significant personal satisfaction in building your own airplane.

The advantages and disadvantages of buying a used factory aircraft are shown in Table 3-1, and a similar table for kitplanes is included as Table 3-2.

Table 3-1. Advantages and disadvantages of buying a used factory airplane instead of a kit airplane

Advantages	Disadvantages
Easier to fly/no surprises	Expensive maintenance
Flyable on purchase	Older
Loans easily available	Little selection on options
Wide selection of types	Ordinary appearance
Wide range of prices	High-cost certified parts
Known resale value	
Parts availability	
No operations restrictions	

Table 3-2. Advantages and disadvantages of kit airplanes

Advantages	Disadvantages
New aircraft for used price	Danger of poor construction
Better performance	Takes long time to build
Lower maintenance costs	Harder to finance and insure
Cheaper parts	Liability on resale
Phased purchase (subkits)	Operations restrictions
Low-cost inactivation	
Attention-getting	
Personal satisfaction	

There is one major breakpoint to determine whether you should buy or build: If your sole intent is to obtain an airplane to fly, you're better off buying a used production plane. You'll have it immediately, and if your tastes change, you can sell it and buy another type.

Building a kitplane is a challenge—many consider the actual flying of the aircraft rather anticlimactic. Unless you are prepared for the grind and mess of construction, and unless you are able to look at the building process as an end unto itself, you're better off shopping in *Trade-A-Plane*.

USED HOMEBUILTS

If you can't face the construction process, but the appearance or performance of a particular homebuilt is still attractive, consider buying a used homebuilt. Prices can be quite reasonable. Some can be bought for little more than the total cost of construction.

There are a number of reasons. First, some folks like the building process more than flying. Some have built five or more aircraft and quite often are craftsmen with excellent workmanship. Second, the occasional builder doesn't do his homework before starting construction, and the resulting airplane doesn't meet his needs. Finally, a builder could own his creation long enough to get tired of it and wish to move on to something else.

Aviation writer Budd Davisson has a saying: "You're better off buying a used snake than a used homebuilt."

Obviously, significant problems could exist with the aircraft. A detailed prepurchase inspection is vital.

Inspections

First, find someone with experience on the type of aircraft, preferably someone who has built the same model. They should pass judgment on the workmanship and adherence to the construction plans. While some deviations are minor, changes to the control system, rigging, or basic structure should set the alarm bells ringing.

An inspection by a licensed airframe and powerplant (A&P) mechanic also would be a good idea because they're experienced in rapid assessment of airframe and engine condition. However, for two-stroke Rotax-powered aircraft such as the Murphy Maverick (Fig. 3-3), check at the local ultralight center for someone who knows two-stroke engines.

Both you and the experienced builder you brought with you should test fly the aircraft. You can judge if the plane is right for you; the experienced builder will determine if the plane flies like it should. Examine the logbooks. An aircraft that is flown regularly probably flies well; a plane that sits a lot might have problems.

Single-seat airplanes such as the most World War I replicas (Fig. 3-4) present a problem. Will the owner let you test fly it? There are a number of cases of prospective buyers crashing single-seaters. If your experienced builder/friend arrives in his own version of the same model aircraft, the owner shouldn't have too many worries. Or if the aircraft is otherwise acceptable, you may be able to buy the aircraft contingent on an acceptable test flight.

Inspections and Maintenance

Maintenance is another issue when buying a used homebuilt. You cannot receive a repairman certificate because you didn't build the aircraft. The original builder can perform the annual condition inspection if she still retains her certificate and is willing. Neither is guaranteed. If you are buying the plane from the original builder, a fresh annual inspection should be part of the deal. Have an outside mechanic inspect the aircraft as well, but licensed mechanics charge more for annuals than for prepurchase inspections.

Fig. 3-3. *A used kitplane such as this Murphy Maverick can be a good deal, as long as you can verify proper construction and maintenance. However, finding a mechanic qualified to check out the Rotax engine might be a problem.* Murphy Aircraft Manufacturing, Ltd.

Fig. 3-4. *If you owned a cool single-seater like this Aerodrome Aeroplanes Fokker Triplane replica, would you let a prospective buyer test fly it?* Aerodrome Aeroplanes.

Don't fault the builder if she is unwilling to continue inspecting the aircraft after you buy it. The builder probably isn't a professional mechanic, and the legal and other repulsing pressures are strong. If you buy from a friend, though, he may be willing to continue.

Any A&P mechanic is allowed to maintain and inspect homebuilt aircraft. One difference between homebuilts and production aircraft is that the A&P doesn't have to have an inspection authorization (IA) to perform an annual on a homebuilt. That'll save a bit. If the plane has a Rotax or converted auto engine, however, it might be hard to find an A&P mechanic willing to be legally responsible for its maintenance and continued operation. Check around before you buy the aircraft.

As mentioned in Chap. 2, the FAA maintenance regulations do not apply to homebuilts. The owner of a homebuilt aircraft is allowed to maintain his airplane as long as an A&P mechanic signs off on the work within the next twelve months. If you can find a mechanic willing to work on this basis, you've got the best of both worlds—authorizations equivalent to a repairman certificate with an experienced eye to keep the aircraft safe.

The final analysis

As you're discussing the deal with the seller, one of the primary things you should determine is *why* the aircraft is for sale.

Does the plane just have a few hours on it? That should trigger a few red lights. Maybe it handles so strangely that the owner is scared of it. Maybe she discovered some major mistakes made during construction.

Examine the aircraft logs. Has the plane flown regularly? If the plane were ten years old, I'd be a lot happier to see a thousand hours in its logbook than a hundred hours. Good-handling, reliable aircraft are *flown*. A plane airborne for only a few hours per year may have problems.

Probably your safest route is to know the seller in advance. I've seen a number of homebuilts come up for sale within my local Experimental Aircraft Association (EAA) chapters. With my knowledge of the builders, I would have had no hesitation about buying most of the aircraft. In fact, I bought a completed homebuilt from a fellow EAA chapter member almost 10 years ago and have been quite happy with it.

With the boom in kitplanes, more and more homebuilts are available on the used market. They're a definite option for those looking for something more than a typical factory plane but without the patience or inclination to build their own.

PARTIALLY COMPLETED HOMEBUILTS

There's a thriving market in flyable homebuilts. Uncompleted kitplanes are another kettle of worms. Buying and selling unfinished homebuilts is nothing new. It used to be said that 30 percent of homebuilt aircraft projects were flown eventually—10 percent by the first builder, 10 percent by the second owner, and 10 percent by the third or subsequent owners. One local Fly Baby ran through six owners until it was completed—20 years after construction started.

A flying homebuilt has passed its most critical test, and one can use its flying characteristics to judge how well it's built. An uncompleted project? Who knows what might be wrong with it? The sixth owner of that uncompleted Fly Baby found a critical problem with the wing bracing system. The *sixth* owner. Would the previous owner have detected the problem before it was too late?

Still, there are some advantages to buying partially completed planes. The primary one is the work you *won't* have to do. Wouldn't you like to pick up a completed empennage for the cost of the tail kit alone?

In addition, buying an uncompleted plans-type homebuilt can be the equivalent of buying a kit. Would you like to build a Sonerai (Fig. 3-5)? You might be able to pick up most of an airframe for a few thousand dollars and get it flying for a couple hundred hours and the price of a Volkswagen engine.

Fig. 3-5. *Buying a partially completed homebuilt is like obtaining a "kit" for a normally plans-built aircraft.*

Buying a partially completed homebuilt is a gamble. It might hide a deadly flaw, or it might be the best aircraft deal since the $650 Jenny. The problem, of course, is telling the difference. How can you make sure that you're not getting taken?

Do your research

To begin with, know what you're buying. Check the specifications. Read pilot reports. If the aircraft type never became popular, there's often a good reason, usually rooted in flying qualities or being difficult to build. Avoid partially completed custom designs. Stick with designs that have been available commercially.

If the advertised project is a kit, knowing the exact model is vital. Homebuilts have improved dramatically in the last 10 years; an older kit's performance and reliability are reduced compared with the latest model. Today's Airdale Flyer is roomier and faster to build than its 1985-era predecessor, the Avid Flyer. The older kit came with a Cuyuna engine. That's quite a comedown from the four-stroke Rotax 912 or Subaru on the Airdale.

Once you understand the aircraft, find someone who's built one. Ask about common problems. What mistakes are typical? Which are the critical operations? Ask the person if he knows the builder on the project for sale; get his impression of the owner's workmanship. Of course, most folks will be reluctant to slur an acquaintance's ability. Rave reviews are a good sign; a shrug and a "Pretty good" can mean anything.

The examination

Thus armed, it's time to examine the merchandise. Don't go alone. Four eyes are better than two, and if the second set belongs to a builder experienced with the design, so much the better.

Scope out the builder's shop. How much equipment was he working with? Did he have an adequate air compressor to rivet with or some cobbled-up system using a hobby compressor and a small air tank? Does he have a sturdy workbench or a cheap folding table? How did he store the kit materials? Did he lean plywood against the walls or build a flat-storage rack? Does he have just a few discount-store hand tools or a complete machinist's set?

This isn't to say that one can't build an acceptable aircraft with minimal tools and facilities. But you're looking for clues to poor workmanship. Few homebuilders rush out and equip their shop completely from the outset. Someone with a well-equipped shop probably owned the tools prior to beginning the aircraft; such a person's workmanship probably would be good from the start. But if the owner has a less-than-optimal compressor setup, his rivets may have suffered. A flimsy work-table can result in warped components. Poor storage of plywood, steel, or sheet aluminum can ruin good material.

Speaking of the raw materials, check them over for damage and decay. For wood, look for the discoloration of rot and the flaking and checking that results from drying out. The end grain of wood stock should have a coating of paint or varnish to retard the drying process.

Aluminum should be free of corrosion. Look for roughness and dry powdery areas. Sheets often get lightly scratched in normal shop handling, but there is a limit. Deep scratches concentrate stresses and can cause premature failure. Slight bends and wrinkles may not affect the strength, but they make the material awkward to work with.

Any untreated steel is likely to have a patina of rust. Conscientious builders apply a coat of paint as soon as possible. Unlike corrosion on aluminum, rust can be removed safely from steel as long as it isn't deep. If the builder has neglected to prime a steel-tube fuselage, though, derusting it will take considerable time.

You won't have these worries with composite materials. However, epoxies usually have a maximum shelf life. Make sure that it hasn't been exceeded. Some epoxies are shipped in slightly permeable plastic containers. These reduce the shelf life, as well as stink up the shop. It's a positive sign if the builder had transferred the resin to metal or glass containers. Finally, make sure that any fiberglass cloth is dry, clean, and still on the roll.

Checking the aircraft

Of course, your primary item of interest is the aircraft itself. Check its straightness. Crouch down and squint along the edges. Run your hand along surfaces, and check for smoothness. Have the builder remove access panels and other items that hinder a full inspection. You'll soon understand one of the biggest frustrations in buying a partially completed project: The more finished it is, the harder it is to inspect. A closed-up wing may cover a number of sins, yet the buyer will want a higher price owing to the level of completion.

Specific areas of interest vary with the type of construction. For wood, pay careful attention to the glue joints. Find out what type of glue the builder used. The poor filling qualities of traditional glues require tight joints for strength. Modern epoxies such as T-88 do a good job of filling loose joints, but tight joints are still a sign of good workmanship.

One critical item to check is amount of glue in the joints. They all should show signs of glue oozing out when the pieces were clamped. The excess normally is wiped away, but a small fillet should be visible all the way around.

Similarly, composite layups also should be checked for starvation. Completed parts should have a rich amber color; pale patches indicate voids and delamination. An excess of epoxy isn't dangerous, but it's heavy and a sign of sloppy workmanship. The cloth weave should run straight or curve evenly around bends. All parts should be dry and clean of grease and dirt.

Partially completed aluminum aircraft requires careful examination of the riveting. Examine the skin around the manufactured heads. Look for distortion, dents, and other signs of improper procedure. Check both heads for distortion or cracking. Gauge a few shop heads, and make sure that they're at least half the rivet diameter high and one and one-half the diameter wide. Mind the edge margin. The rivets should be spaced at least three times their diameter apart and at least twice the diameter from any edge (see Chap. 9).

Check the straightness of the rivet lines. A little waviness is normal (and for some of us, unavoidable), but you can get a better feel for the builder's level of workmanship. Look for scratches and dents.

Welded structures can be tricky. Factory-welded areas are generally all right, although I've seen a few bad ones. Builder welds are another matter. While it's easy enough to check the surface, it's quite possible that only the surface material had been melted and jointed—that the underlying metal is still separate. *Penetration* of the weld is important but difficult to prove without actually cutting through the joint. Hopefully, the builder made periodic test welds to saw apart and verify penetration.

Before visiting a potential purchase, review the basics of its mode of construction as presented in Chaps. 8 through 11 of this book.

Fig. 3-6. *A nonstandard engine makes buying a partially completed homebuilt a bit more of a gamble. Are you going to be able to pick up where the previous engine modifier left off?*

Engine details

If the builder has been modifying the aircraft for a nonstandard engine (Fig. 3-6), consider whether you wish to travel down the same path. Are you willing to continue with the same alternate engine? Has the airframe been modified to the point where it would take too much work to go back to a conventional powerplant? Is the engine mount included, or will you have to get one custom built?

If the engine is included, see whether it has the required accessories, such as magnetos, alternator, starter, and carburetor. These little components are surprisingly expensive.

If a used aircraft engine is included in the deal, scan through its logs. The engine should have been "pickled" for storage. Special oil coats the crankcase and sump; desiccator plugs are installed in the cylinders, exhaust, and breather; and the engine should be wrapped and stored in a dry place. Otherwise, it could be just a 250-pound block of rust.

Other goodies

In addition to the aircraft and engine, don't forget to examine any other equipment included in the sale. Radios, navigation gear, and other electrical components significantly sweeten the deal.

The deal should include a complete set of plans. Make sure that they're all there. Glance over the inventory of kit components. If the kit originally included parts such as wheels, tires, and instruments, find out if the owner has included them in the sale. Ask to see the owner's builder's log—the more detail, the better. For example, it should list exactly what brand of primer was used. That'll help you to avoid compatibility problems when it comes time to paint. If he's got a collection of newsletters or similar publications, ensure they're included in the deal.

Ask the owner about construction variations. Most builders make small modifications; few of these affect complexity or airworthiness. But be leery of major modifications, such as conversion to a nonstandard engine or a scratch-built retractable gear. There's nothing wrong with people making changes like this; however, you don't want to be stuck with someone else's poorly thought-out modification.

The decision

Finally comes decision time. Has the builder done a safe job so far? Homebuilts can be remarkably tolerant of bad workmanship. As long as the structure is sound, the cosmetics don't affect the airworthiness. But it's your peace of mind as well as your life at stake. It's no fun flying a plane that you have to worry about all the time. And everyone's going to assume that those bad cosmetics are your fault.

If you decide that the kit is acceptable, it's time to haggle over price. Unfortunately, there are no guidelines, no "blue book" for unfinished kits. Their value is determined by too many factors, from the workmanship to completeness to the kit's popularity.

Don't be thrown off by the "current" kit price. The original Glasair kit sold for half today's price but had more sensitive flight characteristics and was markedly harder to build. So the starting point should be around the original cost, not what a brand-new kit might run.

How much to "pay" for the work already completed is going to be the thorniest part. If you have to redo shoddy work, the project is of less value—although it might not be politic to tell that to the owner. It boils down to how much the work is worth to you. Treat the seller's description of "x percent completed" with a ton of salt.

For a starting point, take the original kit price (or the cost of the materials), and then add deltas up or down depending on workmanship and additional equipment. The actual market value is probably between 80 to 120 percent of this amount.

What to offer? This depends on the situation. Some sellers are just tired of the bother and are willing to take a much lower price. Others are in no hurry and are willing to wait for what they feel is a fair price. I know someone who sold a project and then bought it back later for half what he originally sold it for. Each case is different. It never hurts to make a low offer and see how the owner reacts.

Unless you pay peanuts, though, it's probably less of a deal than you think. Talking to local buyers of partially completed homebuilts reveals one common theme: They had to do more work than they thought would be necessary. As everyone will tell you, the last 10 percent of the aircraft takes 90 percent of the time.

Finally, remember that you must perform at least 51 percent of the work on the aircraft to receive a repairman certificate. A partially completed project may not allow you to qualify to do your own maintenance. Get a ruling from an FAA inspector prior to buying an unfinished project.

LIGHT SPORT AIRCRAFT

In the year 2004, a revolution occurred in general aviation. The FAA instituted two programs that heralded a major revolution in the American regulatory environment: the *sport pilot license* and the new *light sport aircraft category*. Not only do these new programs expand the horizons for general aviation, but they also may have a major impact on homebuilding.

There are three major aspects of the new system: the *light sport definition*, the *special light sport certification category*, and the *experimental light sport aircraft category*.

The definition

The centerpiece of the new rules is a legal definition of a new type of aircraft: the light sport aircraft. These are airplanes with no more than a single engine and a gross weight of 1,320 pounds, no more than two seats, fixed gear, a fixed-pitch (or ground-adjustable) propeller, a stall speed of less than 52 mph, and a maximum level-flight speed of 138 mph. A summary of these characteristics is shown in Fig. 3-7.

Any aircraft that meets the basic requirements can be flown by a person on sport pilot privileges. The aircraft does not need to be certified in either of the two new categories. Classic production planes such as Piper Cubs and Aeronca Champs qualify. Many homebuilts, both older and newer designs, are sport pilot–eligible as well (Fig. 3-8).

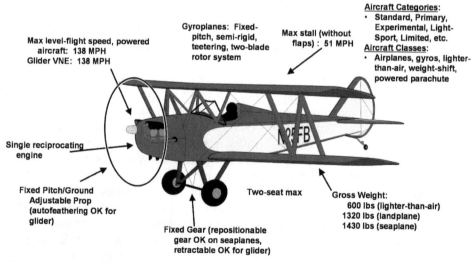

Fig. 3-7. *Basic light sport aircraft requirements.*

Fig. 3-8. *Both existing airplanes such as this Circa Nieuport replica and newly certified special or experimental light sport aircraft can be flown with just a sport pilot license.*

A person can execute sport pilot privileges in two ways. First, he can obtain the new sport pilot license. This license requires less training than the conventional private license, and operators are not required to hold a current FAA medical certificate. A valid driver's license is the only medical qualification necessary.

Second, such a person can possess a recreational or private license (or higher) and merely not renew his FAA third class medical certificate. Again, only a valid state driver's is license is necessary.

The main thing to be aware of is this: Just because an airplane meets the light sport definition does *not* affect how it must be maintained or who can maintain it. A normal category airplane such as a Champ still must be inspected annually by an A&P mechanic with an IA rating, even though a sport pilot can legally fly it.

Special light sport aircraft

The light sport aircraft definition was the basis for a new aircraft certification category: special light sport aircraft (SLSA).

The "special" denotes that the aircraft do not receive standard certification. In other words, the aircraft has not been either designed to or evaluated against the FAA's regulations governing aircraft, engine, and component design; testing; performance; handling; and durability. To this extent, the special light sport aircraft category is similar to the experimental category (see Fig. 1-2).

However, unlike experimentals, SLSA manufacturers are still governed by design standards. Instead of the traditional Federal Aviation Regulations (FARs), though, the design of these planes is governed by guidelines developed *within the aircraft industry.* The FAA does not dictate the limit loads for airframes or the requirements for SLSA engine certification; the *consensus standards* are developed jointly within the aviation industry and published by the American Society for Testing and Materials (ASTM) International, a standardizations organization (ASTM started as an American organization, but now it's an international organization).

These new standards are the equivalent of 14CFR Parts 21 ("Certification Procedures for Products and Parts"), 23 ("Airworthiness Standards: Normal Category"), 27 ("Airworthiness Standards: Rotorcraft"), and 33 ("Airworthiness Standards: Engines"). My commercial copy of these regulations is over 300 pages long. The ASTM standards book (Fig. 3-9) is 80 pages long and includes some elements (such as powered parachutes) that the FARs don't cover.

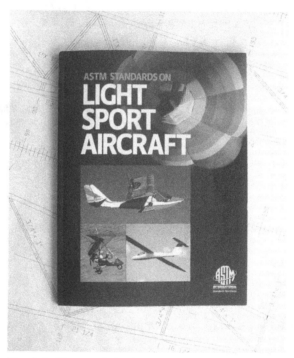

Fig. 3-9. *The regulations governing standard-category aircraft, engine, and parts certification exceed 300 pages, whereas the consensus standard for light sport aircraft is a bit more than a quarter the size.*

The consensus standards establish design, manufacturing, quality control, and testing criteria. They also specify the required level of documentation, such as the airplane flight manual and maintenance procedures, and how aircraft companies must track safety-of-flight issues and how the company will ensure that SLSAs in the field continue to conform to the consensus standards. As with homebuilts, SLSAs are required to carry a placard indicating that they do not meet standard airworthiness requirements.

When a manufacturer is ready to produce a normal-category airplane, the company submits paperwork showing its analyses and the results of its flight-test program to the FAA. There, government specialists examine the data to verify that the new airplane does, indeed, meet Part 23 requirements. Government pilots then evaluate the aircraft.

The light sport aircraft (LSA) process is quite a bit simpler. The manufacturer submits the aircraft's operating instructions, maintenance and inspection procedures, flight training supplement data, and an affidavit that the airplane meets the LSA consensus standard. The prototype aircraft undergoes a safety inspection by the FAA, but otherwise, there is no FAA verification of the manufacturer's analysis or test information. The manufacturer merely agrees to allow the FAA unrestricted access to its facilities if the government wishes to verify that the required information exists.

Once certified, the airplane is similar, legally, to a standard-category plane. SLSAs can be leased or rented. They require annual inspections and, for planes operated commercially, 100-hour inspections.

A&P mechanics are allowed to perform all maintenance, inspections, and repairs on light sport aircraft. Similar authority is granted to those who gain the new light sport maintenance repairman (LS-M) certificate. The LS-M certificate can be earned after an 80- to 120-hour course. An LS-M certificate allows the holder to perform only those procedures that are spelled out in the manufacturer's maintenance manuals and in which he has received the necessary training.

As part of the consensus standard, SLSA manufacturers specify the allowable replacement parts and any allowed alterations. If unauthorized parts are installed or alterations are performed other than those approved by the manufacturer, the aircraft is considered in violation of the consensus standard and is not airworthy. Technically, this applies to *everything*. One can change radio types in a standard-category airplane with the FAA's field approval process, but an SLSA *must* remain in the manufacturer-specified configuration.

As with homebuilts, SLSAs do not fall within the FAA's conventional airworthiness directive system. However, the same level of authority is vested in the manufacturers of SLSAs. They are required to monitor the reliability of their products and, if necessary, issue safety directives. Compliance is mandatory, although an owner can appeal. Also, if an FAA-certified part is used on the aircraft, it is subject to airworthiness directives.

For those looking for a small, fun-flying aircraft, a special light sport aircraft is a good alternative to a homebuilt kit. Costs should be much lower than those of "conventional" new aircraft, and the simple certification requirements are designed to entice new manufacturers.

Experimental light sport aircraft

The experimental light sport aircraft (ELSA) category was created to allow kit versions of light sport aircraft. Unlike traditional homebuilders, ELSA kit buyers are not limited by the 51 percent rule. ELSA kit manufacturers are free to sell their products at whatever degree of completion they wish, from 0 percent (plans-built) to 99 percent.

Of course, there's a catch. To qualify as an ELSA, an aircraft design first must be certified as a special light sport aircraft. In other words, the manufacturer must design and manufacture a prototype to the consensus standard and develop all the flight and maintenance instructions the standards require. Once the manufacturer receives the SLSA certification on this prototype, the company can build ready-to-fly aircraft, ELSA kits, or both.

While builders of "51 percent rule" homebuilts can make whatever changes they wish, an ELSA builder must follow the manufacturer's assembly instructions strictly. The final FAA inspection prior to certification will verify compliance.

There are two other routes to ELSA. First, if an owner of an SLSA doesn't want to comply with a manufacturer's mandatory safety directive, she can recertify the aircraft as an ELSA.

Second, owners of two-seat ultralights operating under the Part 103 trainer exemption and "fat" ultralights (those that do not meet Part 103 requirements) can convert to ELSA through August 2007.

Maintenance

As with other categories of aircraft, the owner of an SLSA or ELSA is allowed to perform specific preventive maintenance tasks. ELSA owners are allowed to perform all maintenance on their aircraft.

ELSA annual inspections may be performed by an A&P mechanic, an LS-M repairman, or someone holding the other new repairman certificate, light sport inspection (LS-I). The LS-I certificate allows the holder to perform the annual inspection *only* on an ELSA that he owns. The certificate requires just 16 hours of training. Unlike the repairman certificates for experimental/amateur-built aircraft, the LS-I certificate doesn't require the holder be the builder of the aircraft.

The decision

Is a light sport aircraft the right pick for you? As with standard-category production aircraft, LSAs provide an immediate flight capability versus the multiyear construction process of most homebuilts. However, the prices for the ready-to-fly aircraft are going to be higher than those of equivalent kit aircraft.

If you're considering buying an already-flying homebuilt, a used experimental light sport aircraft may be a better choice than a traditional experimental/amateur-built plane. You'll be able to take a short training course to be able to legally annual an ELSA, but an A&P mechanic will be necessary for the traditional homebuilt. The ownership differences are illustrated in Fig. 3-10.

If the performance limits of light sport aircraft fit with your needs, either an SLSA or ELSA would be a good alternative to an experimental/amateur-built aircraft.

ULTRALIGHTS

Ultralights, in the United States, are small, very light personal aircraft that are governed by a special set of FAA regulations. They began with the hang gliders of the early 1970s, which hardy souls carried to the tops of hillsides, strapped on, and glided back down again. Hang glider development had been triggered by a number of factors, such as the increasing availability of lightweight aluminum tubing and some experimentation with low-complexity gliding flight. For instance, one popular hang glider design of the era was the Rogallo wing, a sort of "paper airplane" design

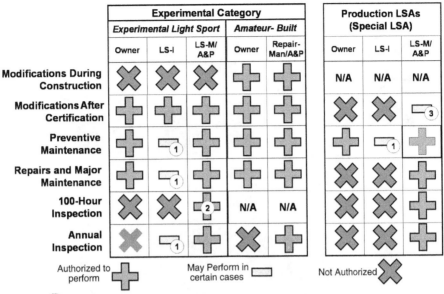

	Experimental Category					Production LSAs (Special LSA)		
	Experimental Light Sport			Amateur-Built				
	Owner	LS-I	LS-M/A&P	Owner	Repair-Man/A&P	Owner	LS-I	LS-M/A&P
Modifications During Construction	X	X	X	+	+	N/A	N/A	N/A
Modifications After Certification	+	+	+	+	+	X	X	☐ (3)
Preventive Maintenance	+	☐ (1)	+	+	+	+	☐ (1)	+
Repairs and Major Maintenance	+	☐ (1)	+	+	+	X	X	+
100-Hour Inspection	X	X	☐ (2)	N/A	N/A	X	X	+
Annual Inspection	X	☐ (1)	+	X	+	X	X	+

+ Authorized to perform ☐ May Perform in certain cases X Not Authorized

(1) Can perform if owner of aircraft

(2) Former Part 103 Two-Seat exemption aircraft transferred to ELSA and used for training

(3) Only modifications specifically authorized by the aircraft manufacturer can be made

Fig. 3-10. *Ownership differences between ELSAs, SLSAs, and amateur-built aircraft.*

in nylon and aluminum. Yet the Rogallo wing was designed originally as a fold-up glider for Gemini spacecraft!

One big drawback with the hang glider: Once you landed at the bottom of the hill, you had to carry the glider back up again.

John Moody demonstrated the solution in 1975: He bolted a go-cart engine and propeller to his Icarus II hang glider—and the ultralight was born. Buzzing contraptions soon arose from fields all across the country. The design and construction of these aircraft were so simple that new designs sprang up practically overnight. Soon, ultralights were the hottest thing in aviation.

The FAA was caught somewhat flat-footed. They agency originally had allowed the basic hang gliders to operate outside the regulations because it was apparent that they threatened no one but the pilot. When the powered versions came out, the engines still had such low thrust that the pilot had to run to get the airplane up to flying speed. Thus the FAA announced such machines were still considered hang gliders if they were "foot launchable."

Soon, however, developers such as ultralight pioneer Chuck Slusarczyk (Fig. 3-11) produced lightweight propeller-speed-reduction units that greatly increased the usable thrust from the small, fast-turning two-stroke engines. Wheels started appearing on the machines. So did "pods" to enclose the pilot (which often included two holes in the bottom so that the ultralights still technically were "foot launchable").

Fig. 3-11. *EAA Hall of Fame member Chuck Slusarczyk (right) pioneered propeller reduction drives that revolutionized early ultralights. In addition to developing the CGS Hawk ultralight, he also appeared in several episodes of the "Junkyard Wars" TV series.*

Other, more serious problems arose. Two-seat ultralights made their appearance. Citizens complained about the noise from the shrieking two-stroke engines just a couple of hundred feet overhead. Ultralight operators with no formal pilot training interfered with the traffic patterns of nearby airports. Experienced pilots attempting to transition were sometimes (fatally) confused by unusual handling characteristics and oddball control systems (such as using foot pedals to control roll).

These problems and the lack of solid regulations soon led the FAA to establish regulations governing ultralights. FAR Part 103 (soon to be renamed 14CFR Part 103) established the regulations governing what the FAA dubbed, "Ultralight vehicles." They are not considered "aircraft" by the FAA; hence most of the FARs don't apply to them.

The ultralight boom went bust around 1985—partly owing to a saturated market, partly owing to a surge in accidents involving a small subset of substandard aircraft and hot-dog pilots, and partly owing to an unflattering story on a TV news magazine that still rankles the ultralight industry 20 years later.

But you can't keep a good idea down. While most ultralight manufacturers went bankrupt, several of the better companies and designs later returned. Most of these companies feature quality engineering and produce ultralights with conventional handling.

Definition of an "ultralight vehicle"

To qualify as an "ultralight vehicle" under 14CFR Part 103, a powered vehicle must:

- Carry no more than one occupant
- Have an empty weight of less than 254 pounds (plus the weight of floats and safety devices such as ballistic parachutes)
- Have a fuel capacity of 5 gallons or less
- Not be capable of exceeding 55 knots at full throttle in level flight
- Have a power-off stall speed of 24 knots or less
- If unpowered, still carry only one person and weigh less than 155 pounds

These requirements are summarized in Fig. 3-12.

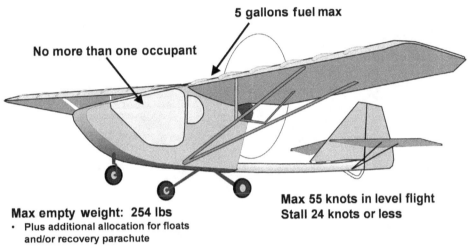

5 gallons fuel max

No more than one occupant

Max empty weight: 254 lbs
- **Plus additional allocation for floats and/or recovery parachute**

Max 55 knots in level flight
Stall 24 knots or less

Fig. 3-12. *Basic qualifications for an ultralight under FAR 103.*

When a vehicle qualifies under Part 103, the FAA does not get involved with either the pilots or the ultralights themselves. No license is needed to fly one; the planes need not be registered nor receive an airworthiness certificate. Anyone can build or maintain an ultralight, and the only inspections are those that the owner feel are necessary.

Of course, this degree of freedom isn't granted without some restrictions. In addition to the preceding weight, speed, and capacity limitations, the FAA does not allow ultralights to fly over congested airspace or within class A, B, C, D, or E airspace without prior permission. Ultralights are limited to day Visual Flight Rules (VFR) flight only. Ultralights must yield the right of way to all other aircraft. Many airports don't allow ultralight operations.

Previously, the FAA granted exemptions for two-seat ultralights to allow them to be used only for training purposes. These were eliminated with approval of the light sport aircraft/sport pilot regulations. New two-seat ultralight-type vehicles will have to be certified as light sport aircraft. Existing two-seat ultralights can receive an experimental light sport aircraft certification through early 2008.

The decision

Ultralights have the most stringent limitations of any personal aircraft. They're too slow for cross-country trips and don't carry enough fuel even if you're willing to take the time. You can't take your friends flying. If there isn't a dedicated ultralight airport in your vicinity, you'll probably have a hard time finding somewhere to fly from.

Owing to the weight limitations, these planes are fairly primitive (Fig. 3-13). Creature comforts are few, depending on how much you can afford to buy. Heaters are rare; a snowmobile suit is a must for year-round flying in much of the country.

Fig. 3-13. *Creature comforts are few in some basic ultralights, although most have some sort of partial cockpit enclosure.*

But the freedom! If all you want to do is *fly*, this is the way to go. The skeletal configurations give the pilot an unsurpassed view of the terrain sliding (slowly) below. No FAA hassles, no pilot's license needed, and no aircraft registrations or inspections. The planes may not be suited for flying cross-country trips, but they are eminently suited to being driven in a small trailer and taken for impromptu flights on a long road trip. And with no 51 percent rule to contend with, the kits are usually fairly complete. Many are available as ready-to-fly aircraft.

Used ultralights can be found at low prices, but there are a couple of reasons to be cautious. First, some early ultralights were developed by persons with no piloting background. Often these designers invented their own control methods, so you

Fig. 3-14. *While early ultralights sometimes had unusual control systems, modern ultralights usually handle like conventional aircraft.*

might have an airplane in which the control response is confusing to a "conventional" pilot. Most designs currently being manufactured, such as the CGS Hawk and the Sky Raider (Fig. 3-14), feature normal handling, but be a bit leery of "bargains" on old ultralights.

Second, ultralights are *not* toys. They are aircraft that require respect and appropriate training. Just because you earned your license on the local FBO's Piper Warrior doesn't mean that you can hop in an ultralight and fly it. Appropriate training is vital.

So there you have it—several alternate routes to airplane ownership. Depending on how much you're willing to spend, you can gain many of the freedoms of the experimental/amateur-built category without having to spend years in your shop.

Still with me? Well, that settles it. I'm no longer going to try to talk you out of building a kitplane. Let's look at how to determine what's the best kit for your needs.

4

Selection

HOW DOES THE AIR FORCE PICK OUT A NEW AIRCRAFT? First, the Air Force makes a careful analysis of the projected mission: What top speed is necessary? How fast should it cruise? How much payload? Maneuverability?

These requirements are then included in a request for proposals (RFP) released to the aerospace industry. Interested companies, such as Lockheed-Martin and Boeing, analyze the requirements, and each submits a design proposal. These proposals show how the company's design will meet or exceed the RFP requirements and how much the aircraft will cost.

The Air Force System Program Office (SPO) in charge of the project analyzes the submissions. The SPO critically examines each proposal in light of the mission requirements, the total cost, the development time before the aircraft is operational, and whether the manufacturer actually can perform to the level promised.

In the government world, this process is slow and expensive.

Things are simultaneously easier and harder for a private citizen choosing a kit-plane. While you don't have to satisfy a Pentagon functionary, family finances don't stretch as far as the military budget. And while kitplane manufacturers strive to please, it isn't cost-effective for them to make a design change to please a single customer.

But the basic process is sound. There are certain performance levels the aircraft must meet, your budget is not unlimited, and you'd like to be flying within a certain amount of time. And you don't want to send your money to a bankrupt or larcenous company.

This chapter will help you pick the plane that's right for you. Even if you've already picked out the kitplane you want to build, you'll learn how much it'll cost to build and other interesting details.

DETERMINING YOUR MISSION

It's easy enough to pick a kitplane—the "eenie, meenie, miney, moe" method, for example. Or the prospective builder can page through a favorite magazine until a design catches his eye. After all, an airplane is an airplane, right?

Too often a kitplane is picked for the wrong reason. A builder thinks that he needs a fast airplane and picks the fastest one advertised. It is discovered too late that the builder's private strip is too short for the hot little homebuilt. Or he finds that the designer made the plane go fast by minimizing the cockpit size, and it's too small for a king-sized builder.

The result is a half-completed project junked or a low-flight-time kitplane up for sale.

To avoid such problems, first determine your basic mission, and then derive the performance needs and other requirements.

The mission statement

There are as many "missions" as there are pilots. Some pilots just like to fly for the pure joy of it—sometimes called "cutting holes in the sky." These folks could get by with a small, single-seat, simple aircraft with good handling characteristics. The 5151 Mustang is an example. However, many pilots want to take spouses and friends flying, so two-seaters such as the Kitfox and the Hevle Classic (Fig. 4-1) are their choice.

Such basic airplanes aren't enough for people who want room for radios and baggage, higher cruise speeds, and longer range. Some pilots still need the capability to operate from short grass fields, but others intend to use their kitplanes for serious travel between large cities. The RV series, the Glasair, and the Lancair IV-P are aimed at these folks.

Fig. 4-1. *Many of those interested in the single-seat Fly Baby chose another aircraft because they wanted to carry a passenger. The Hevle Classic hopes to tap this unfulfilled desire.*

Fig. 4-2. *The popular Skystar Kitfox also can operate as a waterbird with a set of floats.*

Other needs exist. While many kitplanes are capable of light aerobatics, a few builders need competition-level performance. For them, a One Design would work. Or some pilots want to splash around a local lake in a homebuilt amphibian (Fig. 4-2).

But is the mission really right for you? Will you outgrow a simple little "putt-putt"? Are you truly interested in aerobatics? Do some serious thinking about how your needs might grow. It might take five years to build your airplane; it would be a pity if you outgrew it before it flies.

Sometimes it's tough to determine a single mission. For example, you'd like a fun little knock-around aircraft but still require fast cross-country travel. Keep in mind that good cross-country aircraft can be rented at practically any FBO. Buy the little Zenith just for fun, and count on renting a Piper Arrow or equivalent for heavy-duty traveling.

To start, come up with a single sentence expressing your basic mission: "Carry one passenger on long cross-countries between major airports." "Solo competition aerobatics." "Nostalgic-looking biplane for local hops." "Fast-building kit for moderate cross-countries with light aerobatic capability." "Operate from short private airstrip, with high-speed travel to nearby destinations." "Simple to fly for kids' lessons."

Determining your mission might seem trivial, yet it is an important step in figuring out which homebuilt fits your requirements for the least money.

A few notes on irrationality

"Mission?" some of you are wondering. "I just want to own a Finkerbean Special." In other words, you're not making a logical, rational decision based on careful analysis of needs and features.

There's nothing wrong with that!

Few of us *have* to own an airplane. The kitplane is being bought with your discretionary income and isn't expected to "pay its way." Some people might buy a sports car; others, a boat. We just prefer airplanes.

However, there are things you expect to do with your Finkerbean Special. Maybe your fantasy is flitting to that little mountain strip near your favorite fishing hole. Or packing the wife and kids aboard and winging your way to Grandma's at 250 knots.

Before you spring the bucks, shouldn't you at least make sure that the plane matches your needs?

One of the rites of adulthood is the buying of the first car. Considerable effort goes toward finding *exactly* the right set of "wheels" to fit one's transpubescent self-image. When found, it is bought, whether Mom and Dad approve or not.

Reality soon intrudes. The engine leaks oil like a grounded tanker. The chassis is sprung, the shocks are shot, and the electrical system shorts out. Have you ever seen a shiny older car at the side of the road with a stricken teenager tenderly stroking the steering wheel waiting for Dad and a tow rope? Too often that first car is one's introduction to the need for rational decisions instead of emotional ones.

It's bad enough with a $500 car. Do you want to go through the same process on a $35,000 kitplane? With so much money and thousands of hours of construction time at stake, shouldn't you make sure that the plane you get is the one you need and not just the one you want?

But hey, it's your money and your decision. If your analysis finds that Brand X better meets your needs than the Finkerbean Special, but you still prefer the Finkerbean, go for it!

PERFORMANCE AND FEATURES

Let's break the simple mission statement into hard requirements. In addition, let's discuss how your personal preferences might affect your final selection.

Performance requirements

First, let's look at how much runway you have. If your mission calls for operation from "standard" airports only, there's no real takeoff/landing requirement. By standard, I mean a 3,000- to 4,000-foot runway, a fairly clear approach, and the like.

But if you plan to operate from smaller strips, then the takeoff/landing distance requirement is set by the shortest runway you anticipate flying from. To account for variations in technique and density altitude, subtract 10 to 20 percent from this distance to set the required runway length. Check the runway dimensions of likely destinations, too.

While stall speed usually has some relationship to handling, it's generally directly proportional to the kitplane's short-field capability. Therefore, there's little need to set a requirement for maximum stall speed. Be advised, though, that if you're looking for a fast airplane, the stall speed is going to be high.

Once the aircraft has taken off, rate of climb becomes paramount. However, it's hard to set a required value—*climb gradient* is more important than rate but is rarely stated. If a 500-foot hill is located one mile from the end of the runway, a 60-mph plane with a 500-fpm climb rate will clear the hill, whereas a 180-mph plane with a 1,200-fpm rate won't. Set a rate of climb minimum if it means that much to you.

Four interrelated parameters are cruise speed, range, fuel tankage, and fuel consumption. Refueling operations severely affect total time en route; therefore, set the endurance requirement (usable fuel divided by fuel consumption at normal cruise power) first. Some kits have only two or three hours of onboard fuel, which is fine for local pleasure flights, but serious cross-country travel requires more. Again, set the minimum value based on your mission. Be realistic. How long do you honestly want to fly without a rest break? A typical value might be three hours; add another 30 or 45 minutes for reserves, and round it up to four hours.

Now state the desired cruise speed. Note that *cruise speed* multiplied by *endurance* equals *range,* so no separate requirement for range is necessary. Local sport flying, of course, doesn't need a cruise-speed requirement.

Compare your required performance against the advertised performance of available kits. Eliminate the kits whose performance is completely out of the ballpark.

Believing the numbers

Casting a jaundiced eye toward the performance figures supplied by the kitplane manufacturers might be wise. How well can the figures be believed?

Few manufacturers actually lie, if the word "lie" is defined: "Published performance values known to be untrue." But exaggerating the qualities of homebuilt aircraft is almost as old as aviation itself, as Fig. 4-3 demonstrates. Let's take a look at one performance factor, cruise speed.

This seems pretty easy: Take off, set cruise power, and read the gauge.

An airspeed indicator works by measuring the pressure differential between ram air and a source of neutral pressure. Most pilots consider the pitot tube the primary source of airspeed data, but the static vent is just as important. The makers of factory airplanes go through considerable effort to find the most accurate position for the static vent.

The easiest way to install a static system is to not install one at all—in other words, just vent the airspeed indicator to the cockpit. This approach appeals to most homebuilders and many kit manufacturers. However, the external airstream

AT LAST!

An Aeroplane for Everybody!

FLIES WITH MOTORCYCLE ENGINE

Think of flying with an ordinary twin cylinder motorcycle engine! This is the only aeroplane that will do it. It is the smallest and most efficient of all aircraft. No longer is flying the sport of acrobats and millionaires. Every man and boy in the world can build one of these remarkable aeroplanes with ordinary tools in a few weeks and learn to fly at home with safety. No shop is needed. If you can use a hammer, saw and a pair of pliers and have a shed, barn, a basement or a back yard you can build one of these remarkable flyers for a few dollars and in spare time if necessary. Costs less than 1/50 the cost of the average aeroplane and can be built for less than a fifth the cost of going to a flying school. It is the smallest, simplest, safest and most successful aeroplane in the world. The wonderful

WHITE MONOPLANE

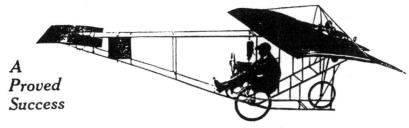

*A
Proved
Success*

YOU CAN BUILD IT

Remember, this is not a toy or an experiment, but a thoroughly perfected man-carrying aeroplane with 18 foot spread and a speed of 30—60 m. p. h. Lifts 190-pound man with twin cylinder motorcycle engine. Hundreds are already in use in the United States, Canada, Mexico, South America, England, Australia and the Philippine Islands. Hundreds of amateurs are building and flying these aeroplanes, men and boys who knew absolutely nothing about aeroplanes and flying before. It is the simplest and safest flyer in the world. Any make of motorcycle engine can be used.

WHAT OTHERS SAY:

POPULAR MECHANICS says:
 "Undoubtedly the smallest successful monoplane in the world."

AERIAL AGE says:
 "One of the most interesting machines yet to be developed."

SCIENTIFIC AMERICAN says:
 "A unique type of monoplane."

L. A. EVENING HERALD says:
 "Has solved the problem of producing small machines at a cheap price for universal service."

Working Drawings $2

Send $2 at once for a complete set of working drawings of this wonderful monoplane showing all details and dimensions in a simple manner so you can easily understand everything. Here is your chance to get into the greatest of all industries. Thousands of experienced flyers and builders are needed.

Don't Miss This Chance! Send $2 Now!

GEORGE D. WHITE
DESIGNER AND OWNER OF SOLE RIGHTS

3832 South Main Street, LOS ANGELES, CALIF.

Fig. 4-3. *"An Aeroplane for Everybody"? Hype and exaggeration for homebuilt sales isn't new. This ad dates from 1917.*

often creates a slight vacuum inside the cockpit. The lowered static pressure will cause the airspeed indicator to read higher than actual. The result is a "faster" kitplane, at least on paper.

And why should the manufacturer disbelieve it?

Next, take a look at that cruise power setting. How accurate is the tachometer? If it indicates 100 rpm low, the engine is running faster than indicated, and the cruise speed inches higher.

Finally, the airplane used to determine the performance figures is probably optimized. It wasn't built by an amateur builder in a garage shop. It may mount a tuned and blueprinted engine. Gaps are nonexistent, surfaces are mirror-smooth, and the weight is shaved. There is no reason to carry radios or extra instruments on performance test runs, either.

The average owner-built kitplane can't match it. Once the first blush excitement of the first flight wears off, most kitbuilders have two regrets: heaviness and slowness.

How can the prospective buyer tell if the performance figures are realistic? The only sure way to find out is to fly formation with another aircraft and compare instrument readings. One aircraft I flew beside indicated 10 mph faster than mine did. But you're not likely to have the chance to do so until it's too late.

The lesson is to take published performance figures with a grain of salt. We are stuck with the published figures; rarely does any other source provide such data. But don't select one plane over another merely because the manufacturer claims slightly higher performance.

Wing choices

A couple of questions about the aircraft's configuration should be settled at some point, too.

Do you prefer a high-wing or a low-wing aircraft? Generally, a low-winger gives better flight visibility for maneuvering in crowded airspace. A high-wing aircraft is better for sightseeing. Biplanes combine the disadvantages of both; you can't see what you're turning into or away from unless the wings are mounted aft of the cockpit.

What about storage and transport? If you expect to keep the plane at home and trailer it to the airport, folding wings or at least quickly removable wings will be necessary.

However, ask around—you'll find that very few people actually keep their plane at home. For one thing, it adds a lot of hassle to the flying. The owner can't just drive out to the airport and fly; she's got to go home, hook up the airplane, tow it, unhook it, unfold the wings, check them carefully, and only then she can fly. After landing, the entire process is repeated in reverse. Do you want to go through this every time the skies call?

A more common use of folding wings is to store more than one aircraft in a hangar. This brings considerable savings, although a little bit of time is lost folding and unfolding.

Even if you don't use it often, though, it's a useful feature. Some areas have unflyable seasons; the aircraft easily can be brought home for several months. In these cases, removable wings are almost as handy as folders. When possible, pick a kitplane with the landing gear attached to the fuselage or a stub wing section. That way, the aircraft still can be rolled about with the wings removed.

Passengers?

How many seats? A simple little local cruiser just needs a single seat. But then, you'd like to give your friends a ride in your pride and joy. Besides, single-seat homebuilts are hard to sell when you wish to move on to something else (Fig. 4-4).

Fig. 4-4. *Single-seat homebuilts are fun to fly and own, but they are slow movers when it comes to sell.*

A multiseat kit is more complex than a single-seater. You'll install dual controls, extra seats and support structure, and a pitch trim system. Because the plane is heavier, the structure is more complex. Larger, more costly engines are needed. It all adds up to more building time and greater expense. But you still build only one airframe and install one engine, no matter how many seats.

All in all, it's better to buy a two-seater than a single-seater, just for the increased resale value, if nothing else.

The seat configuration plays a significant role in performance. A tandem-seating aircraft (such as a Cub or Champ) puts the pilot on the aircraft centerline equally between the left and right windows. It results in good visibility to both sides. Because the aircraft can be narrower (only one person wide), tandem-seaters theoretically are faster. The person in back, though, usually can't see forward worth a darn. Spouses, especially spouses with pilot licenses, usually dislike that.

Side-by-side seated aircraft are more civilized. Both occupants can see the instruments and have good visibility forward. Because aircraft engines are as wide as two people anyway, the performance cost is often negligible. The Van's RV-4 and RV-6 are good examples. With the same wing and engine, the side-by-side RV-6 and tandem-seat RV-4 cruise at about the same speed.

Side-by-side seated planes have more panel space, too. If Instrument Flight Rules (IFR) capability is part of your mission statement, make sure that candidate kitplanes have enough panel space for the necessary instruments. Is it a stable instrument platform? Read the reviews of each kit, and take a thorough test flight.

One of the biggest changes in the kitplane market in the past 20 years has been the vast increase in the number of four-seat (and larger) homebuilts. The Velocity (Fig. 4-5), the Express, the Zenith CH801, the RV-10, and the Lancair ES and IV all illustrate this trend. Just like stepping up from a solo job to a two-seater, a four-seat plane takes longer to build, probably needs a larger engine, and is more expensive. But at least you won't have to install controls for the rear seats.

Fig. 4-5. *The Velocity was one of the first four-seat composite kits. The company now offers several variants of the basic design.* Credit: Velocity Aircraft

Landing gear

One last configuration consideration is the landing gear. Many kits are offered with the choice of *conventional (taildragger)* or *tricycle* landing gear (Fig. 4-6). Conventional gear is more rugged and better suited to rough and short fields. It's easier to build—tailwheels simply can be attached to a leaf spring, bolted to the tail post, and connected

Fig. 4-6. *Many kit companies offer the choice between conventional ("Taildragger") and tricycle gear. All of Van's two-seaters offer the option.* Credit: Ed Hicks/Van's Aircraft, Inc.

to the rudder system via cables and springs. Nosewheel support structure is far more complex. Again, conventional-geared aircraft theoretically are faster than trigeared types because tailwheels cause less drag than nosewheels.

There is a third type of landing gear: the unicycle type used by the Europa. This is really a taildragger configuration with a single large center-mounted main gear and small outriggers near the wingtips, like the U-2 reconnaissance aircraft. If you're worried about the configuration, make sure to get a test flight before you decide. In the Europa's case, a tricycle-gear version is also available.

Why are most new homebuilts tricycle-geared? Market pressure—most pilots are more comfortable with trigear. There's no question that trigears are easier to operate and safer. A taildragger checkout is sometimes difficult to arrange, too. The theoretical performance disadvantage is minor to nonexistent. Looking at the RV-7 (taildragger) versus the RV-7A (trigear), the nosewheel costs only 2 mph in cruise and 50 fpm in climb rate.

The cumulative effect is interesting. In going from the RV-4 tandem-seating taildragger to the RV-7A side-by-side seated tricycle gear, 4 mph in cruise and 300 fpm of climb are lost—all on the same wing and engine. The RV-7A is 300 pounds heavier but has the same useful load and has almost 200 miles more range.

Of course, to really gain some speed, you'll need to retract the gear, right?

Poppycock. Several homebuilts are made with both fixed and retractable gear. Take the Glasair SII FT with fixed gear versus its retractable stablemate. The straight-leg version is only 11 mph slower at cruise and has 75 pounds more useful load. And will *never* land gear-up. Your insurance agent will love it.

Retractable gear adds weight and needless complexity. Only the very fastest planes really need it.

For planes with cruise speeds of less than 200 mph, it's generally a cosmetic issue: The planes look better with the gear gone.

It's one of those emotional issues, so I won't dwell much longer on it. Consider this, though: Fixed or retractable, you can't see the gear from the cockpit. Add a non-functional gear switch to your straight-leg bird, and use a little imagination (Fig. 4-7).

Fig. 4-7. *The biggest improvement when retractable gear is added to most homebuilt is in cosmetics, not performance. With the help of a photo editor, the author's Fly Baby retracts its wheels.*

Performance and features summary

At this point, you should know the approximate performance figures for your dream kitplane, as well as the preferred configuration and features. Use these parameters to narrow down the list of potential candidates.

Meeting a requirement isn't necessarily a pass-fail situation. Chances are that no single kit meets all your criteria, unless you have your heart set on one and have biased the numbers in its favor. Shame, shame. In any case, the last section of this chapter includes a method to help decide close calls.

Other factors are involved in the building decision, though. It's time for the tawdry issue of money in this conversation.

CASE STUDY: FINDING THE RIGHT MATCH

When potential builders think about what they want to build, their minds often go on flights of fancy. Sleek fiberglass rockets streaking along at flight-level altitudes behind a shrieking turbocharged recip or the howl of a turboprop.

When they take the time to figure out what they *really* need—well, a lot of folks turn to planes like the Murphy Rebel. This Canadian kitplane is almost the antithesis of the whole homebuilt image: It's not fast, it's not sleek, and it's not fancy.

Why is it so successful? Probably because it offers two features many builders secretly want but won't often publicly admit: utility and a conventional design.

"I wanted the type of airplane to match the type of flying I like to do," says Chuck Bailey (Fig. 4-8) of his decision to go with the Murphy product. "The Rebel is perfect for a Sunday flyer."

Bailey completed his Franklin-powered Rebel (Fig. 4-9) after almost eight years of construction and now has over 200 hours on the aircraft.

Fig. 4-8. *Chuck Bailey took almost eight years to build his all-metal Murphy Rebel.*

Fig. 4-9. *Bailey's Rebel is powered by a Franklin engine license-built in Poland.* Credit: Charles Bailey.

Selection and delivery

Since he planned to leave the plane tied down outside, he preferred to avoid a fabric-covered aircraft. Manufacturer support also was important. Says Bailey, "I wanted a dealer I could get to." It would give him an opportunity to examine the aircraft closely and get a feel for the company.

Bailey's home is a three-hour drive from the Murphy factory located in Sardis, British Columbia. As it turns out, he *also* lives about the same distance from the Van's Aircraft factory in Oregon.

Thus the final decision came down to the Rebel or the RV-6. He made the final choice after an honest look at his past experience building boats and model aircraft. "I looked and found where I generally make mistakes," he says. "Measurements." The Murphy kit requires the builder make fewer measurements because many of the pilot holes come predrilled. The RV-6, then Van's only side-by-side seating aircraft, did not (the RV-7 and RV-9 now include pilot-drilled holes).

Bailey ordered the full kit from Murphy and drove to the company's British Columbia facility from his home near Seattle. The entire kit came in a crate about 12 feet long, 4 feet wide, and 2 feet high. The 700-pound box rode nicely in the back of his pickup truck, though it stuck out the back a bit.

Building experience

Bailey's workshop was typical of that available to most homebuilders—a two-car garage attached to his house. The walls were insulated and sheet-rocked, and a natural-gas furnace provided heat. He didn't take these actions specifically because of his homebuilding project, though. The area had always been his shop. As usual when an airplane is constructed in a garage, Bailey's airplane project shared space with household items such as freezers and lawn mowers.

The garage was 21 feet long from the doors to the back wall. Unfortunately, the Rebel fuselage is *21 feet, 4 inches long* once the engine and spinner is installed. Without the engine, the fuselage fit snugly in the garage (Fig. 4-10). Bailey made his first

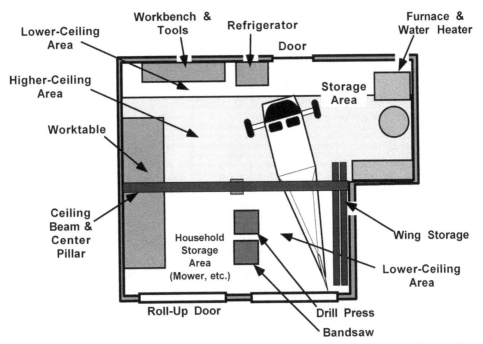

Lower-Ceiling Area

Higher-Ceiling Area

Worktable

Ceiling Beam & Center Pillar

Workbench & Tools

Refrigerator

Door

Furnace & Water Heater

Storage Area

Wing Storage

Lower-Ceiling Area

Household Storage Area (Mower, etc.)

Roll-Up Door

Drill Press

Bandsaw

Fig. 4-10. *Chuck Bailey's garage is a typical homebuilt workshop layout. It was too short to allow the engine to be installed in the garage, so the trial installation was done in the yard.*

engine test-fit outdoors and transported the aircraft to a hangar at a local private field for final assembly.

While the Rebel's 6061T-6 aluminum has excellent corrosion-resistance capabilities, Bailey elected to corrosion-proof the entire airplane by acid-etching, anodization, and priming each part prior to assembly. "Now I would just clean the joint, prime, and rivet, just like Murphy calls out," he says. "The way I did it may have added a year or two to the construction time."

Bailey praised the support provided by the Murphy company. "They're very good people to work with."

As far as tools go, Bailey says, "I wouldn't build any airplane without a bandsaw." For Murphy builders, the tool he recommends most highly is a pneumatic rivet puller. The Rebel kit includes 20,000 Avex rivets, which are of the pulled ("pop") variety rather than the driven type. When riveting two panels together, he was able to insert all the rivets and go down the line quickly.

Like many builders, Bailey finds the small Dremel Moto-Tool a great aid. He also recommends using a small razor saw to cut angles and channels of aluminum. Aviation snips tend to flatten out the channel when going around the corners, so he cuts the corners first with the razor saw.

First flight

Bailey's first flight came about seven and three-quarters years after picking up the kit. Construction had been delayed during a two-and-a-half-year period where he had to commute two hours each way to work. "I was out of the house for over 12 hours per day."

While his first engine run was at a nearby private airport, his flight advisor convinced him to take it to another field about 50 miles away. The runway at the nearby field was just 3,000 feet long and completely surrounded by trees. Bailey removed the wings and trailered the plane to the other field, which has several runways and plenty of emergency landing fields.

The move was a stroke of luck, really. "There I meet Steve Sloan, who has a Rebel, and he took me under his wing." Sloan gave Bailey some checkout time and rode along during the taxi tests of Bailey's plane. " Best thing that could have happened," Bailey told me. "If you put this in the book, say, 'Thanks Steve.'"

Finally, the day came. A friend at the airport called at 6 a.m. and told him that it would be a good day for his first flight. He flew Sloan's Rebel around the pattern a couple of times and then climbed into his own plane. Sloan and another friend flew chase. "My wife thought it to be just another flight, until she spotted the axe and large medical pack in the back of the truck. Then she got a little concerned!"

The takeoff was normal, and the chase plane reported that all was well. Airspeeds between the two Rebels were nearly identical. Engine instruments were okay, and slow flight, turns, and climbs were dead normal. "I set up for a normal landing on 34 . . . one of the best landing I have done," says Bailey.

"This first flight turned out to be just another 20-minute flight. Guess that is the way it should be."

Since the first flight, he has had to replace a valve guide at 35 hours owing to excessive oil leakage. Also, he had a blade crack on his noncertified adjustable-pitch propeller about a hundred hours later. He replaced the prop with a standard Sensenich fixed-pitch metal unit.

Advice

"Don't work when you're tired," is Chuck Bailey's common-sense advice. "Measure once, cut twice still bites me."

Bailey's final advice to new builders: "Patience. Just keep at it, It's worth it."

COST OF CONSTRUCTION

That's *cost of construction,* not *cost of kit.* There's a considerable difference. To understand the construction costs involved, let's first look at what's included in the kit.

Types of kits

Kits can be divided into four types: materials kits, subkits, complete kits, and quick-build kits.

The materials kit is the cheapest. It contains the raw materials necessary for a plans-built aircraft. No work is performed by the kit supplier; in fact, the supplier often has no connection with the aircraft designer. The kit provides aluminum sheets, plywood, metal extrusions, and other materials that require considerable

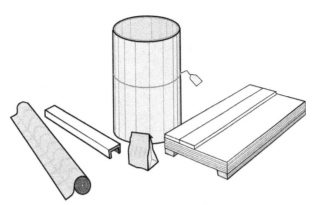

Fig. 4-11. *A materials kit could consist of some rolled-up aluminum sheets, plywood, rolls of fiberglass, extrusions, and sacks of small hardware. The only advantage is the ability to order a single kit instead of listing pages of parts numbers and quantities.*

work to become aircraft parts (Fig. 4-11). Materials kits don't exist for molded composite airplanes.

If the materials kit doesn't do any work for the builder, what's the advantage? Basically, it's a painless way of obtaining the necessary hardware. The aircraft plans probably include a total list of materials, but rather than filling out an order form with countless lines, you can just order a single part number. There's always a price break for buying material in quantity, and a materials kit usually qualifies.

One has to watch out, though. Check the price list of materials, and then compare with local prices. I know of at least one supplier whose aluminum prices are well above those I pay local distributors—and I pocket the shipping charges.

Obviously, pilots looking for a way to build an aircraft quickly should pass up materials kits. While cowlings and other fiberglass parts are often available off the shelf, it's certainly not what most people consider a "kit."

The next step upward is those designs that offer subkits to hasten construction. Usually, the aircraft is one that can be build directly from plans as well. This is the most flexible method—the builder can buy whatever subkits the budget allows and build the rest of the airplane from scratch (Fig. 4-12). Or a builder unsure of his skills can buy subkits for critical or complex components. For example, a predrilled wing spar reduces the chance of error and helps to ensure full strength of the final product.

Another advantage of the plans-and-subkits method is that the builder can still finish the aircraft if the kit manufacturer goes out of business. If the manufacturer of a molded-composite kitplane goes under and hasn't shipped a major component such as the fuselage or wing, you're out of luck.

Some designs offer subkits only for those items that need welding or other special work. Other companies sell the entire aircraft in subkit form. Some molded composite designs can be bought as subkits, but all the subkits are required, and the builder can't finish by plans alone. There's generally a small savings in buying all the subkits at once, in the form of a lower crating charge.

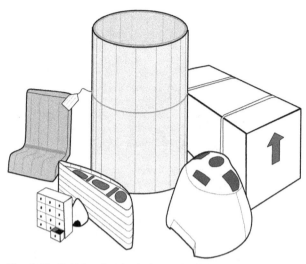

Fig. 4-12. *Subkits often include much of the same stock items as the materials kit, but completed parts such as ribs and cowlings are often available. If the budget permits, the builder can greatly reduce his building time by buying additional subkits.*

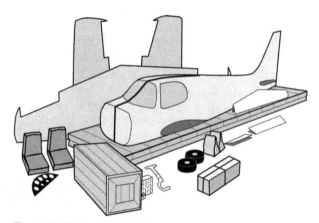

Fig. 4-13. *The complete kit is everyone's dream, but they are the most expensive of the lot. Complete kits for metal airplanes often consist of merely all the subkits shipped at once.*

However, not all subkit aircraft can be built from plans. Subkits are a popular method for spreading the cost of the kit out over the period of construction.

Of course, when the entire aircraft can be ordered as a single item, it becomes the third category—a complete kit (Fig. 4-13). Few are truly complete; even the best kits don't include paint or batteries.

The final evolution of the complete kit is the quick-build kit. These are generally complete kits that are partially built by the manufacturer. For instance, for RV kits, the wings and fuselage are available with nearly all basic assembly already accomplished.

The fundamental question arises, though: What about the 51 percent rule?

As mentioned in Chap. 2, the Federal Aviation Administration (FAA) has transitioned to a process where the builder is expected to perform at least 51 percent of the tasks, not the specific labor elements. The quick-build kit makers work with the FAA to ensure that their kits receive approval under the 51 percent rule. If in doubt about a particular kit, consult the "Revised Listing of Amateur-Built Aircraft Kits," published online by the FAA. Search for "Amateur-built aircraft kits" using your favorite Internet search engine. Just because a kit isn't listed doesn't mean that it fails the 51 percent rule, but if it's on the list, your local FAA office won't make it an issue.

Reducing the kit cost

The cost of a kit depends on two factors: the cost of materials included in the kit and the amount of work already done. Materials kits are cheap because the kit "assembler" (which can hardly be called a "manufacturer") performs no work other than preparing parts for shipment.

Composite kits are the most expensive, but it's not because fiberglass is more expensive than aluminum or wood. Producing the molded components is a labor-intensive affair that is reflected in the kit price. Numerically controlled production machinery has reduced the costs of aluminum kits significantly because the material can be cut to size, bent, and predrilled with little human intervention.

One nice feature of the subkit method is the ability to optimize construction time based on budget. The plans for the Murphy Renegade tube-and-fabric biplane, for instance, show how to build the fuselage from aluminum tubing and the company's custom extrusions. However, if the builder's budget allows, she can save a considerable amount of time by buying a prebuilt fuselage directly from the factory (Fig. 4-14).

This has always been a big advantage of homebuilding—the builder can save money by doing more work herself. Complete kits are great but expensive. If money is tight, pick a design that can be built solely from plans, and buy only the most necessary subkits.

Figuring out the cost of the kit is easy; knowing how much you'll spend before the aircraft flies is another thing.

What the kit doesn't include

The more goodies the kit provides, the more the kit costs. This is perfectly obvious, isn't it? The problem is, How do you tell what the kit provides?

The most complete kits include descriptions like, "Everything but the paint and battery." That's not bad—it'll still cost a pretty penny more to build the plane, but at least you aren't stuck with any big-ticket items.

The biggest of the big-ticket items, of course, is the engine. A few years back, most smaller homebuilts used air-cooled two-stroke engines such as the Rotax 503; these often were included in the kit. Times have changed to some extent. Many of these airplanes have standardized on the larger, liquid-cooled Rotaxes such as the 582 or even the four-stroke 912. These engines are quite a bit more expensive; the kit price

Fig. 4-14. *A factory-built fuselage saves a lot of construction time. The factory probably can build them lighter as well. But budget builders can scratch-build for far less.*

may not include them. Few planes using conventional aircraft engines include the powerplant, and the additional cost is frightening

Engine prices can range from a few hundred dollars for a used two-stroke to well over *$60,000* for the larger certified engines. The kit for the Van's Aircraft RV-7 sells for about $17,000 and includes "just about everything needed except the engine, prop, instruments, and tires."

A new Lycoming O-320 engine will cost about $25,000 (and a lot more if bought directly from the factory). But a remanufactured, rebuilt, or used engine will reduce the bite dramatically. If you insist on a new engine, several kit makers have arranged special discounts with Lycoming and Continental. Chapter 3 discusses engine options.

Let's look at what additional items might cost. Propeller? A Rotax prop costs $250; thousands for a constant-speed model. Instruments? At least $1,000 for basic Visual Flight Rules (VFR) if you're buying new. Avionics? No kit includes radios or a transponder. Simple VFR birds can get by with just a hand-held transceiver, but even these sell for more than $300. By the time you add a nav receiver, global positioning system (GPS), and a transponder, avionics cost can pass $3,500.

Most kits don't include these components and a great many others besides: Brakes? Paint? Upholstery? Epoxy and brushes? Crating and shipping? Find out what more you'll need before you order. The information package offered by the manufacturer should list what isn't included. If it doesn't, ask. If the manufacturer's answer is broad, such as "the fuel system" or "the interior," building will take considerably more money and time.

Ground money

Everything in the preceding section is "air money" for items that will fly. However, don't forget "ground money," which is the money paid for items necessary to build the airplane.

How well equipped is your workshop? The tools you already own probably aren't enough. Most builders agree that a bandsaw is a necessity, as is a drill press. Other tool costs depend on the kit materials. For instance, if you are building a metal-monocoque airplane, you'll be doing lots of riveting. Do you own a compressor? That'll probably run between $300 and $600. A rivet gun will cost another $250 at least; add bucking bars, air hoses, rivet cutters, and so on. Again, sometimes these items can be bought used for considerable savings. Table saws, belt sanders, and grinders—they all add up. Many kits require a solid and level workbench at least 10 feet long, which'll probably cost at least another $100.

Most kits require the same tools. Some of the smaller kits "can be built with simple hand tools," but you'd be better off getting a drill press at least. The construction material greatly affects the tools required; wood seems to take the most, primarily in the saw category, of course. Metal is next, and composite kits seem to get by requiring the fewest tools. If you're building a composite kit in the cold country, adding heat to a garage might be a one-time and continual expense.

One advantage of the subkit system is the ability to bypass operations that require specialized tools. Bending a Wittman-type spring-steel landing gear is beyond the capability of the typical home shop, so the availability of a prebent gear is a definite plus. By purchase of the proper subkit, you might be able to save money on tools as well as decrease overall construction time.

Some tool companies off "builder's tool kits" for popular homebuilt airplanes. If you're building an RV-10, for instance, you could buy all the necessary tools in one package. While it makes the starting process a lot easier, it does tend to highlight how much the construction tools will cost! They can range from $500 to over $2,000 and usually don't include the common shop tools such as benchtop sanders and bandsaws.

One last item: Don't forget to keep your flying skills current during the building process. Too often an aircraft's first flight is the builder's first flight of the past two years. Currency is a must; allocate a certain amount of money each month toward flight time. Or drop flying altogether, and plan on going through a thorough refresher just prior to the first flight. Count on needing at least 10 hours in the preceding month if you go this route. Consider refresher training in an airplane similar to the one you are building: conventional gear or high performance.

Adding it up

It would be nice if kits included everything from engine to screwdrivers, right down to a warm, comfortable, completely equipped workshop. But they don't. Where most prospective kitplane buyers have only a general idea of the performance and features desired, all seem to know the limits of their budget. It's no fun to stretch your budget to buy the kit and then find out that you'll need another $1,000 just to start building.

Van's Aircraft (*www.vansaircraft.com*) has an online calculator on its Web page to help determine the total cost of construction of an RV. Table 4-1 presents some of the costs the company predicts for various elements. Chapter 5 guides you through the selection process for the engine, instruments, and avionics, and Chapter 6 covers tools. Determine the total cost for your configuration from the information in these chapters.

Table 4-1. Cost of typical options for an RV-7

RV-7A kit

Standard	$17,840	
Quick-build	$25,865	

Engine (O-320) and accessories

Used (run-out)	$10,000	Includes $3,000 for engine accessories
Used (midtime)	$15,000	Includes $3,000 for engine accessories
Used (low-time)	$21,000	Includes $3,000 for engine accessories
New (OEM)	$25,500	Includes $3,000 for engine accessories

Propeller

Wood, fixed-pitch	$500–$1,500	
Metal, fixed-patch	$2,000	
Constant speed	$7,000	Includes accessories

Engine instruments

Analog	$1,000
Electronic	$4,000

Flight Instruments

VFR	$1,100
IFR	$2,200

Avionics

VFR	$1,000–$2,500
IFR	$4,000–$12,000

Other

Basic interior	$1,000	
Electrical	$2,000–$3,000	Including aircraft lights
Painting (professional)	$2,000	
Painting (builder)	$6,000	

Plus

$300–$600+	for shipping
$550–$2,000	for tools, workbench, etc.

Source: From Van's Aircraft's Web page.

Spending more money—voluntarily

As I've gotten older, I sometimes look at money a different way: It exists to make my life easier.

Sure, I can park for free in the far recesses of the mall parking lot and walk a half mile through the driving rain. Or I can shell out a couple of dollars and use the valet parking right at the entrance.

Twenty-five years ago, I would have walked. Nowadays, well, let's just say my shoe leather is lasting longer.

Like all airplanes, homebuilt aircraft are compromises—even in the financial department. For instance, take a typical kitplane that can be built from plans for about $20,000, including a used engine. If a kit is available, you'd take less time to build, but your total cost might rise to $30,000. Using a quick-build kit might add another $10,000 to that, and a completed example (built by someone else) might cost $60,000 total.

But one doesn't have to be limited by those "even" steps. Many people "farm out" the instrument panel or hire a professional to paint the aircraft. These kinds of activities are permitted by the FAA and do not affect the 51 percent evaluation. They'll cost you more money, but they'll save a lot of time. FAA Advisory Circular 20-139, "Commercial Assistance during Construction of Amateur-Built Aircraft," covers this issue. It is available online. Just enter "FAA AC 20-139" on your favorite Internet search site.

There are also businesses that will help you build your aircraft. Some are operated in conjunction with the kit makers; some are independent operations. The kit maker programs generally lead you through most of the critical structural operations and allow you to finish the plane at home. This gives you access to specialized tools and jigs.

The independent operations may operate the same way, or they may merely provide shop space, tools, and on-site expertise to help you build your plane.

Care must be taken whenever you have outside help with your project: *You* have to perform 51 percent of the tasks. If the FAA thinks that you hired someone to *do* the work, the aircraft may have trouble getting certified. Several builder-assistance centers have come under scrutiny.

The best bet would be to discuss the situation with your local FAA office prior to engaging outside assistance with the routine construction of your plane.

CONSTRUCTION TIME

Which naturally leads us into another prime topic: What is the total time required to build your airplane?

Kit type

Obviously, materials kits take the longest time to build, and quick-build kits, the least. The degree of completeness has a serious impact. Unless you order the non-included part in advance, you might not notice its absence until the part is needed.

If there isn't another task ready, work stops until the shipment arrives. Is the engine included? Because of the high cost of aircraft engines, builders often wait until a "good deal" is found. That "good deal" might delay completion for months.

Does the kit include the engine mount? Getting a mount made might take weeks. Many kits don't include anything forward of the firewall. Kits that include the engine take far less time to build because all the nitty-gritty of powerplant installation (mount, fuel strainer, throttle linkages, and so on) usually is included as well.

Other construction time factors

Beyond the kit type, there are four main factors that affect construction time: type of construction, complexity of the aircraft, detailed work included, and the amount of thinking the builder is required to do.

For equivalent aircraft, composite kits take the least time, followed by metal kits and, finally, wooden kits. This is part of the cost trade; composite kits generally cost more because a more completed aircraft is supplied. A Lancair Legacy costs more than an RV-7, yet its primary structure assembles far faster.

However, the complexity of the aircraft has a significant impact. The RV-7 has a simple, rugged, fixed gear. The Lancair comes in both fixed-gear and retractable models. If a builder opts for the retractable version, the time he gains through composite construction might well be lost when it comes time for gear installation.

After all, one cannot expect to build a Glasair faster than a simple tube-and-fabric sport plane. Let's look at just one small element, cockpit entry. The sport plane might have only an open cockpit. But the fast glass bird's sliding window/door is necessary to allow the Glasair's high cruise speed. It'll take a number of hours to fabricate individual components, prepare the opening in the fuselage, assemble the structure, install, and adjust.

It's not a flaw of the Glasair design; it's just a reflection on the performance ranges. If your wish is for blistering cruise speeds, you'll pay for it with both a higher purchase cost and longer construction time. The Glasair has numerous systems; installing flaps, trim, retractable gear, and controllable-pitch prop will take additional time.

Of course, the kitplane manufacturer can reduce the effect of additional complexity by including many prefabricated components. If an assembled gear linkage can be supplied, the time impact of retractable gear is reduced. A Pulsar can fly rings around a Baby Ace, but because of its modern kit design, the Pulsar is finished far faster.

If you've got the money, go with the quick-build kits—they'll save a lot of time. Otherwise, trade building time for dollars.

The fourth factor is the amount of thinking the builder has to do. Imagine two kits that supply basic VFR instruments. Kit X includes a blank instrument panel. Builder X can lay out the panel however he chooses. Kit Y specifies the mounting location for each instrument.

Does kit X sound better? Yes, in some ways. After all, we're building a custom aircraft, and we'd prefer to configure things exactly the way we want. But the time builder X takes to figure out a panel layout is time builder Y uses to actually install instruments.

Kit Y eliminates measurement time, too. Kit X includes instructions like, "Cut a wire long enough to go from the instrument to the electrical bus." Kit Y says, "Cut a wire 23 inches long, and connect the instrument to the bus." For that matter, the maker of kit Y could sell a precut wire harness subkit and save even more building time.

One major function requiring builder decisions is engine installation. If the kit manufacturer specifies a particular engine, the kit design can be optimized, and detailed installation instructions can be included. Otherwise, or if a nonstandard engine is chosen, the builder is on his own.

Builders buy kits for one of two reasons: They are not qualified to design an airplane or they want to complete the airplane as fast as possible. If you're essentially paying someone to design an airplane that can be quickly built, you want detailed instructions. Every time you have to wonder, "Gee, how should I do this?" is time wasted.

This isn't to say that thinking is bad. Nor am I telling you to place your brain in park whenever you enter the shop. No kit matches the ideal; none is perfect. You must beware of inconsistencies and outright errors by questioning any step that doesn't seem right.

Fewer decisions means flying sooner.

The manufacturer's estimate

On the surface, this seems the easiest part of the process. After all, each kit manufacturer lists an estimated construction time. But if you thought verifying performance figures was complicated . . .

Cynical experience calls for doubling the published construction time. If you've never built an aircraft before, this is optimistic.

How does the manufacturer estimate construction time, anyway? In some cases, it's just a ballpark guess. Judging how long something will take to build is a science; few manufacturers have industrial engineers on the payroll. After all, building time is as much a marketing ploy as cruise speed. As mentioned in Chap. 1, the published construction time for the BD-5 was 300 hours. Those BD-5s completed by private builders generally required more than 2,000 hours.

Again, the concept of "average skills" comes into play. To pilots, the ability to land an aircraft without breaking anything is considered an average skill. Yet to the common citizen, such a skill is quite extraordinary. It's that way with kitplane building as well. Someone with 10 years of riveting experience can't understand how a newcomer might take a while to get the knack.

The manufacturer's time estimate doesn't include correcting mistakes. New builders will make errors; some errors might not be costly in terms of dollars, but all increase construction time—especially if the builder has to write the factory for a component to replace the ruined one.

Another item the factory estimate might not include is preparation, jig building, and cleanup time. You can't just glue two pieces of wood or composite structure together or just drill and rivet aluminum; the surfaces first must be prepared and positioned properly. The jigs that hold the pieces together during joining must be

Fig. 4-15. *An RV-4 fuel tank in its jig. More proof that wood-working skills are needed even on a metal airplane.*

constructed accurately before a component can be built (Fig. 4-15). Some kits incorporate the jigs within the shipping crate, which is a good reason for reading the "How-to-Open" instructions.

Finally, your shop will become extremely messy. Cleanups will range from an occasional squirt with an air hose to full-fledged scrubdown.

The final analysis

Again, the kits that generally take the least time are composite. They are followed by the metal aircraft (either monocoque or tube) and, finally, by wood. Complex features of the aircraft (folding wings and retractable gear) add to construction time.

To get a feel for comparing the construction times between kits, look for the following:

- Completeness of kit
- Fiberglass parts such as cowlings and wingtips
- Predrilled parts
- Covering envelopes (ragwings only)
- Use/inclusion of standard latches and hardware
- Prebuilt wire harnesses
- Firewall-forward packages (engine, mount, firewall, and accessories)
- Upholstery and interior trim
- Machined and welded parts
- Special tools and building accessories, such as rubber gloves and mixing cups

- Prebuilt jigs or jigless construction
- Precut instrument holes

The best way to find out how long a kit takes to build is to talk to someone who actually built one. Magazine articles are one source, but these aircraft are usually trophy winners that the builder spends an inordinate amount of time on. Talk to the run-of-the-mill builders, instead.

Failing that, though, your best bet is to double the manufacturer's estimate. Of course, I'm being unfair to those companies who strive to advertise accurate times. My apologies. However, with no common basis to compare kits, it's best to assume the worst-case situation. If you finish in less time, great. Prevent heartburn by allowing a generous period for construction rather than basing the schedule on suspect estimates.

THE PLANS

"Don't buy a pig in a poke."

In the kitplane world, this means, "Don't buy a kitplane unless you have a chance to look at the complete set of instructions."

The instructions for any given kitplane can be obtained two ways. First, they can be borrowed from an acquaintance who is building the same aircraft. Second, they sometimes can be bought separately from the manufacturer. Often the instructions' purchase price will be deducted from the kit's price should you decide to order.

Remember, though, it's a no-no to make a photocopy without permission of the copyright holder. It's especially wrong to build a plane from copied plans. I've never understood how some people can spend $20,000 building an airplane and yet stiff the designer the lousy hundred bucks or so for the cost of the plans.

With the plans in hand, let's check 'em out.

Clarity

Do you understand them? It takes a special person to write kitplane plans. A brilliant designer doesn't necessarily have the knack for writing kitplane instructions. Some famous inventors had trouble with spelling and grammar.

Engineers often have particular trouble writing kitplane instructions. It's not that they can't use English; it's just tough for an engineer to write for a nontechnical audience. Technical language is the engineer's universe, and specific terms within the universe have clear-cut, unambiguous meanings. Unfortunately, the terms are completely incomprehensible to the layperson. (The author is an engineer with a major U.S. aerospace firm.)

The instructions for each operation should include a complete parts list. Components assembled from previous operations should be identified (such as "Assembly A112"), as should the part number, quantity, and description for each individual part.

Diagrams or pictures should provide unequivocal illustration of the construction of each component (Fig. 4-16). I prefer drawings to photos—cluttered backgrounds sometimes obscure the point being made. That's just a personal preference, though. The reality is that it costs less for a kit maker to shoot photos than to hire an artist to make the drawings.

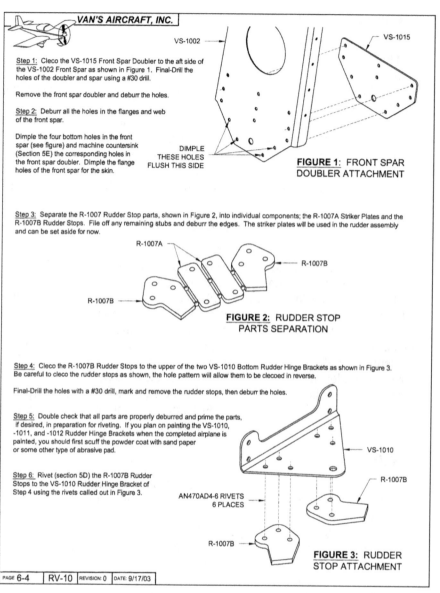

VAN'S AIRCRAFT, INC.

VS-1002

VS-1015

Step 1: Cleco the VS-1015 Front Spar Doubler to the aft side of the VS-1002 Front Spar as shown in Figure 1. Final-Drill the holes of the doubler and spar using a #30 drill.

Remove the front spar doubler and deburr the holes.

Step 2: Deburr all the holes in the flanges and web of the front spar.

Dimple the four bottom holes in the front spar (see figure) and machine countersink (Section 5E) the corresponding holes in the front spar doubler. Dimple the flange holes of the front spar for the skin.

DIMPLE
THESE HOLES
FLUSH THIS SIDE

FIGURE 1: FRONT SPAR DOUBLER ATTACHMENT

Step 3: Separate the R-1007 Rudder Stop parts, shown in Figure 2, into individual components; the R-1007A Striker Plates and the R-1007B Rudder Stops. File off any remaining stubs and deburr the edges. The striker plates will be used in the rudder assembly and can be set aside for now.

R-1007A

R-1007B

R-1007B

FIGURE 2: RUDDER STOP PARTS SEPARATION

Step 4: Cleco the R-1007B Rudder Stops to the upper of the two VS-1010 Bottom Rudder Hinge Brackets as shown in Figure 3. Be careful to cleco the rudder stops as shown, the hole pattern will allow them to be clecoed in reverse.

Final-Drill the holes with a #30 drill, mark and remove the rudder stops, then deburr the holes.

Step 5: Double check that all parts are properly deburred and prime the parts, if desired, in preparation for riveting. If you plan on painting the VS-1010, -1011, and -1012 Rudder Hinge Brackets when the completed airplane is painted, you should first scuff the powder coat with sand paper or some other type of abrasive pad.

Step 6: Rivet (section 5D) the R-1007B Rudder Stops to the VS-1010 Rudder Hinge Bracket of Step 4 using the rivets called out in Figure 3.

VS-1010

R-1007B

AN470AD4-6 RIVETS
6 PLACES

R-1007B

FIGURE 3: RUDDER STOP ATTACHMENT

PAGE 6-4   RV-10   REVISION: 0   DATE: 9/17/03

Fig. 4-16. *The most important thing is whether the plans are clear enough to allow you to build the airplane with a minimum degree of puzzlement.* Credit: Van's Aircraft

Builder's skills required

Make no mistake about it: To build an airplane, you will *have* to learn new skills. Be it riveting, ripping capstrips on a table saw, or bonding fiberglass parts, there are going to be things you've never done before.

However, one should establish exactly those skills that will be necessary to build the aircraft. For example, look for the following:

- *Machining* ("Turn the rod down to 0.875 inch." "Line-bore a half-inch center hole.") implies access to a lathe, and appropriate skills. While some of these processes might be done on other, simpler tools, it would be best to chase down a machinist.
- *Welding* ("Tack-weld the rod in place." "Anneal the metal under low flame.") should not be considered an average skill. A typical area requiring welding is the landing gear.
- *Brazing* ("Braze the washer in place.") is arguable because it's not that difficult, and a propane torch might suffice for small jobs. However, it does involve waving an open flame around the shop, which does require a modicum of attention.
- *Fiberglass mold making* ("Glue foam in place, and carve the shape shown in the figure." "Protect the engine with plastic; then cover it with chicken wire and build up the cowling shape with plaster of paris."). These days, no one should be surprised if a kit requires some fiberglass work. However, much more time is required to make the mold than to lay the glass. Making a fiberglass cowling from scratch might take 20 to 50 hours. Pick a kit that has major fiberglass parts (cowling, wingtips, fairings) supplied.

Again, subkit options allow bypassing operations that require a skill you don't have. If only one or two parts require welding, buying a welded-parts subkit is certainly faster than learning aircraft-quality welding.

It's not unusual for a homebuilt aircraft to require any of these processes. But kitplanes shouldn't require welding, machining, or making fiberglass wingtips from scratch. They aren't grounds for rejection of a design. But find out how much they'll affect your construction time and total cost.

A welding side note

By mentioning that welding should not be a part of the builder operations of a normal kitplane, I don't wish to discourage those who wish to give it a try. Welding courses are taught at many community colleges; many a homebuilder started on her way to a low-cost airplane by taking a class three evenings a week.

If you want a quality airplane at the lowest cost, a plans-built welded-fuselage model should be just the ticket. For some of us, though, welding is a skill that we don't care to learn or don't have the facilities for.

THE MANUFACTURER

You're about to drop between $6,000 and $25,000 or more on an aircraft kit. What do you know about the company that takes your money? Depending on the attitude and solvency of the manufacturer, your building time could go quick and smooth, or . . .

Kitplane manufacturers come and go. A few have become successful and prosperous. Others rose to the surface and then disappeared again with nary a ripple. A few surged aloft and then dropped back with a resounding splash.

A prospective buyer can prevent problems with the manufacturer.

Don't be the first

Getting caught in the undertow of a dying manufacturer is depressing; builder support vanishes with the source of major components. Of all the BD-5 kits sold during the 1970s, only a few have flown.

This is not just a long-ago historical phenomenon. When I was working an earlier edition of this book, a local company was producing a fast composite kit. The company was apparently successful; its prototype was on all the covers of the homebuilt magazines.

Then I got a phone call about a curious phenomenon in relation to this company: The company had laid off all their workers *except the sales staff.* Money was tight. They were using the company's remaining resources to gain new customers, not produce kit parts for the existing ones.

Sure enough, the company went under a few months later. A number of buyers were left with partial kits they couldn't finish.

Why had these people been caught? It wasn't really their fault. There had been no real indication of a coming bust. They were hurt by being too fast—too eager to buy a well-marketed product of a hot new company. They wanted to be the first in the air with the latest aerial dreamboat.

How do you keep the same thing from happening to you?

In the first place, do not buy a kit produced by a brand-new company. The temptation is overwhelming. The company just announced the production of the Wiltfang 400, the slickest little job you ever saw. You want to buy a kitplane to be different, not to build yet another Kitfox or Glasair. You want to turn heads, not run with the herd.

But buying a kitplane early in the production run can turn your stomach. Five problems arise:

1. As mentioned earlier, the manufacturer can go out of business. At best, you'll be left without any support. At worst, you might get stuck with a partial kit and have no way to finish it.

2. Nobody gets things right at first. Aircraft factories don't; hence there are airworthiness directives. Writers don't either. No matter how many times this text is edited, no one would be surprised to find a minor typo.

   Kitplane manufacturers make mistakes, too. Sometimes the instructions are wrong. Sometimes two parts don't go together right. Builders might interpret instructions differently. How are these problems found?

   Simple. They depend on the first few builders pointing them out. Recall the impact of thinking on building time, and imagine how much slower progress will be if every problem you encounter might be a design problem and not due to you misunderstanding the instructions. Can you imagine the dead ends? The wait for redesigned parts? The frustration?

3. The builder's support network isn't in place. One of the most invaluable aids to building is looking at a completed example or having someone with experience drop in to give you advice. If you're the first, no one can help you but the factory, and it might be thousands of miles away.

4. Engine details probably aren't finalized yet. Take the Murphy Rebel. The prototype, shown in Fig. 4-17, flew with an 85-hp Rotax 912.

   The Rebel's not unusual. Most of today's homebuilts started out with smaller engines. The original Glasairs and Lancairs flew with 115-hp O-235s, for instance.

   If you buy a kit based on the recommended engine at the time the plane is introduced, you may end up changing engines in midstream. If you haven't bought one, that's no big deal. But if you intended to install a 50-hp VW and the kit company decides that at least an 85-hp engine is necessary, you'll be out a lot of money.

5. Things can only get better if the kit company remains in business. Problems are discovered and corrected, and more prefabbed components are included in the kits. Of course, the price might rise quite a bit, too.

Again, do not buy a kit from a newly organized company. Wait a year or two while the kitplane is debugged by someone else.

Of course, there's a flaw in my argument: If no one buys the initial model, the company won't stay in business.

My recommendation is aimed at those who have never built an aircraft before, especially those with limited mechanical backgrounds. Some builders are A&P

Fig. 4-17. *This Murphy Rebel prototype was powered by an 85-hp Rotax 912. The current model has an 180-hp Lycoming. Such an increase is not common as a kitplane design matures and is another good reason not to buy a brand-new design.*

mechanics; others have restored older aircraft or are inveterate tinkerers. These folks can identify problems faster and can come up with sound solutions.

There has never been a shortage of those willing to buy an early production of anything *good*, be it kitplane, car, or electronic gadget. They know they might lose factory support.

But for pilots without the experience and pilots who want to minimize the hassles of construction, wait. Things can only get better. If the airplane is good, it will succeed. If it isn't, you don't want to be involved.

Problems with existing manufacturers

It isn't only new homebuilt companies that go under. Many of the pioneers have gone out of business—mostly without a lot of warning. Most industry watchers were shocked when Stoddard-Hamilton aircraft, the developers of the Glasair and the GlaStar, suddenly closed its doors in 2000. One report at the time said that the total loss to kit purchasers exceeded a million dollars.

Often, a purchaser steps forward to buy the rights to the aircraft, and it returns to the market. The four-seat composite Express, for instance, has had about four owners since its introduction in the late 1980s.

When a new owner purchases a company, the new company is still obligated to deliver previously ordered parts. Unfortunately, the buyers of bankrupt companies usually are under no such legal obligation. At best, one might receive a significant discount on components ordered from the new company. At worst—which is, sadly, more typical—the customers are completely out of luck.

In a few cases, companies have announced their imminent closure and offered a last chance for existing kit buyers to purchase their remaining components. It's better than the company folding unexpectedly, but it does mean that a builder will have to pay for everything at once rather than stretch out the kit buying over a longer time period.

The kit aircraft business is a tough one. A plans seller's only business expense is an occasional trip to "Copy-Co" to run off another set of plans. A kitplane manufacturer needs to invest a significant amount of money in shop equipment, jigs, and computer-controlled milling machines, as well as hiring people to actually build the parts and ship them out. Rumors of unstable companies are rampant. Here are some that I'm familiar with:

- Company A is skipping attendance at major fly-ins (these events are the major single source of orders).
- Company B has defaulted a on government loan.
- Company C has stopped answering its e-mail.
- Company D has dismissed some long-time employees.
- Company E's Web page hasn't been updated for over a year, and some of the links no longer work.

How, then, to protect yourself?

The best way is by involvement with current customers. Talk to local builders. Are they getting their parts in time? Is the company still providing good customer service when questions pop up?

Find the mailing lists and forum pages that discuss the airplane you're interested in, and scan the messages. Complaints about slow deliveries? Reports of surly answers to questions? You can easily find these sorts of Web pages and mailing lists by using the aircraft type for a Google or Yahoo search term. You might have to add "homebuilt" at the end in case the aircraft's name brings in some nonaviation results. Look for the sites that host forum pages.

E-mail lists can be found similarly. Two of the most common hosts for homebuilt aircraft e-mail list are *groups.yahoo.com* and *www.matronics.com*. Homebuilt companies themselves host e-mail lists and forum pages, but of course, these might be subject to deletion of uncomplimentary information.

Evaluating manufacturer support

You should evaluate the manufacturer as well as the kitplane itself. Here are some of the criteria:

1. What is the pricing strategy and delivery time for replacement parts? You undoubtedly will make some mistakes during construction. Make sure that the network exists to buy replacements without waiting forever.

2. Do they supply detailed performance charts, like production aircraft? Some kitplanes give no further data than that supplied in promotional brochures. A detailed owner's manual is a good indication of thorough flight testing. You should be able to obtain or look at a copy before deciding on a kit.

3. Does the manufacturer offer a newsletter, either hard copy or online? It's a positive sign if it does, but the newsletter should be more than a public relations sheet. It should answer common construction questions, provide insight into modifications and recommended changes, and assist in contacting other builders in your area.

4. Does anyone else outside the kit manufacturer offer a newsletter or e-mail list? This is an even better sign—a builder support network is in place.

5. Is there a help line or special e-mail address to answer your questions? Most problems occur during evenings and weekends, of course, and the best help lines will be open during these periods. Don't be surprised if they aren't— kit manufacturers are human too and like weekends off. Sometimes the help line is only available during limited hours, such as 9 a.m. to 4 p.m. on Saturdays. Similarly, some kit manufacturers only e-mail on a specific day of the week.

6. Does the maker have an online repository of support information (Fig 4-18)? We're not talking about just a fancy Web page; you want to find detailed information and resource material.

7. As mentioned earlier, some kits have systems that allow the builder to do some of the construction in the factory before the kit is shipped. This is

Fig. 4-18. *A manufacturer Web site should be more than just pretty pictures—it should feature additional construction details and updates to the instructions.*

great—the factory lets you use the labor-saving tools and the correct jigs. Help is instantly available from the factory reps. Because these programs concentrate on the most critical sections, it helps to ensure a strong and straight aircraft. However, it isn't free and often must be arranged at the factory's convenience. Don't forget your hotel and food bills either.

8. Ask for the names and phone numbers of those building the aircraft in your area. Find out their opinion of the factory's support.

THE AIRCRAFT

We've looked at a lot of factors. You're just about ready to decide. But isn't it time to look at the airplane itself?

Construction material

Do you have any biases toward or against working with particular materials? If you've got a shop full of woodworking tools and enjoy building furniture and cabinets, a wooden airplane kit actually might take less time to build. Otherwise, if you decide on a metal airplane, you'll be on the bottom of the learning curve. Similarly, if you already suffer from chemical allergies, a composite kit might make matters worse.

Don't have any preferences? Subsequent chapters discuss the techniques of working with each material. Study them; see if you start leaning in one direction or the other.

Mission selection also might affect type of material. Fast airplanes are invariably composite, metal-monocoque, or wood; slow knock-around planes are usually tube-and-fabric.

Shop requirements

The Sequoia Falco has a one-piece wing that is 26 feet long. If your garage is only 24 feet wide, you're in trouble. Make sure that you have the space to build the kit before buying it. A two-car garage is sufficient for most; however, it's best to compare the size of the aircraft versus the space available. Yes, the Falco wing could be built diagonally across the garage, but you won't have enough room left to build the fuselage.

Another important factor is handling the project. We've discussed the advantages of folding or removable wings when it comes time to carry the plane to the airport. Imagine the problems that arise when the plane does not disassemble. How are you going to haul a complete airplane, 20 feet long with 30-foot wingspan, from your workshop to the airport?

The easiest solution to this problem is to plan on doing final assembly at the airport. This requires hangar rental, approval of the airport manager, and the like. It is not an ideal or cheap way to solve the problem.

I recommend picking a kitplane that allows the wings to be unbolted and the project legally carried on a standard trailer. Further information on shop size and arrangements is contained in Chap. 6.

Maintainability

With your brand-new repairman's certificate, you'll be able to do all the maintenance yourself. But can you reach the components? Does the cowling remove quickly for access to the engine? Or do you have to remove the propeller to completely expose the engine? If you have to replace rudder cables, how easy is it to get access to the pedals? Are there access panels, or will you have to dive head first under the instrument panel?

It's a serious consideration. Any moving component that you can't inspect is a snake, ready to bite. And should it be necessary to replace a component, you want wide access panels that allow removal and replacement. If there aren't too many visible on a particular model, ask the manufacturer to identify allowable locations.

Cabin size and comfort

Hopefully, you'll find a kitplane that fits comfortably in your workshop. Now, do *you* fit comfortably inside the airplane? Have you tried one out for size? Is the plane easy to get into, or does it require contortions your body just can't handle? Once you're in the pilot's seat, can you move the controls through their entire range? Can you see out?

Until recently, homebuilt designers have ignored *anthropometry*, the science of measuring the human body and compiling statistics on the population.

Consider the last time you drove a car. Your feet reached the pedals (with the help of an adjustable seat), your head wasn't crammed into the ceiling, and you were able to swing your gut behind the steering wheel without honking the horn.

How did the designer know how much room to allow? From statistics developed through anthropometry. Researchers take measurements of volunteers and develop tables that indicate what ranges will accommodate a given percentage of customers.

In my day job, I'm a space systems engineer with a major aerospace company. Years ago, I was involved in the early days of NASA's space station project. We were required to design space station living spaces to accommodate all but the smallest 5 percent (which statisticians call the 5th percentile) of women and up to the largest 95 percent of men (95th percentile). We needed test subjects to try out the facilities, people who fell right at the limits NASA had set. A number of us volunteers were placed on a special jig that measured our maximum arm reach, hip width, and other standard anthropometric items.

I wasn't picked as a test subject. A look at the data gave the answer: While my height was just 6 feet (about 88th percentile), I was well above 95 percent in a lot of areas, such as shoulder width.

I then realized why a lot of homebuilts were just too small for me. Most of the homebuilt industry is in the dark ages of anthropometry. Throughout homebuilding's history, one factor has remained nearly constant: The original designer sizes the cockpit to fit himself. Too many designers have been little people.

A comfortable cockpit is important. The Wizzbang 400 that seems just a mite snug on a trial fit on the ground can get mighty irritating on a five-hour cross-country. You'd hate to build a whole airplane only to find that your legs can't take the cramped position required in the short cockpit.

The eventual solution is obvious: You have to find an example of the aircraft you want to build and try it on for size. Until then, help yourself to the results of my own measurements.

For several years, I carried a tape measure to local airports, taking a set of what I consider to be critical cockpit dimensions for every type homebuilt I could get access to. The measurements I took included

- *Head room*. Distance from the seat cushion to inside the canopy or cockpit ceiling along the approximate line of the pilot's back. This takes seat-back rake into account better than a straight vertical measurement.

- *Hip room*. Width available at the seat bottom between either the cockpit sides or one side and the center console.
- *Leg length*. Distance from the seatback (at the bottom, right above the cushion) to the top of the rudder pedal. The ruler was allowed to bend at the lip of the seat.

The results of my measurements are shown in Figs. 4-19 through 4-21. When taking measurements, the results were rounded up to the nearest half inch. The results for similar measurements on a Cessna 152 and a Cessna 172 are included.

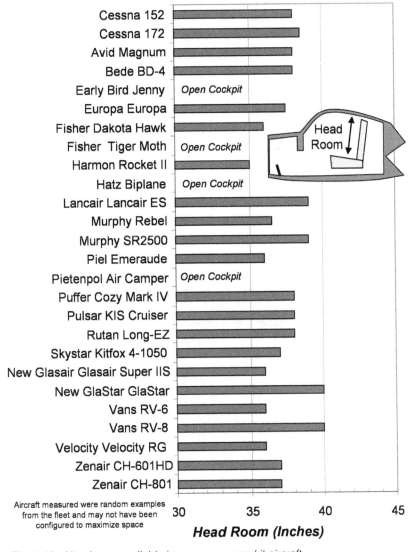

Fig. 4-19. *Headroom available in many common kit aircraft.*

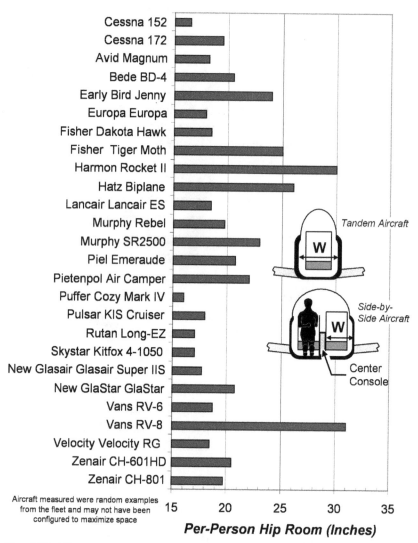

Fig. 4-20. *Hip room.*

These numbers aren't perfect because the aircraft were picked at random and probably were optimized for the size of their builders. But they'll give you a general idea of which planes are roomier.

There's no use spending years building a plane if you won't be comfortable inside it. Before you commit, try it on for size.

How does it fly?

While you're trying out the cockpit, arrange a test flight. Kit manufacturers give rides regularly; there should be no problem as long as the plane isn't a single-seater.

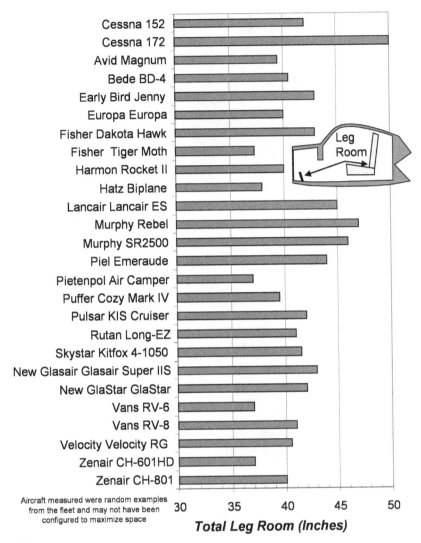

Fig. 4-21. *Leg room.*

Keep in mind that some companies charge for demo rides. The charge is usually refunded if you buy a kit.

Often, the factory is thousands of miles away. Obtain the company's demonstrator airshow attendance plan. If you can't match the schedule, pay the airfare and fly to the company's location. Would you pay $40,000 for a car without so much as a test drive? Call first—some manufacturers have scheduled test ride dates.

A cheaper and easier option, of course, is to cadge a ride from somebody local. It's a little delicate; they've likely been deluged with requests and have no obligation to give anyone a ride.

Fig. 4-22. *This Zenith Zodiac's stick is mounted between the seats for access by both occupants. It's best to try out such an arrangement before committing to buy the kit.*

Eventually, you'll find yourself at the controls. You don't have to play "Mr. Test Pilot"; just fly the airplane to see if it's agreeable. Would your feet be comfortable on a long trip? If the plane uses a sidestick or other unusual control arrangement (Fig. 4-22), do you think you can adjust?

Most people have different ideas as to what makes a good-flying aircraft. Be honest with yourself during the flight test: Do you like the way the plane flies? There have been several planes I've flown that I've felt at home with almost immediately. But there have been one or two whose handling I've absolutely detested. The only way to know is to try it out.

How much extra instruction will be needed before your first flight? Be honest; if you've flown nothing but 150s, there's no way you can step right into a Lancair IV. Besides, experience in unfamiliar aircraft is excellent training for your coming test-piloting chores.

The buzz

Finally, we come to the last, and perhaps biggest, factor involved in your selection. What do builders of that aircraft think? How do they rate the company? What problems are they finding with the aircraft?

The true answers can *only* be found by talking to actual builders. Why trust anyone who might just have a personal ax to grind unless that person is a *builder* who has personally dealt with the company involved? Magazine reviews aren't immune to personal bias or advertiser pressure.

How do you find builders? Ask the kitplane manufacturer. The company should be able to provide you with a list of local purchasers of its kit. If the company declines for some reason, contact your local Experimental Aviation Association (EAA) chapters, and ask around. And, as mentioned earlier, check the online forum pages and e-mail lists.

Additional reading

You're facing a big decision, all right. Before risking $20,000 on a kit, why not spend $30 or $50 more on a couple of other books that'll help you make the big decision?

The first is the *Aerocrafter Homebuilt Aircraft Sourcebook*. It's the "wish book" for homebuilders—it contains photos, drawings, specifications, and information on almost five hundred plans-built and kit aircraft. A chapter on engines presents the basic data for over 60 engines suited for homebuilders. Individual chapters give advice on such details as propeller selection and flight-test operations. Finally, the book wraps up with a complete directory of companies providing components and services. It's a "goodies catalog" that's fun to curl up with after a long evening in the shop. It's now published by the EAA.

If you haven't got a lot of experience, you may be a bit leery of evaluating the flying qualities of your potential purchase. Ken Armstrong's *Choosing Your Homebuilt—The One You'll Finish and Fly* is the best reference available on homebuilt flight characteristics. Armstrong was a multiengine instructor in the Canadian Air Force and is the primary inspector of amateur-built aircraft for western Canada. He knows the aircraft; he's tough and unbiased.

Both these books are available through aviation book suppliers or through the EAA.

THE BIG DECISION

By now, you probably know which airplane best fits your requirements. But what if you can't make up your mind? What if you can't decide between two similar airplanes? One is faster, but the other promises a shorter construction time. How do you pick which one is best for you?

One way is by "weighting and grading." Make a table of requirement categories, such as cruising speed, range, wing position, gear position, and so on. List *every* category that is important to you; add an "appearance" category if styling is important.

Alongside each category, list your desired values or configurations. Then assign a weighting factor for each category on a scale of 1 to 10. A weighting factor of 10 means that meeting your requirements in that category is vital, whereas a factor of 1 means that the category isn't very important.

Now, for each kit examined, assign a grade to each category depending on how well it meets the requirement. Assign a 5 if the requirement is met, going higher or lower depending on how well the kit does. For instance, if your requirement is for a 200-mph cruise speed, and the kit cruises at 210, you might assign a grade of 6. A 120-mph cruise might get a grade of 1.

BASIC MISSION STATEMENT: Long range aircraft that can operate from 2,000 foot
grass strip

MUST-HAVES: Certified engine, wingspan less than 25' to fit in shed,
at least two seats

KIT EVALUATED: *Ullrich 2000*

CRITERIA	PREFERRED	WEIGHTING (1 to 10)	THIS KIT	SCORE	WEIGHT X SCORE
CONFIGURATION:					
Wing Position	Low	3	*Low*	5	15
Gear	Trigear	8	*Taildr.*	2	16
Seating	Side by Side	10	*Tandem*	2	20
Seats	2	5	*2*	5	25
Styling		8	*Nice!*	10	80
PERFORMANCE:					
Cruise Speed	150+	5	*120*	3	15
Endurance	4 hours	7	*5*	8	56
Takeoff Dist.	1500 feet	9	*800*	10	90
Landing Dist.	1500 feet	9	*500*	10	90
CONFIGURATION:					
Building Time	100 hours	4	*1000*	5	20
Type of Material	Metal	2	*Tube*	4	8
Total Cost	$35,000	8	*15,000*	8	64
Completeness of Plans		6	*Poor*	3	18
OTHER FACTORS:					
Type of Engine	Lyc./Cont.	10	*Cont.*	5	50
Cabin Comfort		8	*Good*	6	48
Flying Qualities		9	*Great*	10	90
Aerobatic		1	*Yes*	5	5

TOTAL SCORE: **710**

Fig. 4-23. *Sample kit evaluation sheet.*

The grades don't have to be proportional to the amount the kit misses or exceeds the requirement, but they must be consistent. If two planes cruise at the same speed, they should get the same score. If one is "uglier" than the other, it should be reflected in the "appearance" category, not under "cruise speed."

For each category, multiply the weighting by the grade. Then add all the category scores to determine the total score for the kitplane. Figure 4-23 shows a sample score sheet using this process.

Was that any help? Or had you already made up your mind? No matter. But you're not quite ready to send in that order sheet. Let's look into some of the choices that might affect your decision.

5

Decisions

EVEN THOUGH YOU'VE PICKED OUT A KITPLANE, the decision process isn't over. How are you going to pay for your dream ship? What avionics and instruments should be installed? If the kit doesn't come with an engine, how do you find the right one at a reasonable price?

The material in this chapter, for the most part, shouldn't affect your choice of kit. But there are a few things to settle before making the chips fly or even before signing the purchase order.

CAN YOU AFFORD TO OWN AN AIRPLANE?

People build airplanes for a variety of reasons. Economy is one; homebuilts are cheaper to operate than new or used production aircraft. But owning a kitplane isn't free by any means.

It's quite possible to end up in over your head even if you paid cash for the kit. The purchase money might have come from some special nest egg, careful scrimping over a number of years, an inheritance, or whatever.

So the nest egg is gone, but the plane still must be supported: fuel, insurance, hangar rent, and so on. Unless you've owned an airplane before, you're likely to be surprised at how much they cost to own—even homebuilts.

Before you go much further, figure out if you can afford to operate the homebuilt you've selected. Let's add up the ownership costs for a typical kitplane.

There are two types of expenses, *direct* and *fixed*. Direct costs are those that accrue directly from flying the aircraft: fuel and oil bills, for the most part. Fixed costs are those you must pay even if the airplane doesn't fly. Tiedown or hangar rental and insurance are the major elements. If you finance the aircraft, the monthly payments also must be added to the fixed costs.

Fixed costs include the maintenance expenses—those required to keep the aircraft in a flyable condition. Typical maintenance expenses are the annual inspection and routine repairs. The base cost for a Cessna 150 annual inspection can be up to around $800 for the labor, plus parts. A homebuilder with a repairman certificate saves the labor charges and saves on those replacement parts that he can build. The kitplane owner is likely to pay more for an "aircraft" part than a certified mechanic because the mechanic probably gets a discount. But the owner of that certified aircraft is not likely to see any difference by the time the part ends up on the bill.

Accountants identify one other expense of owning an aircraft: lost interest. If the builder spent $40,000 to build the airplane, that money could instead have been invested and earned interest. This lost opportunity is lost income; hence it's the same as an expense.

As far as I'm concerned, this "expense" is balanced by the simple joys of owning an aircraft. The freedom a personal plane grants is impossible to put a value on, but I'm willing to stipulate that it's worth at least as much as the lost interest. Let's look at those expenses that come out of actual pockets instead of out of an economics text.

Fixed costs

Fixed costs are the most irritating aspect of aircraft ownership because they continue whether the plane flies or not. It might snow 30 days out of the month, and on the thirty-first day, the driveway has to be shoveled, but you still have to pay the hangar, insurance, and other fixed costs. For the average owner who flies a hundred hours or so a year, fixed costs *are* the major expense of ownership.

Storage, in the form of hangar or tiedown rents, hogs the lion's share. Tiedowns are cheaper, but hangars provide better protection. No matter what kind of kitplane, one element is common: The builder is loathe to tie it down outside. After spending thousands of hours and dollars, you'll want to hangar it as well.

Have you priced a hangar lately? The closed hangars at my airport rent for $240 a month, which is a steal compared with most major metropolitan airports. Hangars eight miles away at a controlled field run $500 on up; in a year, that's $6,000. If you fly 100 hours a year, the hangar rent alone amounts to $60 an hour.

Reduce the bite by hangaring with another aircraft and sharing the rent. Home-builts are small; hangars generally are designed for the Cessna 210 or larger class of airplanes. If your kitplane will slip under another plane's wing, hangar cost is halved (Fig. 5-1).

Outside tiedowns are cheaper. Rent varies depending on the desirability of the location. As with hangars, tiedowns cost more at close-in major airfields with good security. A tiedown might run $120 a month at these fields, but a grass strip 20 miles away might have spots for $35. Whether this works for you depends on the severity of the local weather and the type of kitplane: composite, wood, and so on.

Don't be too eager to see a folding-wing aircraft stored in your garage as the solution of the hangar woes (Fig. 5-2). Not only does it become a hassle (discussed in Chap. 4), but most airplanes can't take a steady diet of being dragged over public roads. The highway vibration is far worse than takeoffs and landings, and a continuous 50 mph is far different from an occasional acceleration to takeoff speed. It causes premature wear and is hard on the instruments. Consider all the alternatives before deciding on keeping an airplane at home.

However, folding wings do make it easier to share a hangar. Several kitplanes require only a few minutes to unfold the wings and prepare for flight.

Fig. 5-1. *Homebuilts often are small enough to share hangars, hence cutting the rent in half. Here, a Thorp T-18 snuggles nicely under the wing of the Wickam Bluebird, an all-metal four-seat homebuilt from the 1950s.*

Fig. 5-2. *While keeping a folding-wing homebuilt at home is impractical in most cases, they make hangar sharing easier.*

You can easily find out the hangar and tiedown rates at local airports. But the cost of renting a hangar might be moot if none are available. The $240 closed hangars mentioned earlier have a waiting list at least five years long. Larger hangars have a shorter waiting list, but they rent for almost $150 a month more. If you hope to keep your kitplane in a hangar, get on the waiting lists immediately. If your name comes up and your plane isn't ready, sublease the hangar to someone else.

Insurance

As mentioned in Chap. 2, insurance can be difficult to obtain for some high-performance homebuilt aircraft. Coverage is a bit easier for the more run-of-the-mill kitplane. Premiums vary with coverage, pilot experience, and the type of aircraft.

It's easy enough to find out what your homebuilt will cost to insure; you usually can obtain a rate quote online. But let's look at some ballpark figures and familiarize ourselves with some common terminology.

The basic policy covers *liability* only. Liability coverage insures you for damage or injuries caused by your aircraft. Coverage limits are usually about $100,000 per person and $1 million per accident. Note that a liability policy does not cover *your own* medical expenses or damage to *your aircraft*. However, it should include the recovery expenses if you force land in a field somewhere. Expect liability-only coverage to cost $500 to $1,000 a year for a typical homebuilt. Some friends of mine operate "bare," with no coverage at all. In this litigious age, I really don't recommend it.

Liability won't pay for any damage to your aircraft. If you wish to insure the aircraft for repair costs, you'll need *hull* coverage.

Hull coverage comes in two flavors: *not in flight* and *in flight*. Not-in-flight coverage only insures your plane from damage that might occur while it's parked or hangared. If the hangar roof falls in or your plane is vandalized, you're covered. But the moment the engine starts, you aren't. To have insurance in an accident, you'll need a policy that provides in-flight coverage. In-flight coverage is the equivalent of automobile "collision" coverage, whereas not-in-flight coverage is about the same as an automotive "comprehensive" policy.

Hull coverage can be expensive. One major insurance company quoted yearly premium rates of 10 percent of the kitplane's value. An owner of a $20,000 Early Bird Jenny would pay $2,000 a year for hull insurance. If you fly 100 hours a year, that's $20 an hour just for insurance!

But it pays to shop. The company I selected for my aircraft quoted a premium closer to 4 percent a year.

As the saying goes, "Your mileage may vary." Higher deductibles reduce the premium, as will thousands of flight hours in your logbook. Features such as retractable gear drive the rates higher. Amphibious aircraft (Fig. 5-3) and helicopters seem to have the highest premiums.

Many homebuilders don't carry hull coverage at all. They prefer to self-insure; if they break the airplane, they'll fix it out of pocket. Deciding to self-insure is like betting with the insurance company. Fly 10 years without "totaling," and you win. It's not for the faint of heart or those with expensive kitplanes. Also, if the airplane acts as loan collateral, the bank will require full insurance.

Self-insurance is more feasible with homebuilts than with production aircraft because labor is a major factor in the cost of repair. When the owner has a repairman certificate, the labor for any work, whether damage repair, routine maintenance, or annual inspection, is free. The accountant might argue that the time spent doing the annual inspection could be better spent working for minimum wage at the local burger emporium, but we've already decided to ignore her.

Fig. 5-3. *Homebuilt amphibians such as this Thurston Seafire can be very expensive to insure.*

Consequential expenses

Even if the labor's free, though, you'll have to make or buy replacement parts. Many parts have to be replaced at regular intervals. For instance, many emergency locator transmitters (ELTs) must have the battery replaced every two years, and other parts such as hoses have a finite life and must be replaced on a fixed schedule.

Otherwise, parts cost predictions are difficult because cost varies among types. A hydraulic failure in a retractable-gear aircraft might cost many hundreds of dollars, whereas torn fabric on a simple ragwing could be fixed for nearly nothing. While these costs are related directly to operation of the aircraft, I lump them with fixed costs. A good term for them is "consequential expenses" because they are only related indirectly to actually flying the aircraft.

Consequential expenses vary in an odd relationship with the aircraft's total hours. The first hundred or so hours will be costly because problems will be discovered and repaired. And getting used to a new aircraft might result in a few minor dings.

After the bugs are ironed out, aircraft are relatively trouble-free for awhile. Then, as they age, repairs start creeping up again.

How much these costs run during the test period is anyone's guess. During the "steady as she goes" midlife of my aircraft (a simple homebuilt), I averaged about $100 a year for engine and airframe parts. The amount rises with the aircraft's complexity: radios, retractable gear, constant-speed prop, and the like. Not to mention luck! I'd guess that consequential expenses for an average homebuilt might run $500 per year.

This isn't adequate for major expenses, but we'll establish a cash reserve under the direct costs.

If you've taken out a loan to buy your kitplane, don't forget to include the monthly payments in the fixed costs. Other fixed costs are rather minor. Most states require annual aircraft registration. My home state charges homebuilts a flat $50. Some state governments find ways to clap "luxury" taxes on aircraft, too.

A homebuilt has a major advantage over a production aircraft: Fixed costs can be eliminated completely during financial crises. If necessary, the aircraft can be disassembled and brought home for storage. Surrender the hangar and cancel the insurance. The kitplane can't be flown, but that saves direct costs as well. Mothball the plane until you can afford to fly it again.

But if it'll have to be sold, don't bring the airplane home. You'll get a lot more for a flying aircraft than one in pieces.

Direct costs

Direct costs are easy to compute. Take the engine's per-hour fuel burn and multiply by the cost of fuel. Add a buck or so for oil, and there's the direct costs.

There is one major variable in the direct cost equation: cost of gas. The solution is to simply install an engine capable of running on 80 octane aviation gasoline (avgas). Of course, 80 octane avgas doesn't exist anymore, but automobile gasoline (autogas) does. Most engines designed to operate on 80 octane avgas will happily burn 87 octane autogas or 92 octane unleaded.

The autogas solution isn't perfect. For one thing, aircraft engines designed to use 80 octane aren't made anymore. If you want a new certified engine in your dream ship, count on feeding the engine 100LL. Most of the new engines intended for homebuilts, such as the Rotec radial and the Jabiru, can run on autogas.

If you're building a cross-country machine, forget the 80 octane engine because only 100LL is available at most airports. This can be an important issue with auto-engine conversions. If the conversion uses the oxygen sensor and computer of the car setup, the sensor will be quickly ruined by the lead in 100LL.

Autogas aircraft encounter many of the same problems as aircraft kept at home. It's a bit of a hassle. Feeding VW engines isn't so bad, with their 3.5-gph fuel consumption. But imagine keeping a 50-gallon tank filled; five-gallon jerry cans aren't gonna' do it, buddy.

If you do use autogas, use *good* autogas. Buy major brands from dealers with rapid turnover. The gas refiners optimize the mix for the season, and cut-rate dealers get special deals by buying leftover winter gas in the spring. Avoid all gas suspected of adulteration. Two-stroke engine users should be especially wary; those engines are very sensitive to gasoline quality. Autogas also has stability problems in long-term storage. Many users, such as myself, switch to 100LL in the winter when the aircraft won't be flown very often. Aviation fuel is a lot more stable.

Another problem in some areas is the addition of alcohol to the gasoline. This is often indicated on the pump, but not always. For instance, some states, such as California, mandate alcohol year-round. Auto engines and fuel systems are designed to handle alcohol, but aircraft systems are often vulnerable. The seals in some aircraft carburetors can swell if they are exposed to alcohol. Most other oxygenating additives, such as MTBE, are OK.

To check for alcohol, use a small gradated cylinder such as a test tube or a syringe. Fill it 20 percent full with water; then fill it the rest of the way with gas. Shake for 30 seconds, and then let the water separate from the gas. The water will "swell" as it absorbs alcohol; if the tube is now more filled with water than before, don't use the gas in your airplane.

As far as oil is concerned, engines typically burn a quart every 5 to 10 hours. The oil should be changed every 25 hours for engines without oil filters or at 50-hour intervals if an oil filter (not an oil screen) is installed. Aviation oil costs almost three bucks a quart, hence the oil costs come to a bit over a dollar an hour.

Routine fix-up expenses are included under consequential costs, discussed earlier. However, occasionally, things go bad. Very bad. A friend had to replace two engine cylinders in his Tri-Pacer. The total cost came to almost $2,500.

The best way to handle these problems is by establishing a *maintenance reserve*, where you set aside a fixed amount of money every flight hour. The basic purpose of this fund is to pay for the eventual overhaul of the engine. It also comes in handy for other expensive repairs.

Determining the rate to "charge" yourself for the maintenance reserve is relatively easy. It's based on the *time between overhauls* (TBO) of the engine and the cost of overhaul. Any engine is going to need overhauling sooner or later. In the case of the Rotax, it's sooner but cheaper—400 hours or so between overhauls, but a full overhaul costs only a few hundred dollars. A production aircraft engine overhaul comes later (1,600 to 2,400 hours) and is much more expensive.

To determine the hourly rate for the maintenance reserve, take the overhaul cost and divide by the hours until overhaul (at the time of engine installation). It's as simple as that. If you install a new Lycoming O-320 (TBO 2,000 hours), count on an overhaul cost of $8,000, and build your maintenance reserve at the rate of $4 a flight hour. If you had installed a used Lycoming with 1,000 flight hours, the rate jumps to $8 an hour—the overhaul charge must be based on only 1,000 flight hours.

The maintenance reserve should be deposited into a separate account to keep it isolated from other needs. Of course, it is available for worthier causes, especially if your spouse simply must have some "essential" home improvements.

Alternatively, you might decide to ignore the maintenance reserve. Most owners don't keep their aircraft that long; you'll probably sell the plane long before overhaul time approaches. Being more grasshopper than ant, my preference is dig up the money from somewhere when the time comes. But when I owned a Stinson in partnership, my partner and I maintained separate maintenance reserves based on our flight hours.

Keep in mind, though, that TBO is a rather nebulous term. Engines are discussed further later in this chapter.

Financing

As of this writing, only one major finance company offers loans for those interested in building their own airplanes:

NAFCO
P.O. Box 7050
Lakeland, FL 33807
 (800) 999-3712
http://www.airloans.com

There are limitations, of course. The company won't insure a project unless at least 25 examples have been completed and flying. The purchaser must have a good credit rating and a fairly low debt ratio. A down payment of at least 10 percent is necessary. Loan rates depend on the components being insured; the company will cover buying the engine (which has a definitive value) at a lower rate than the aircraft structure. The total loan period can be five years or longer.

In addition to NAFCO, a few kit companies offer loans to purchase their aircraft. Other avenues exist. A lower-rate loan might be available from your local bank or credit union. Most traditional loan institutions require collateral: home equity, property, or other valuables. If you have something that can act as collateral, a loan shouldn't be too difficult. The bank merely wants to be sure of recovering its losses should the loan go into default.

Signature and *unsecured line-of-credit loans* are offered by many financial institutions. They're granted on the basis of a good credit history and the ability to afford payments. The loan officer bets whether you are a trustworthy character who pays your debts. Maximum loan values generally are 10 to 20 percent of your annual income (more if you're a third-world country).

Probably the best way to finance is a loan against a 401(k) retirement plan. You are essentially borrowing money from yourself, so the interest rate is quite low. I bought half-interest in a Stinson with a loan against my 401(k) at a rate less than two-thirds the going rate for signature loans.

However, the cheapest way of financing a kitplane is to buy subkits as your budget allows. This frees you of interest payments, although crating and shipping charges are higher than if the entire kit is purchased at once. As you might expect, the monthly payments play hob with the per-hour cost as well.

Adding it up

Table 5-1 shows typical costs based on various annual utilization and the autogas/avgas decision. Option 1 is a typical "performance" kitplane, whereas option 2 is a light, inexpensive fun airplane. Fixed costs are the major expenses until utilization reaches about 100 hours per year.

One hundred hours per year works out to about two hours a week, about eight hours a month. That's quite a bit of flying, especially for knock-around fun planes like the Kitfoxes and CGS Hawks. Your actual utilization probably will lie between 25 and 75 hours a year unless you plan on making specific, regular flights. The first year will be greater, of course.

At these low utilization rates, renting from the local FBO is sometimes cheaper. Exceptions are fast ships such as Falcos and Ventures. Equivalent rental planes such as Arrows or 182RGs are quite a bit more expensive; the breakeven point between renting and flying a homebuilt is fairly low in these cases.

Table 5-1. Typical ownership costs

Fixed Costs

Hangar:

    Option 1: $400/month (closed hangar, metro area)

    Option 2: $125/month (shared hangar, suburban airport)

Insurance:

    Option 1: $2,000/year (liability + hull coverage on $40K value)

    Option 2: $700/year (liability only)

Loan payments:

    Option 1: $270/month ($20K loan, 6%, 8 years)

    Option 2: $0 (no loan)

Consequentials:

    Option 1: $100/month (more complex A/C)

    Option 2: $25/month

Total fixed costs:

    Option 1: $11,240 per year

    Option 2: $2,500 per year

Total Ownership Costs (Fixed Plus Direct)

Annual ownership costs

	25 Flight Hours/Year		50 Flight Hours/Year		100 Flight Hours/Year		200 Flight Hours/Year	
Fuel Used	*Auto*	*Avgas*	*Auto*	*Avgas*	*Auto*	*Avgas*	*Auto*	*Avgas*
Option 1 (8 gph)	$12,812	$13,072	$13,192	$13,712	$13,952	$14,992	$15,472	$27,552
Option 2 (4 gph)	$2,880	$3,140	$3,260	$3,780	$4,020	$5,060	$5,540	$7,620

Per flight hour

	25 Flight Hours/Year		50 Flight Hours/Year		100 Flight Hours/Year		200 Flight Hours/Year	
Fuel Used	*Auto*	*Avgas*	*Auto*	*Avgas*	*Auto*	*Avgas*	*Auto*	*Avgas*
Option 1 (8 gph)	$512	$523	$264	$274	$140	$150	$77	$88
Option 2 (4 gph)	$115	$126	$65	$76	$40	$51	$28	$38

Note: Autofuel is assumed to cost $1.90/gallon; avgas $3.20/gallon.

However, at such a low yearly utilization, the per-hour rate is about the same for simple or complex homebuilts. Owning a Lancair is cheaper than renting an Arrow, but a Fisher Dakota Hawk will cost more than renting a Cessna 152. Our resident accountant sees no reason for owning an airplane at all.

But how important is convenience to you? If you wake up Saturday to blue skies, can you schedule a plane for the afternoon? For as long as you want it? Probably not.

How safe are the planes you're renting? Does the FBO fix every problem immediately? Or are they bandaged together until the next 100-hour? At least if you own it, you'll know the exact maintenance status. And rental aircraft are subject to considerable abuse.

Yes, in some cases, owning your own plane costs more, but we own airplanes because we *like* airplanes. We own them so that we can fly them when we feel like it, not when an FBO's schedule allows.

Sure, ownership is sometimes a hassle. But renting a plane is like riding the bus: You're locked into someone else's schedule, and you never know what the last rider did on the seat.

Don't base your decision solely on the per-hour cost. Work out your budget, and if you can afford ownership, go for it. The convenience and safety are worth it.

ENGINE SELECTION

If you're out to get the plane flying with minimum difficulty and the least amount of time, stick with the engine recommended by the kit manufacturer. As mentioned previously, homebuilders often regard the engine as another "black box"—an item to be installed and not necessarily understood. Mounting an engine other than the recommended one causes more problems than just a change in engine mounts. Some possible problem areas include

Center of gravity (CG). Any change in weight or distribution will change the CG. As you learned in pilot training, out-of-limit conditions can affect flyability dramatically.

Fuel usage. If the same size tanks are used, installing a bigger engine actually might shorten the aircraft's range. You could install larger tanks, but the affect on CG must be considered, and the extra fuel weight comes out of the useful load.

Cooling. Bigger engines produce more heat—it's simple physics. The cooling inlets and outlets on the cowling might be too small.

Access. A larger engine will take up more space under the cowling, making it harder to perform maintenance.

The following sections discuss the engine options open to the homebuilder.

CERTIFIED AIRCRAFT ENGINES

The number of certified aircraft engine manufacturers has expanded slightly over the past 10 years. Lycoming and Continental are the traditional sources. License-built versions of the Franklin engine are being built in Poland and are appearing in such planes as the Velocity. Reciprocity agreements with European countries have facilitated the introduction of a certified version of the Rotax 912.

Once you get away from small, low-cost airplanes, most kitplane manufacturers specify certified aircraft engines for their products. This is a trifle odd from one point of view. These light aircraft engines have evolved from engines introduced in the 1930s and 1940s. And they put these old-fashioned engines in our twenty-first-century kitplanes?

Certified aircraft engines have one great advantage: proven reliability in an airborne installation. Auto-engine converters claim equal or greater reliability. They produce reams of calculations and stacks of test-bed results.

But these engines haven't been flying for 50 years. The Federal Aviation Administration (FAA) and the National Transportation Safety Board (NTSB) track failures of certified aircraft engines; the resulting airworthiness directives (ADs) and service bulletins continually improve the breed.

Aircraft engines are recommended because they are a known quantity, not because they are the most efficient or the most powerful. Their faults and quirks are well known, and if a problem arises, any competent airframe and powerplant (A&P) mechanic can track it down. When you experience a problem with an auto-engine conversion, it's up to you. The best you can expect is advice by long distance.

Aircraft engines aren't cheap. Prices can range from merely breathtaking to exoatmospheric. Depending on your requirements, you might be able to buy an adequate engine for a third of the cost of a new one.

Engine types

By far the most common type of certified aircraft engines are the Continental and the Lycoming (Fig. 5-4). Both sides have their adherents, and both manufacturers have produced "lemon" engines.

One fast way of telling a Lycoming from a Continental is Lycoming's preference for locating the starter ring on the front of the engine, just behind the prop. Look into the front of the cowling; if you can see the crankcase, it's either a Continental or an older Lycoming without a starter.

If the cowling is off, the manufacturer's name is stamped into the valve covers, although older Lycomings show an "L" in a hexagon. The final obvious difference is in the oil tank—small Continentals use a separate kidney-shaped tank, and larger Continentals have an attached oil pan that's smooth and rounded. Lycoming's oil tank is an angular pan bolted directly to the bottom of the engine. In fact, the residual heat from the oil tank to the carburetor is one reason Lycoming engines are less prone to develop carburetor ice than Continentals.

The manufacturer is about the only thing you can determine with a casual glance. For further information, check the engine data plate.

The type, configuration, and selected accessories are indicated by the engine's designation, such as O-360-C2A, TOL-200-A, and so on. The designation's prefix shows the engine configuration, the numbers in the middle are the engine's displacement

Textron Lycoming.

Fig. 5-4. *Lycoming produces the O-320 and the O-360, which are the two most popular certified engines for kitplanes.*

in cubic inches, and the suffix is used by the manufacturer to indicate small changes in configuration.

Common letters used in the prefix are

O   Opposed cylinders

L   Inline cylinders

I   Fuel injected

T   Turbocharged

S   Supercharged

L   Left-hand rotation

G   Geared

Although different aircraft might use the same basic engine, slight differences are seen. Changes in magnetos, the addition of a controllable-pitch prop, and the like result in a unique suffix to identify the variant. For example, the Cessna 152 mounts a Lycoming O-235-L2C, which is an opposed 235-cubic-inch engine with Slick magnetos. The nearly identical Cessna 150 mounts a Continental O-200A.

Certain Lycoming engines also feature the *dynofocal engine mount* (Fig. 5-5). The mounting holes for other engines are horizontal and parallel; the dynofocal mount reduces vibration by "aiming" all mounting holes toward the propeller mounting flange. There are two kinds, *type 1* and *type 2*. Type 2 is needed when prop extensions

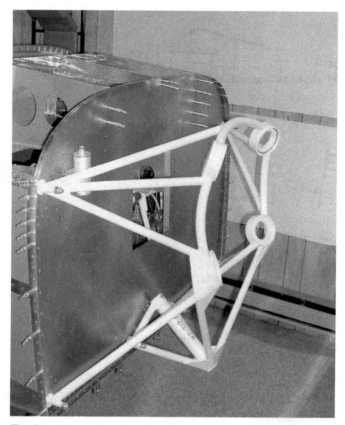

Fig. 5-5. *A type 1 Dynofocal engine mount for an RV-7A.*

require a slight variation in the pointing of the mounting holes. Make sure that you have the correct mount.

It's important to know the differences between engines of the same series, especially when looking for a good deal in used engines. The question is, How exactly must you match the plan's engine requirements? If the plans specify a Lycoming O-320-E1J, can you substitute a Lycoming O-320-H2AD instead?

In this case, the -H2AD uses a dual magneto drive (two mags are essentially installed in one case). This has some reliability impacts: If a single magneto drive fails, it kills both mags. However, there's another, bigger problem. The -E1J engine includes a governor for a controllable-pitch prop; the -H2AD doesn't.

Hence the substitution probably would be all right so long as you weren't using a controllable-pitch prop and were willing to trust the two-in-one magneto.

This is typical of engines considered the same basic "type": While the core section of the engine remains the same, the accessories change.

How will this affect your choice? By changing one or two of the specific dimensions of the engine. A change in the induction system, for example, may require more room below the engine; a cowling modification may be necessary.

Before starting your engine search, talk to the kitplane manufacturer and get a list of alternative engines. The manufacturer should be able to tell you what (if any) changes would have to be made to use a particular engine.

The point made in Chap. 4 still applies: The more brain work you have to put into installing a different engine, the longer the construction time. But sometimes the extra work is worth it if you can find an engine at a good price.

Check the ADs issued for candidate engines. The Lycoming O-320-H2AD, for example, was the subject of many ADs. By now, the problems should be fixed. All Lycoming O-320s aren't tarred by the same brush—the suffix makes all the difference. The O-320-E2D (used in Cessna Cardinals and selected 172s) would be an acceptable alternative.

There are several ways to check the ADs on a particular engine model. An online search is easiest. The FAA maintains ADs back to the 1940s on its Web page. Enter "FAA Airworthiness Directives" on your favorite Internet search engine to find the page; then do a search on your engine type.

In addition, every A&P mechanic with *inspection authorization* (IA) is required to maintain AD lists. If you have an "in" with an A&P mechanic, that would be a good place to start. For that matter, any competent A&P mechanic should have some idea of the ADs affecting certain aircraft and engines. Your Experimental Aircraft Association (EAA) chapter probably has at least one A&P mechanic who can give you some good advice.

If you don't have the right contacts, a book entitled, *The Aviation Consumer Used Aircraft Guide,* is a good start. Some libraries carry it, but it's available via mail order from several places. The *Guide* gives a no-nonsense evaluation of most used production (and several homebuilt) aircraft and usually includes an AD summary. Find the aircraft that used the engine(s) you're looking for, and check out the ADs.

Now that we know what we are looking for, let's look at the engine-buying process. There are three classifications: new/remanufactured, overhauled, and used.

Nontraditional certified engines

I tend to think Lycoming or Continental when the issue of certified aircraft engines arises, but in reality, there are a number of foreign-manufactured engines that also have type certificates. As mentioned earlier, the Franklin engines used on light aircraft of the 1940s and 1950s have been license-built for quite a while. The four- and six-cylinder inline LOM engines are imported from the Czech Republic. The newly approved light sport aircraft (LSA) category undoubtedly will generate at least a trickle of new engines certified under the simpler LSA standards.

Finally, the Russian M-14 radial engine is a production model, but it does not possess a U.S. type certificate. But that doesn't stop it from being used on planes such as the Murphy Super Rebel.

A salient point when using any of these engines is the availability of replacement parts. Distribution networks for foreign-based certified engines tend to fade away, making it difficult to keep the engines in repair. If you do decide to go for this option, ensure that you'll have a steady source of parts.

Buying a new or remanufactured engine

The most expensive option is to buy a brand-new engine. They're offered direct from the factory and through dealers. The same engines are found at various prices depending on how much the dealer is willing to discount.

Often, the kitplane manufacturer arranges an *original equipment manufacturer* (OEM) deal with the engine companies, which allows you to purchase new engines at a substantial discount. Check with the kit manufacturer.

Once you get by the considerable obstacle of price, a new engine has several advantages. Lycoming's warranty pays parts and labor during the first year and pro-rates parts costs during the second year.

Remanufactured (or *zero-timed*) engines are probably the best deal. The manufacturer takes a used engine, disassembles it, and replaces major components such as cylinder heads, cylinder barrels, valves, and pistons. The remaining parts are checked, and those that don't meet the tolerances required for new parts are also replaced.

Although the engine might have had thousands of hours on it on arrival at the factory, it is now considered the same as a new engine, hence the term "zero-timed," and it receives the full new-engine warranty. It gets a new logbook, with no mention of the previous flight hours.

Zero-timed engines are as good as new but cost around 20 percent less. However, you must supply a "core" engine. Unless you find an old junker cheap, this might add several thousand dollars to the total price.

Like new engines, remanufactured ones are offered through both the factory and dealers.

Buying quasi-new

An interesting trend has popped up. A couple of companies that make aftermarket parts for Lycoming engines (legitimately) have begun selling complete kits for these engines. The engines are nearly identical to those Lycoming sells, except that they are not certified and cannot be used on a standard-category aircraft.

But they're just fine on homebuilts. These kit engines are a good way to obtain a low-cost new powerplant if you are up to building the entire engine from scratch. These companies sometimes offer programs where you can build your engine under supervision at their facility: a very good option.

Buying an overhauled engine

Engines can be overhauled to two standards: *new limits* or *service limits*. Overhauling to new standards is identical to the remanufacturing process, but because only the original manufacturer can zero-time an engine, any third-party work is only an overhaul—even if the rebuilder exactly duplicates the factory operations.

An overhaul to service limits means that the engine is completely disassembled and all parts checked. Those parts that don't meet the service limits established by the manufacturer are junked and replaced. These limits are looser than the new limits but still should provide adequate margin to allow the engine to run to TBO.

In both cases, the engine is returned to service with the original logbook, carrying the same number of hours it had before the overhaul. This is why you'll see engines advertised as "2,500 TT (*total time*), 250 SMOH (*since major overhaul*)."

As an interesting side note, engine manufacturers have begun overhauling engines as well as remanufacturing them. These engines are rebuilt to service limits. While the prices are higher than you'll find at local shops, they are quite competitive with the quality rebuilds.

Most engine difficulties occur around the cylinders, not within the block. Problems with the cylinder head, valves, or even the cylinders themselves can result in a *top overhaul*. A top overhaul is similar to a regular overhaul, except that the engine casing isn't disassembled; therefore, the condition of the main bearings and other interior parts is unchanged.

Overhaul costs may vary widely, even for the same engine. Quality also varies. Some companies specialize in precision overhauls to like-new limits. These professional shops automatically replace some parts whether in limits or not. They produce high-quality engines, but sometimes the price approaches the cost of a factory remanufacture.

Don't be fooled into accepting the factory list prices, for that matter. You can by those same factory-remanufactured or factory-rebuilt engines from independent vendors for discounts of up to 20 percent from the factory price. Shop around!

But prices can get even lower. Any A&P mechanic can overhaul an aircraft engine. There are a number of small shops performing overhauls, with quite a variation in prices. Shops range from FBOs to A&P mechanic schools to retired mechanics who like to tinker with engines.

These smaller shops usually overhaul to service limits, which can reduce the number of parts that must be replaced. You don't end up paying for the replacement of components that might have thousands of hours of life remaining. However, sometimes this can go too far. The man that overhauled my Stinson's Franklin reused the old spark plugs—an example of excessive frugality.

One other option is to rebuild the engine yourself—it's legal for a homebuilt. Find an experienced mechanic to watch over your shoulder and give advice when you need it. Rebuilding manuals are available through most homebuilder's supply companies.

Overhaulers don't sell engines; they sell rebuilding services. Therefore, you will have to supply a *rebuildable core*. This is an engine without major damage—generally, it means at least an intact crankshaft. Some of the larger outfits might sell you a rebuilt engine outright, but a core charge will be tacked on. If you send them your old engine, portions of the core charge will be returned if the components are rebuildable. This isn't unique to the aviation world, as you know if you've ever bought rebuilt car parts.

As mentioned earlier, the engine manufacturers themselves offer overhaul services. However, they are getting picky about previous work done on an engine. If the engine block has been repaired by another company, they may refuse to accept it and keep your core charge. This has been happening even when the original repair company was an FAA-approved repair station. If this occurs, request return of the unacceptable parts, and try to sell them yourself.

Finding the large professional overhaul shops is no problem. They advertise extensively in *Trade-A-Plane* and other publications or can be found with an Internet search on the term "Aircraft engine overhaul services." Talk to local FBOs; some act as agents for the large rebuilders. The price will be higher than dealing direct, but the FBO also handles crating and shipping your core.

The FBO might offer to do the job for less, or you might find an independent A&P mechanic. Ask for references, and check with other members of your EAA chapter. All rebuilds are not alike, even if they pass FAA certification.

There is one trap when having an engine rebuilt. Occasionally, the rebuilder will say, "Y'know, I can save you even more money. Since this is going on a homebuilt, we don't need all the FAA paperwork. I'll cut the price by x dollars if we don't make this an official overhaul."

The paperwork certifies that all parts meet the tolerances of the service limits. Eliminating this guarantee is no problem with an honest and conscientious A&P mechanic. But bad apples grow in all orchards.

A friend of mine bought a rebuilt Lycoming on this basis for $10,000. He mounted it on his homebuilt; it ran fine. However, he made a mistake with a hoist and bent a couple of pushrods. He took it to another FBO for repair.

The FBO refused to reassemble the engine. Many parts were beyond tolerance, and the interior of the engine was filthy. My friend paid $3,000 more to get the engine into legal condition, besides the $1,000 for repairing the hoist damage.

The engine is the second-most-important component on the aircraft. You and your passengers are first. Rebuilding an engine is more feasible than raising the dead. No matter how much they cost, quality rebuilds are cheaper than tombstones.

Buying a used engine

Buying a used engine has its advantages, primarily in the cost department. Midtime used engines can be bought for half the cost of a remanufactured engine. A used engine is ready to roar on installation, whereas new engines have to be babied until they're broken in. Tight new/remanufactured/rebuilt engines run hot, a problem compounded by the extensive ground testing needed before a homebuilt's first flight. The builder can easily overheat a tight engine during the static runups and taxi tests.

Used engines run cooler. In an ideal world, used engines would be for ground and flight testing, and a new engine would be installed when the plane has proven itself. Otherwise, low-time used engines are sometimes the best compromise.

Used engines become available for a variety of reasons. An owner might be upgrading to a larger engine, or an accident may leave the engine still usable. The best way to buy a used engine is running—that is, mounted either on an airplane or on a test stand.

Failing that, you'll have to take a careful look at the engine's history. Two key things to look for are prop strikes and sudden stoppage. A prop strike might occur during a hard landing or taxiing into something medium soft.

A sudden stoppage is the king of prop strikes—it hits something hard enough to stop the engine right now. All that rotational energy has to go somewhere. Typically, it goes into the crankshaft, which is the single most expensive part and the most difficult to repair.

After a prop strike of any sort, the engine should be checked for internal damage. Lycoming, in fact, requires a complete teardown. Such a check should be included in the engine log. You might need to play detective; if there is an entry stating that a damaged propeller was replaced, check into how it was damaged. Someone might have backed a pickup into it while the engine was off, or a prop strike might have been ignored.

Another factor to consider when buying any used engine is how long it's been sitting. When an engine doesn't run, rust begins to form on the inside. It doesn't do the engine any good.

Each logbook entry should include the total time on the engine and the date of service. At minimum, each annual inspection should be indicated. If the last signoff was five years ago, there might be significant internal corrosion.

The logbook includes other information as well, such as results of compression checks performed for the annual, airworthiness directives complied with, and histories of all other work on the engine. Logs aren't perfect because they indicate that certain actions were taken, but not necessarily why. However, logbooks are all we have to go on.

That is, assuming they're available. You'll occasionally see an engine advertised as "no logs." This is sticky. The logs could have been lost honestly. Or the engine may have some significant damage history that the owner would rather not reveal. Again, the crankshaft *is* the engine. The owner may have discovered a crack and decided to dump the engine for what he can get. The engine can't be reinstalled in a factory airplane, so the only possible market is the homebuilt.

Buy it, and turn it in as the core for a rebuilt or remanufactured? Careful, all rebuilding agreements require turning over a rebuildable engine. One with a cracked crankshaft might not apply.

If you know the seller personally and trust him, it might not be a bad deal. Otherwise, you're taking a considerable chance. Don't pay much.

Used engines are found from myriad sources. Aircraft junkyards make a major portion of their income from removing and selling the engines from wrecks. Prospective homebuilders pick up engines with the intention of using them someday and eventually give up. Local FBOs and independent A&P mechanics often have one or two sitting around.

One of the best sources is from air-taxi operators. They cannot legally operate an engine beyond TBO, and the engines are babied by professional pilots and regular expert maintenance. All it's worth to them is the core charge. Well-maintained engines can go significantly beyond TBO.

One drawback to this approach is that you'll find only the larger engines available on this basis—few (if any) air-taxi aircraft use O-320s or other engines suitable for the air recreational vehicle and similar aircraft.

Otherwise, check around, and let it be known which engine you need. See what the grapevine can produce. If the right one doesn't turn up locally, check *Trade-A-Plane* and online sources. If you want a cheap engine, start looking *early*. Have the cash on hand and ready—you don't want to lose a good deal just because it takes a couple of days to get a bank loan.

One option might be to buy a freshly wrecked airplane. Some owners don't carry insurance and might be more willing to sell to a private individual. You can afford to offer a bit more than the junkyards because you'd have to pay an additional markup if you end up buying it through them. In addition, the wreck might have usable radios and other parts that aren't included in your kit.

But the salvagers know how much the wreck is worth, and it's usually worth more to them. You probably don't have a need for the wings or tail feathers; they know where to find buyers. In the old days, homebuilders eagerly sought wrecks as sources of cheap bolts, pulleys, and other small hardware. But this hardware should be included with your kit.

Besides, where are you going to store something the size of a wrecked Cessna 172?

No matter how you find a used engine, have an A&P mechanic check it over. If the engine can be run, so much the better. Otherwise, you'll have to trust the mechanic's judgment. But there are many problems that can be missed without a full teardown.

Know which accessories the seller is including. Any additions or deletions are a starting point for price adjustment. Buying a new alternator, starter, magnetos, or fuel pump can make a cheap used engine no bargain. After all, two new mags and their harnesses will cost $1,000 or more.

Used engine prices vary depending on accessories, condition, and total time since overhaul. An engine near or past its TBO is called a *run-out,* and prices are the lowest at this stage. An engine advertised as a "first run-out" has never been overhauled before and is worth more. Common terms used in engine advertisements are shown in Table 5-2.

Just because an engine is past its TBO doesn't make it junk. The condition is more important; if it runs well, if the compression is good, and if a borescope doesn't turn up anything negative, the engine could last for years with just normal attention.

After all, what does TBO mean?

The TBO myth

Everyone puts great stock in the concept of the manufacturer's recommended TBO. Here's a quote from Textron Lycoming's warranty on new aircraft engines:

> If the engine proves to be defective in material or workmanship during the period until the expiration of . . . recommended TBO, or two (2) years from the date of first operation, whichever occurs first, Textron Lycoming will reimburse you for a pro-rata.

In other words, failures are covered under the warranty for only two years. Since Lycoming engines have TBOs of up to 2,400 hours, you would need to fly 100 hours a month.

If you fly only an hour a month, an 1,800 hour TBO doesn't mean that the engine will go 150 years between overhauls. Engine manufacturers recommend a minimum usage of at least 10 hours a month. In a service bulletin, Lycoming says, ". . . because of the variations in operation and maintenance, there can be no assurance that an individual operator will achieve the recommended TBO."

Table 5-2. Common used-engine abbreviations

SCOH	Since chrome major overhaul (chrome increases the hardness of internal components and makes them last longer)
SFREM	Since factory remanufacture
SFRM	Since factory remanufacture (same as SFREM)
SN	Since new (same as TSN)
SMOH	Since major overhaul—usually accompanied with the engine's total time: "200 SMOH TT3040"
SOH	Since overhaul (same as SMOH)
SNEW	Since new (same as TSN)
STOH	Since top overhaul—usually accompanied with the engine's total time: "250 STOH 1500 TT"
TBO	Time between overhauls
TSMO	Time since major overhaul (same as SMOH)
TSN	Time since new
TSOH	Time since overhaul (same as SMOH)
TT	Total TIme
TTSN	Total time since new (same as TSN)
TTSFRM	Total time since factory remanufacture

Other terms of interest

Balanced	Connecting rods and other components have been matched or adjusted to weigh the same—results in a smooth-running engine
First run (out)	Engine has never been overhauled before
FWF	Firewall forward—includes all parts forward of the firewall (mount, muffler, etc.)
Green tagged	Repairable
Red tagged	Unairworthy
Run out	Due for overhaul
Yellow tagged	Ready to run

When an engine doesn't reach operating temperatures on a regular basis, the moisture and acids produced by combustion and condensation collect in the engine instead of being vaporized and eliminated through the exhaust and crankcase breather. These by-products remain and contribute to the formation of rust on the cylinder walls, camshaft, and tappets. When the engine is finally started, this rust becomes a fine abrasive. As the engine components wear, the metal scraped off attacks softer metals such as piston pins.

Keep in mind, TBO isn't a regulatory number. The engine doesn't have to be overhauled at that point; and nothing guarantees that it will last even that long.

Storing the engine

It's too easy to rush out and buy the engine at the same time as the kit. Sometimes you can't help it. If you find exactly the right engine at an incredible price, it's best to go ahead and buy it.

But it can cause problems. Building the plane might take years. Where are you going to store this 300-pound block of metal? As mentioned in the last section, sitting idle does nasty things to the inside of an engine. What's going to keep it from rusting up?

The first order of business is to find a nice dry place to store the engine. The second is preparing the engine for storage, a practice called *pickling*.

The procedure varies among engines, but it generally begins by draining the oil and replacing it with a special pickling oil. The sparkplugs are removed, and pickling oil is squirted into each cylinder. Then the engine crankshaft is turned over by hand to spread the oil on the cylinder walls. Dehydrator plugs are screwed into the sparkplug holes; these plugs contain silica gel to absorb moisture. Similar plugs are inserted into the exhaust pipes. Crankcase vents are sealed off, and the engine is wrapped in plastic (Fig. 5-6). The engine's overhaul manual will give exact instructions.

Pickling isn't a permanent procedure. The silica gel in the dehydrator plugs must be replaced periodically, and the crankshaft should be turned over occasionally to keep pickling oil on the cylinder walls. Count on checking the engine at least once a month.

Fig. 5-6. *A pickled Lycoming O-320.*

Summary

Certified aircraft engines have many advantages to the homebuilder. Reliability has been mentioned. Maintainability is another factor; if you're on a trip and something breaks, most parts will be available right at the airport.

It's hard to pin down engine prices. Here's an example of what might be expected in the mid-1990s when buying an engine such as a Lycoming O-320:

New price (list): $30,000+

Street price (new): $24,200

OEM price (via kitplane manufacturer): $22,500

Manufacturer remanufacture: $20,700 (plus core)

Manufacturer overhaul: $17,300 (plus core)

Low time: $18,000

Midtime: $12,000

Run-out: $6,500

These prices, especially those for the used engines, are just approximations. Shop around to find the best deal.

Determining the cost of operation is easy; after all, there are a number of production aircraft flying the same engine. The most commonly used engines will burn fewer than 10 gph, usually in the range of 6 to 8 gph. As mentioned earlier, picking an engine that can run on autogas cuts direct costs almost in half.

UNCERTIFIED ENGINES

As nice as certified aircraft engines are, they also have several disadvantages. Their basic design dates from the 1930s. High technology, in itself is not an advantage. One can make a considerable case for the "tried and true" traditional designs. However, advances in metallurgy and electronics in the last 50 years have been considerable.

Technological advancement hasn't made inroads in small aircraft engines because development and certification costs aren't justified for the available market. General Motors can develop a new car engine and expect to sell millions of copies. Lycoming would be lucky to sell hundreds.

A certified engine's second disadvantage is the lack of adequate engines in the lower-horsepower ranges. The Continental A-50 and A-65 engines made the J-3 into the first mass-produced RV. Sixty-five horsepower is a pretty good value; it's adequate for sprightly single-seaters and light two-seaters. Yet no in-production fully certified engine is available at anywhere near this power level.

The A-65 engine has been out of production for more than 40 years. It and other small engines (such as the Continental O-200) have become somewhat rare and therefore expensive. At one point, the supply of small Continentals dried up owing to their popularity in homebuilts such as the VariEze. And the pressure from RV builders has made Lycoming O-320s and O-360s somewhat rare on the used market.

The reliability advantages of obsolete certified engines can be disputed. How many times can an engine be rebuilt until it's no longer safe? Some engines and their accessories (magnetos and generators, for instance) are kept operating through stocks of leftover original parts. Can you trust a 30-year-old fuel pump?

The final major disadvantage of the certified engine is cost. A rebuilt O-360 might cost $10,000 or more; then, if a cylinder goes bad, replacement "jugs" run around $1,000 each. Yet new auto engines go for less than $2,000; buy a reduction drive for $3,000, and you'll have a more powerful engine for half the price. Parts and services are available cheaply at auto parts stores. A set of four VW cylinders and pistons sells for less than a hundred bucks.

However, if the engine you select isn't one that is supported by the kit manufacturer, you can have quite a task ahead of you. Most firewall-forward instructions can just as well be thrown out, and you have to work out the mechanical details yourself. And perhaps you will even need to build a new cowling from scratch.

On the positive side, many of the alternate-engine suppliers sell "firewall forward" packages for their designs. These take a lot of the worry out of installing another engine. The drawback, of course, is the additional cost of the package.

Even with these problems, alternate engines are popular. The online forums and mailing lists usually feature several people who have been through the same processes you are contemplating.

Let's take a look at the main contenders among uncertified engines.

Dedicated engines for homebuilts

Homebuilts have become popular enough that several companies have developed brand-new engines for the market. Some were based on engines developed for cars or sports vehicles. Others were developed specifically for aircraft use. Often they are allowed to be used on small production aircraft in the countries where they were developed.

The Aerovee engine (Fig. 5-7) is a twenty-first-century Volkswagen conversion. It's based on the 2,180-cc VW engine and produces 80 hp. The kit sells for $6,000.

In the 1980s, two Australian men developed a small two-seat composite aircraft for certification and production. A month after the new Jabiru received its Australian type certificate, the engine manufacturer announced it was ceasing production. Having experienced problems with two-stroke engines, the two men decided to develop their own 60-hp four-stroke engine.

This engine eventually led to today's Jabiru 2200 (80 hp), Jabiru 3300 (120 hp), and Jabiru 5100 (180 hp). Prices begin at about $10,000 for the 80-hp model. It's currently being used in planes such as the Sonex (Fig. 5-8). The realities of the kitplane business are reflected in the engine's design philosophy; the engines are designed specifically to be manufactured in small-batch quantities by subcontractors using computer numerically controlled (CNC) machine tools.

The availability of CNC machining also came into play with another Australian engine. Rather than small light aircraft powerplants, Paul and Matthew

Fig. 5-7
The Aero Vee engine is derived from the Volkswagen but features improved power and reliabilty for aircraft use.

Sonex.

Chernikeeff went for the old-time look with their Rotec R2800 radial engine (Fig. 5-9). This engine, appearing very much like a three-quarter-scale Warbird radial, is designed for homebuilts with a nostalgic bent. This 110-hp engine, ready to run, sells for about $12,500.

Rotax made its name in aviation with its two-stroke engines, but the company has expanded into the mainstream with its four-stroke Rotax 912 and Rotax 914. These engines are almost perfect for homebuilts needing 100 hp or so, with prices starting about $13,000. These engines seem to have achieved a better reputation than their two-stroke brethren. In fact, one model is now certified and powers some standard-category trainers.

In addition to these engines, the advent of light sport aircraft should encourage other companies to enter the market. These will present greater choices for the homebuilder, as well as make used engines of these types available.

The same warnings should be considered as I mentioned for any new aviation company. Business failures are common; this could leave the purchaser without a source of spare parts. There was a "turf war" among distributors of one new engine in

Fig. 5-8. *While the Sonex was designed originally around the Aero-Vee engine, the company also offers a Jabiru option.*

Fig. 5-9. *The Rotex R2800 radial not only appeals to the nostalgia buffs but also is a pretty good airplane engine in its own right.*

which people who bought engines from another, then-defunct distributor were denied support.

These new engines don't have 50 years of in-flight experience like Lycomings and Continentals do. New engines generally have some "teething problems" that irritate or even endanger initial buyers. As with homebuilt designs, *not* being the first customer for a new engine design is a good idea. The engine developers don't always pay for the changes required.

Liquid-cooled auto engines

Conversions of liquid-cooled auto engines have been popular since the 1920s. Some have even been certified, the Ford Model B engine in the Funk for one. Quite a number of Pietenpol homebuilts have flown with Ford Model A engines as well.

Traditionally, auto engines have been used mainly to fill the gap when a properly sized aero engine isn't available. This is changing now that the prices for new Lycomings have passed the $25,000 level.

The biggest attraction for auto engine conversions is their lower price. A Ford V-6 engine can be picked up at a junkyard and completely overhauled for $2,000 or less. The biggest problem with auto engines is their horsepower curves: Auto engines are designed to produce their maximum horsepower at relatively high speeds—around 5,000 rpm.

Aircraft propellers aren't efficient at those speeds. The tips are turning at supersonic speeds, causing excess drag, noise, and lack of efficiency. Aircraft engines have their peak power at around 2,500 rpm to match the most efficient propeller speeds.

Most auto engine conversions, then, include a reduction drive (Fig. 5-10) to convert revolutions per minute into thrust. A 2:1 reduction ratio changes a 5,000-rpm horsepower peak into 2,500 rpm at the prop. These units are called Propeller Speed Reduction Units, or PSRUs. With a PRSU, the auto engine is more suited for aircraft use.

Power usually is transmitted to the prop via a cog belt drive incorporating the appropriately sized pulleys. While some homebuilders design their own drives, most buy one of the commercially available units. These will cost about $3,500. The engines themselves can come from the junkyard for less than $1,500. With inspection and rebuild, you'll probably spend less than $6,000 on the engine and reduction drive.

Ignition systems are one issue to consider. Most auto-engine conversions use a standard auto ignition system controlled either by the computer used with the car or by an aftermarket unit. Many people argue that an electronic ignition is more reliable than the old-fashioned magneto used on certified engines. If you like ignition redundancy, several companies offer dual-ignition modifications. Sometimes these use separate spark plugs in each cylinder, but often two separate ignition systems fire the same plugs. Specialists claim that the hot sparks produced by modern electronic ignitions prevent the fouling that drove magneto-based ignitions to two plugs per cylinder.

Auto engines are a little heavier than aircraft engines, but liquid-cooled engines are more efficient. Sizewise, they're similar but usually are deeper and narrower. Figure 5-11 shows a size comparison between the Continental O-200 and the CAM-100, a converted Honda Civic engine.

Fig. 5-10. *Most auto-engine conversions require a speed-reduction unit to get the most efficiency out of both propeller and engine. This is a Northwest Aero Products reduction drive mounted on a Chevrolet V-6 engine.*

CAM-100   **O-200**

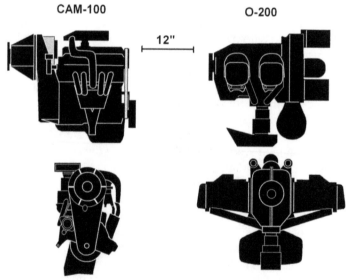

Fig. 5-11. *The CAM-100 is a converted Honda engine. Note that an airplane using this engine might need a deeper but narrower cowling than a Continental O-200.*

Chevrolet and Ford engine aficionados have carried their war from the streets to the air, and you often see aircraft powered by either engine. Various models of the Subaru are also popular, especially since they are sometimes similarly configured to conventional aircraft engines. The Mazda rotary engine springs up on occasion. While it generally has higher fuel consumption and generates a lot of heat, it gets good marks for smoothness and reliability.

Traditionally, the main drawback to selecting an auto conversion is the extra work and construction time. Converting the engine is at least as complex as building the airplane.

This long route is still available. However, complete firewall-forward engine packages mentioned earlier are available for many auto-engine conversions. Such a package does boost the price a bit, though. A ready-to-fly Subaru conversion costs about the same as a good used Continental O-200. However, the total package cost is not as affected by total horsepower. A firewall-forward package for an O-200 equivalent may cost as much as a used O-200, but an auto-engine conversion for a 360-hp Lycoming O-540 is probably going to be quite a bit less.

The cheapest route is to build it all yourself. But you'll spend a lot of time working on it. The camshaft might have to be reground or coolant port locations changed. The reduction drive itself must be assembled and installed. You'll have to build a custom engine mount, design a carburetor heat system, install a coolant header tank, and have a special prop built.

And this completely ignores the work involved in preparing the aircraft to use the converted car engine. Here's an excerpt from *Sport Aviation* magazine concerning an auto-engine conversion in an RV-4:

> The first thing he had to do was remove the aft portions of the cheek cowls . . . rivet in some aluminum angle to make up, structurally, for their absence. . . . He had two long, skinny radiators made and mounted under the bottom of the fuselage. . . . instead of a P-51 type scoop, he built a second bottom into the fuselage. . . . He welded-up his own bed-type engine mount, and hammered out an aluminum nose bowl [not to mention working out routings for throttle cables, fuel and coolant lines, and more].

All of this could have been avoided by installing the O-320 called for in the plans.

If you decide to install an auto engine, take an honest look at your past engine-work experience. If you haven't much experience working on car engines, it'd make more sense to buy a commercially available conversion. Talk to the company before deciding, and determine how much the company will support your effort. After all, the company's only market is homebuilt aircraft. Check on the availability of standard engine mounts and whether anyone else is installing the engine on the same type of airplane.

For further information on auto-engine conversions, see *Contact!* magazine (*http://www.contactmagazine.com/*) or *Converting Auto Engines for Experimental Aircraft,* by Richard Finch.

AUTO-ENGINE SAFETY

Auto-engine conversions have rather spotty reliability. Some conversions have been flying reliably for hundreds of hours, and some don't even make it around the pattern on their first flight. Unlike certified aircraft engines, there is no FAA to track trends and notify users if a problem is found. The builder is generally on his or her own. Sometimes the company that developed the conversion will send out notices if a problem occurs, but of course, this doesn't always happen.

If you want to get in a fight with a bunch of homebuilders, ask them about the suitability of car engines for aircraft use. You'll find some folks who'll claim that the auto conversions are more durable and others who'll swear that parts such as crankshafts will snap if you wink at them.

The truth is somewhere in between. In examining the accident reports for auto-engined aircraft, it is apparent that the basic engines themselves don't fail all that often. What does cause problems are the accessories, such as the cooling system and the ignition and exhaust systems.

It's difficult to develop useful statistics on auto-engine accidents because the FAA aircraft records don't always specify what kind of engine is installed in a homebuilt. Without knowing the total number of auto-engine conversions in the experimental fleet, we can't determine the accident rate for auto-engine homebuilts versus the overall fleet.

However, examining the accident reports themselves provides some insight. Of the planes with aircraft engines involved in accidents, problems with the powerplant were blamed in about 17 percent of the cases. However, about 28 percent of the accidents affecting auto-engined homebuilts were due to engine difficulties.

Air-cooled auto engines

Air-cooled engines have always been a subset of the auto world. They faded even further when the federal government required automakers to control pollution. For a variety of reasons, it was more difficult to institute these controls on air-cooled engines. Thus the engines pretty much faded away from the new-car front, although some companies, such as Subaru, have used a combination of air and liquid cooling for their engines.

While these engines have (mostly) faded from the new-car showrooms, the air-cooled auto engine hasn't faded away. The original VW Beetle was manufactured until recently in Mexico and is still produced in countries such as Brazil. The original Beetles are a hot collector's item in the United States, and parts are plentiful and cheap. Similarly, General Motors' Corvair is kept alive by aficionados.

Let's take a brief look at these engines.

THE MIGHTY VW

Certified engine disadvantage number one (old technology, dating from the 1930s) also applies to the Volkswagen engine. Ferdinand Porsche designed the engine for Hitler's "people's car" concept. Air cooling was picked in the interest of simplicity and reliability.

The people's car was magically transformed into the slabsided bucket car (*Kubelwagen*) for the German army. Wartime pressures refined the engine and drive train. The *Kubelwagen* continued its yeoman service after the war, obligingly being blown up, machine-gunned, and driven off cliffs for the cinematic armies of Warner Brothers, MGM, and Twentieth-Century Fox.

Volkwagenwerk AG recovered quickly after the war, producing almost 2,000 civilian cars by the end of 1945. By the late 1940s, humpbacked autos swarmed from the Wolfsburg factory. The VW Beetle became the most widely recognized car in the world. While the parent factory converted to more modern designs in the 1970s, the Bug and its variants are still produced in Brazil and Mexico.

Experimentation with aircraft began shortly after the war. With a continuing shortage of small aircraft engines, Europe has been the most enthusiastic proponent of VW power. Limbach of Germany even certified a VW-derived engine for use in motorgliders.

Acceptance in the United States was slower. Thousands of small Continental engines were still available. The Volksplane (Fig. 5-12), one of the first simplified homebuilt designs, was one of America's first popular VW-powered designs. Its VW engine was converted for aircraft use by removing the flywheel, starter, and generator, replacing the distributor with a dune buggy magneto, and replacing the crankshaft pulley with a propeller flange.

The VW's biggest boost came with the Q2. The original single-seat Quickie used an 18-hp Onan GPU engine. The two-seater Q2 was designed for VW power, getting almost 200 mph on the Revmaster conversion. While few kits currently list the VW as the standard engine, it's a common builder modification, and the VW engine appears often on plans-built aircraft.

The usual VW aero conversion moves the carburetor underneath the engine (using a Posa or other aircraft-style carb) and installs a standard aircraft magneto at

Fig. 5-12. *The Volksplane was the first popular U.S. homebuilt using a Volkswagen engine. This example sports a closed-canopy arrangement.*

the aft end of the crankshaft. Prop hubs are either the machined-to-fit or shrink-on variety, where the hub is slightly undersized but, when heated, expands enough to slip over the crankshaft.

Horsepower almost can be selected. The standard 1,600-cc engine produces about 45 hp, but larger cylinders and pistons bolt on for higher performance. Only the very largest engines (2,100 cc and up) require machine work, and they produce about 65 hp. VW-derived engines offer 75 hp and up. Ready-to-fly weight for a basic VW engine is about 160 pounds.

The VW engine has four bolt holes on a wide flange located at the rear of the engine. This is perfect for simple and cheap engine mounts; in fact, the Volksplane merely bolts the engine directly to the firewall.

Assembling your own VW engine is the cheapest route. A good used engine can be bought for $500 or so, with another $500 budgeted to replace major components like cylinders and pistons. Complete kits are available, starting around $1,700 for a 1,600-cc model.

However, these prices don't include carburetor, intake manifold, accessory cases, or ignition system. Some builders use automotive carbs on a stock manifold. This places the carb above the engine, which usually requires installation of a fuel pump. A below-engine mount gets enough "fuel head" via gravity.

While the VW engine looks pretty good on the surface, it has some long-standing disadvantages when it comes to aircraft use. As with liquid-cooled auto engines, the VW engine produces its maximum power at an rpm level that is a bit high for efficient propeller operation. However, the VW engine produces pretty good power at a significantly lower rpm level than the liquid-cooled crowd, about 3,400 rpm. A few reduction systems have been developed, but the usual practice is to use a smaller, lower-pitched prop to allow the engine to turn that fast. The main drawback is in acceleration; the VW-powered planes take a while to get up to speed and aren't as suited for short or rough runways. The VW and VW-derived engines work best in low-drag, speedy airplanes.

The VW engine also is hurt by its poor power-to-weight ratio. A 50-hp VW engine weighs about 160 pounds, ready to fly. A Continental A-65 with 15 hp more weighs only 15 pounds more. And the propeller *likes* these horses—the A-65 produces full power at only 2,150 rpm. (At the opposite end of the spectrum, the Rotax 503 engine has the same horsepower and, with a gear-reduction drive, achieves full power at about 2,700 rpm at the prop, and the Rotax is 60 pounds lighter.)

One curious fact becomes apparent when you study VW-powered homebuilts— except in a few cases, the VWs eventually have been replaced with aircraft engines. The Q2 turned into the Q-200, with a Continental. The VariEze went the same way. The Zenith Zodiac transitioned to the Rotax 912 four-stroke engine.

VWs turn in the wrong direction, counterclockwise as viewed from the cockpit. This can cause transition difficulties because left rudder (instead of right) must be held on takeoff. The small prop diameter magnifies the P-factor effect, which means that the plane swings more when power is applied for takeoff.

Opinions are mixed on VW engine reliability issue. Single-ignition engines run their own risks. Some argue that the engine was designed for 1,600 cc and 36 hp; anything more reduces longevity. The stock VW engine probably operates with a narrower margin of safety than a certified engine. Then again, the VW design has stood up to auto and air racing for years, as well as pushing thousands of dune buggies.

VW engines have one recurring maintenance problem. The valves have a tendency to become too tight and must be adjusted at least every 25 hours. This is a minor procedure that takes only 15 minutes or so. Some companies offer hydraulic lifters that eliminate the difficulty.

Drawbacks or no, the VW engine has remained active on the homebuilding front. It's even had a new lease on life, sort of: Some small aircraft use a cut-down VW engine sporting only two cylinders. Surprisingly, much of the horsepower is retained with a significant weight drop.

CORVAIR

Chevrolet introduced the Corvair in 1960. Several years earlier, the company had noticed the rising interest in compact cars, especially the Volkswagen, and decided to produce a family of cars that was cheaper to build, cheaper to own, and offered sporty handling.

The resulting Corvair shared a lot of traits with the VW Beetle—air-cooled, rear-mounted engine, a compact transaxle, and similar suspension. But the car was designed for the American market. It was larger both inside and outside, boasted a 100-hp six-cylinder engine (versus the VW's 40-hp four-cylinder engine) and unitized construction, and was the first production car with an optional turbocharger.

Aircraft builders pounced on the Corvair's air-cooled engine almost immediately. Bernie Pietenpol installed one in an Air Camper the year the car was first introduced. Interest has continued since, with Air Campers and numerous other aircraft types taking to the air on GM's engine.

In its most common aircraft conversion, the Corvair produces 100 hp at about 3,200 rpm for takeoff. The basic auto engine was rated at 180 hp, so the aircraft conversions are pretty conservative. It might be thought that parts would be difficult to find for an engine that went out of production over 40 years ago, but that doesn't seem to be the case. After all, GM produced almost 2 million of them.

The current "guru" of the Corvair world is William Wynne of Port Orange, Florida. He is the author of *Converting Corvair Engines for Experimental Aircraft* and sells complete kits for assembly of Corvair aircraft engines on his Web page at *http://www.flycorvair.com.*

Two-stroke engines

The Rotax two-stroke engines single-handedly saved the ultralight industry and were a shot of fresh air to small homebuilts as well. The ultralighters were buffeted on all sides: TV exposés on safety, noisy engines irritating the public, shoddy manufacturers trying to cash in on a craze, and cobbled-up conversions of two-stroke stationary engines that rolled their eyes up and died at the least provocation.

Along came Rotax, with engines proven in snowmobiles, not just in stationary power units, a wide range of engines, effectively tuned muffler systems, and a bolt-on gearbox that eliminated prop noise as it doubled or tripled efficiency.

Is it any wonder that most ultralights and small homebuilts now use Rotaxes? Traditionally, the bigger the engine, the better the horsepower-per-pound ratio. The standard VW engine has a 1:3 ratio; the O-200, a 1:2.5 ratio; the Lycoming O-360, a 1:1.87 ratio; and so on. The Rotax 582 (Fig. 5-13) has about a 1:1.5 ratio; it produces 65 hp and weighs about 100 pounds. Best of the lot.

Fig. 5-13. *The two-stroke Rotax 582 features dual ignition and oil injection and offers a high power to weight ratio. Those who desire four-stroke reliability tend to opt for the four-cylinder Rotax 912.*

The popularity of the Rotax means good support. A number of companies sell accessories. Propellers are available in a variety of styles and at low prices. The liquid-cooled Rotax 582 is standard equipment on an astounding number of kitplanes.

The prices of these engines, though, have gone up quite a bit. Five years ago, many small homebuilts got along quite happily with the $1,600 fan-cooled single-ignition Rotax 503. Nowadays—well, by the time you get a starter and a heavy-duty gearbox, you'll have $6,500 in your dual-ignition water-cooled Rotax 582.

Other companies offer two-stroke engines with the same advantages of the Rotax, although their dealer networks aren't as extensive. Zenoah and 2SI engines have been used in ultralights for years.

Two-stoke engines usually come with a tuned muffler and exhaust system, which optimizes the performance. The length of the system is critical—don't shorten the pipes. You can bend and position the exhaust in a variety of ways, but the centerline length should stay the same. If you wondered why so many small homebuilts equipped with two-strokes have the ugly muffler and exhaust pipe hanging below the fuselage, now you know.

Most two-stroke engines need oil mixed with the gas. Maintaining the proper ratio is critical; too much oil, and excessive carbon deposits form. Too little, and the engine might overheat and seize. If you are on a cross-country trip, you can't just taxi up to the gas pumps and fill up. Rotax recommends a separate container be used to blend the gas and oil before the mixture is added to the tank. It's difficult to carry such a container in the aircraft, so you end up carefully adding the right measure of oil after a fill-up. But unless the aircraft's fuel tank can be sloshed about, the oil might not mix properly.

The oil will tend to separate if it is not agitated regularly. If the plane will be stored over the winter, drain the gas and use it in the snowblower instead. About a quarter of all two-stroke engine failures are caused by fuel quality problems, so extra attention to fuel mixture and quality is warranted.

This leads to the major objection to two-stroke engines in general: reliability. They aren't certified engines; they were designed for other purposes and adapted to aircraft. The Rotax's TBO is 400 hours, less than a quarter of a typical aircraft engine.

In the August and September 1989 issues of *EAA Experimenter* (reprinted in the June 1990 issue of *Kitplanes*), Hank Fritz of Renton, Washington, presented the results of a survey of air-cooled two-stroke engine failures. The combination showing the best reliability was a Rotax 503 with dual carburetors, fan cooling, and mounted upright in the tractor configuration. Change the plugs and fuel filter every 50 hours, along with checking the timing and fan belt and cleaning the air filter.

However, as discussed earlier, a 400-hour TBO means little to sport flyers. Fly 50 hours per year, and the engine might not need an overhaul for eight years. And Rotax engines are used so widely in nonaviation applications that a complete overhaul costs less than $500. You can do it yourself for even less.

Two-stroke engines have other, minor disadvantages. Vibration is one. They're shaky little devils and require careful attention to shock absorption. The gear-reduction systems often turn the prop counterclockwise, which, like the VW, requires left rudder on takeoff.

Other options

Diesel engines have a long, if not necessary extensive, history in aviation. Junkers equipped its JU-86 twin-engine transport with diesels in the 1930s. Diesel engines offer better fuel efficiency and eliminate the need for an ignition system, but generally they are heavier and more expensive than their gasoline-powered counterparts. After the Junkers JU-86, interest languished for a number of years.

Diesel engines started to rise again as the refineries around the world started to cut back on the production of aircraft gasoline. Airports in the third world were handling mostly jet or turbine aircraft; they tended not to carry the 100LL fuel required by most existing aircraft engines.

Gasoline engines are like thoroughbred horses; they require great attention and the finest feed. Diesels, on the other hand, are mules. They "eat" practically anything.

Several diesel car owners in the United States power their vehicles on the grease left over from cooking french fries at fast-food restaurants.

It's true that you're not likely to find a Burger King at a remote airstrip. But you *are* almost certain to find jet fuel. Or failing that, you can use the ordinary diesel fuel that powers local ground vehicles.

As of this writing, diesel engines have yet to make major inroads into homebuilt aviation. But Thielert Aircraft Engines has certified its Centurion 1.7 diesel, and Deltahawk is not only in the final states of certifying its diesel but also is flying a Velocity as a test bed.

Of course, there's another way to burn jet fuel in aircraft: Install a jet or turbine engine. Several homebuilt designs accommodate either modern small jet engines or older engines that are available at much lower costs. Prototypes are flying, and some customer-built planes are under construction. Since jet aircraft require higher altitudes for efficiency, these planes usually must accommodate oxygen systems and/or pressurization. Not the best project for the garage mechanic.

Turbine engines are a bit more common. The 650-hp Walter M601 engine, produced in the Czech Republic, has found its way onto the noses of homebuilt aircraft such as the Legend, the Aerocomp, and the Lancair Sentry.

On the smaller, side, there has been considerable activity on conversions of the Soloy T62 APU unit to an aircraft powerplant. This results in a very small, very light (112 pounds) package that produces over 150 hp (Fig. 5-14).

Fig. 5-14. *This small turbine engine, converted from a helicopter APU, offers 150 hp in a package weighing just over 100 pounds. However, its fuel consumption is much higher than similar-power reciprocating engines.*

As in any alternative engine, working out the installation details of these engines may take more time than building the airplane itself. Also, while jet fuel may be more common than avgas at third-world airports, that isn't the case in the United States. Something other than a diesel or turbine would be the best pick unless you always plan to operate from major airfields.

Summary

Engine selection depends on a number of factors: the kitplane manufacturer's recommendation, your engine budget, and your mission. A summary of the weight and power statistics for common engines used in homebuilts can be found in Figs. 5-15 (100 hp or less) and 5-16 (over 100 and less than 300 hp).

At present, there are three commonly used engines that produce over 300 hp. These are the Walther 601D turbine (657 hp, 425 pounds), the Lycoming IO-720 (400 hp, 568 pounds), and the Vedeneyev M-14P radial (360 hp, 475 pounds). Derated version of these can be found as well.

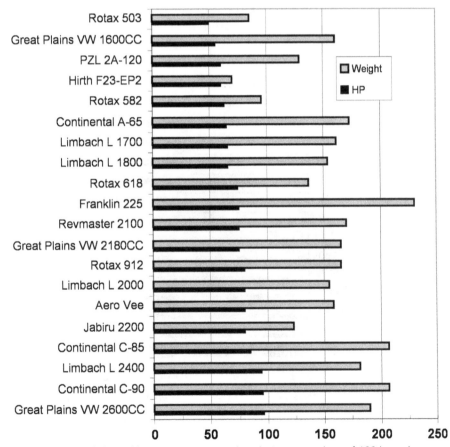

Fig. 5-15. *A weight and horsepower comparison between engines of 100 hp or less.*

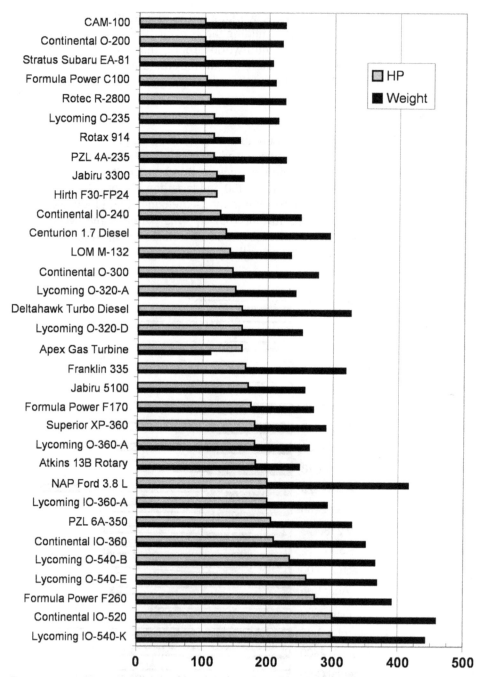

Fig. 5-16. *A weight and horsepower comparison between engines of more than 100 hp and less than 300 hp.*

PROPELLER SELECTION

Prop selection isn't as easy as you might think. Not only must it be matched to the engine but to the aircraft as well. A Zenith CH701 and a Europa (Fig. 5-17) might both use the same engine, but the Europa's mission of high-speed cruising requires a different prop.

Two parameters define the characteristics of a particular propeller. The *diameter* is the distance across the swept circle of the prop disk. The *pitch* describes the angle of the blades. A prop with a 46-inch pitch would pull itself 46 inches forward in a single revolution.

Zenith STOL CH 701
Construction: Aluminum
Gross Weight: 960 lbs
Cruise: 80 MPH
Takeoff: 75 feet
Engine: Rotax 912

Europa XS Mono-Wheel
Construction: Composite
Gross Weight: 1370 lbs
Cruise: 161 MPH
Takeoff: 600 feet
Engine: Rotax 912

Fig. 5-17. *While a STOL airplane such as this Zenair CH701 and an Europa might use the same engine, their propeller requirements are far different.*

Prop options

You'll see four basic types of propellers on homebuilt aircraft:

1. *Fixed-pitch props, either of wood or metal.* These are just like the units you'll see on most small production aircraft. Fixed-pitch props are defined by the diameter and pitch; that is, a 72 × 48 prop would have a 72-inch diameter and a 48-inch pitch.

2. *Ground-adjustable props.* These allow you to change the prop prior to takeoff to match the anticipated needs of the upcoming flight. While such props can't be changed in the air to match particular conditions, they give you the ability to optimize for cruise flight when you're on a cross-country trip.

3. *Constant-speed props.* These are common on aircraft such as the Cessna 182 or Piper Arrow. Using a cockpit control, the pilot selects the desired engine rpm, and the propeller automatically adjusts its pitch to maintain it.

4. *Cockpit-adjustable props.* These allow the pilot to change the propeller pitch to optimize performance. They differ from constant-speed units in that the pilot directly controls the blade pitch rather than selecting the desired engine rpm. These props are cheaper than constant-speed units but do require a bit more work by the pilot.

The pocketbook issue is important. Most aircraft perform best with a constant-speed prop. But we're talking $3,000 and up just for a used one. Many of the smaller planes don't see enough of a performance improvement to justify the cost.

Many builders decline the additional expense and go with a fixed-pitch prop. The other options have a number of proponents as well. The kit manufacturer will have the best handle on the right engine and propeller combination for your aircraft. If you decide to go beyond their recommendation, try to find someone who has installed a similar setup on the type of airplane you're building.

Fixed-pitch propeller parameters

If you decide to go with a fixed-pitch prop, you'll have to select the diameter and pitch. Three items affect selection: mission, aircraft, and engine.

The mission decides your choice of a prop optimized for climb, cruise, or a compromise between the two. A cruise prop will maximize speed at the cost of short-field performance.

One prop maker has an all-wood design in which the twist of the blades actually changes depending on the load on the propeller. A friend installed one on his BD-4 and found that the prop actually did a pretty good job of optimizing itself to various flight conditions. While he was reasonably happy with it, he eventually replaced it with a cockpit-adjustable model.

Since your aircraft already has been optimized for both the engine and the mission, how much can it affect propeller selection? Simple: What ground clearance does it leave for the prop? Small aircraft have short gear legs. A standard prop might be too long; even if it doesn't actually touch the ground, a hard landing might make the prop hit. So some planes use a shorter prop for clearance reasons and compensate with increased pitch.

The engine affects prop selection in two ways: First, there's no single standard prop hub. The bolt circle diameter varies, and between four and eight bolts might be used. Second, the pitch and diameter must be selected based on the mission. The horsepower output of the engine is directly related to its rpm. But the rpm isn't just controlled by the throttle—forward speed decreases the blade's effective angle of attack and allows the engine to turn faster.

Your objective is to select a prop based on the desired aircraft speed at which engine power must be maximum. If you need short-field performance, the engine should be running at its maximum torque point at full throttle and 60- to 80-mph airspeed. This requires less pitch and a larger diameter.

How much difference does it make? Cessna 150s and early Lancairs both mount Continental O-200 engines. A typical cruise prop for the Cessna 150 is 69 inches in diameter with 50-inch pitch, and it gives about 125 mph. The same engine on the Lancair must be propped to allow 200 mph; one prop maker recommends a 58-inch diameter and a pitch of *71 inches.* The Lancair has limited ground clearance as well. Both factors serve to increase pitch.

Metal versus wood

At some point, you'll have to choose between props made of wood or metal. There aren't any metal props made specifically for homebuilts; hence metal props will be expensive certified models. If you must have a new metal fixed-pitch prop, count on spending $2,000 to $3,000. However, used props can be found for less than $1,000. Propeller shops often can make a good homebuilt prop by cutting-down a damaged certified model.

Wood props are cheap in comparison; prices run about $500 or less for most engines. A Rotax prop might cost only $200 or so. There are a number of custom prop-making companies aiming specifically at the homebuilt markets.

Should you pick metal or wood? Like wood airplanes, wood props seem to run smoother. Because they are in production at low cost, you can order a prop specially for your aircraft and mission. Wood props are classier as well.

However, metal props are durable. Bent props can be repaired to a certain extent. Wooden props don't bend; they break (although by breaking, they can prevent engine damage in a prop strike). If you buy a wood prop, you're stuck with the diameter and pitch. Metal props can be shortened and repitched by certified propeller shops. Of course, for the price of a single new metal airscrew, you could buy a wood prop for every season.

Metal can pick up some nicks and dings from gravel and debris, but these can be repaired easily. Flying rocks can break wooden blades. This problem is more severe with pusher installations. However, pusher applications are tougher on props because the blades have to "chop" through the wash from the wing, pylon, or whatever on every revolution. This is why you'll always see wooden or composite blades on Long-EZs and Velocities.

Finally, a metal propeller is impervious to weather. Wood props should be protected from moisture; small cracks in the varnish or epoxy allow water to enter and cause an unbalanced condition. Use a cover or move the blades to horizontal when the plane must be kept outside. Also, flying in the rain can damage the wood's finish.

Wooden props need careful attention to the prop bolt tension. The wood expands and contracts with the weather. After a long dry spell, the hub shrinks and makes the prop bolts relax their grip. Always install a steel crush plate in front of the prop, and periodically retorque the bolts. If you buy a wooden prop well in advance, store it in the stereotypical cool, dry, place. The wall of your den might be a good spot.

Propeller selection

If the kitplane manufacturer recommends a particular engine, it will specify the propeller as well. If it doesn't, picking the right prop can be difficult. If you've decided to go with metal, consider the same propeller as the production aircraft closest to your kitplane with the same engine.

When hunting for propellers and other aviation accessories, you often hear the term "tagged," as in "yellow-tagged" and "green-tagged." When a component is checked by an approved maintenance facility, a tag grades the component based on its condition. In the case of propellers, the facility checks various dimensions of the propeller against the minimum standards published by the propeller manufacturer.

A prop that meets all standards is given a yellow tag, which means, "This component is airworthy and operational." If the prop is damaged, the facility compares the amount of damage to the manufacturer's repair allowances. A prop missing a little from a tip often can be shortened slightly, and bent blades can be straightened to some extent.

If the discrepancies are within specified limits, the prop gets a green tag, meaning, "Unairworthy, but repairable." Once repairs are completed, the tag is replaced by a yellow one. The rest are given red tags, which prohibit use in certified aircraft.

But kitplanes aren't certified. Propeller shops can rework some red-tagged units into perfectly acceptable homebuilt props. These sell for between $500 and $1,000. Be advised, though, that sometimes these reworkings violate the "repairable" ranges specified by the prop manufacturers. There is a danger of harmonic vibration and fatigue problems in cut-down metal props.

Some of the bigger-engined RV-4s and RV-6s are being built with constant-speed propellers—new cost around $4,000. One common Hartzel model is used on selected Mooneys and Arrows. The RVs can take a slightly cut-down version of this prop; prop shops start with a red-tagged Hartzel, cut the blades down a bit, and sell it for about $1,000 less than a new one.

If you buy a green-tagged fixed-pitch prop, count on spending about $500 to get it rebuilt. If you need it repitched, add around $200. Check with local propeller facilities before buying a new prop. You might be able to save 50 percent with a used, refurbished model.

But metal can't do everything. There's a limit to how much a prop can be repitched and shortened. Sometimes it's easier just to buy a custom wooden prop. Talk to the propmaker; see if he's made props for your kitplane before. Props for the more common kitplane-engine combinations often can be bought right off the shelf. It doesn't guarantee a perfect solution. A friend has bought three props for his VariEze and still isn't satisfied.

However, you aren't buying from a factory. Wooden prop shops generally are one-person operations. Many don't make certified props and therefore aren't necessarily familiar with federal standards. Ask around, and look at some examples. A local wooden prop owner found little pieces of epoxy varnish littering the ground one day; flakes were peeling off his brand-new prop.

Selecting and finding a prop isn't a quarter of the battle involved in picking the right engine. But it has almost as great an effect on performance and satisfaction.

INSTRUMENTS

Your instrument decisions are fairly easy: Are you building for Visual Flight Rules (VFR) or Instrument Flight Rules (IFR)?

If VFR, your airplane must include the following:

- Airspeed indicator
- Altimeter
- Magnetic compass
- Tachometer
- Fuel gauge

Other instruments are required depending on what type of engine is installed. For four-stroke engines, you must have oil temperature and oil pressure gauges. If you're using a liquid-cooled engine, add a coolant temperature gauge. Similarly, add gear lights if the plane is equipped with retractable gear and a manifold pressure gauge if the engine is turbocharged or has a constant-speed prop.

What's interesting about this list (from FAR 91.205) is what *isn't* required for VFR flight:

- Vertical speed indicator (VSI)
- Directional gyro (DG)
- Turn and bank/turn coordinator
- Clock
- Artificial horizon

These are required for IFR flight but not for VFR. Many builders install more (Fig. 5-18), but if you're never going to fly into the clouds, why spend the money and waste time on their installation?

Instrument prices aren't that bad; in fact, I think the prices have come down a bit over the last couple of years. You can buy most of the primary instruments for $200 or less each. Overhauled certified gauges run about 75 percent of the new price. If you want special features, such as a 2-inch gauge rather than the standard 3-inch gauge, you'll pay extra for it.

Or buy used instruments and save 50 percent or more. You'd like to find a gauge directly removed from another aircraft or one that's yellow-tagged.

Keep in mind that you're building an experimental aircraft and don't have to use instruments approved under a technical service order (TSO). I've seen automobile

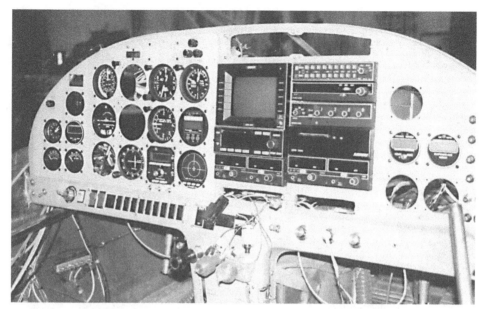

Fig. 5-18. *The level of avionics depends greatly on the type of airplane under construction. This Lancair IVP can only be described as "loaded."*

compasses installed in homebuilts. The airspeed indicator requirement could be met by a spring-loaded calibrated wind vane.

They're legal solutions, but don't get too unusual. Stick with what you're used to. Besides, the aircraft eventually will be inspected by an FAA inspector who might take a dim view of such shenanigans. Inspectors can't legally downcheck an aircraft on the basis of nonstandard instruments, but they can always find some other excuse to reject it.

This doesn't mean that you must use only TSO'd instruments. The homebuilder's supply outfits sell many good-quality noncertified gauges meant for homebuilts. These are perfectly acceptable. In fact, the auto-parts store can stock much of your panel (Fig. 5-19) But if you're building an IFR airplane, use TSO'd instruments only.

Some companies now offer all-in-one electronic units. These include most of the standard VFR and IFR flight instruments and an altitude encoder into a single package for about $2,000 (Fig. 5-20). The power required is minimal, and the total weight barely exceeds that of two of the gauges they replace. Many builders are selecting these units as an option. However, one should consider one or two standard gauges for backup in case the electronic one develops a problem in flight. This is even more important on an IFR aircraft.

Electrical system

In the past, many homebuilts could get by without an electrical system. The airspace was free, towered airports were few, and transponders were only installed on airliners and military planes. Folks navigated by a paper chart spread on their laps.

Fig. 5-19. *Half this Volksplane's instruments came from the auto-parts store, including the compass.*

Things are different, now. If you're building a simple knock-around airplane, you probably still can get by without an electrical system. You'll save 50 to 100 pounds in weight and tens of hours in construction time. If you don't install an electrical system, then you aren't required to have a transponder to operate within the 30-NM-radius class B veil. You still need it to enter class B and C airspace, though.

Most of us these days need the power. Incorporating an electrical system will require installation of a battery, an alternator, a regulator, an ammeter or voltmeter, fuses/circuit breakers, switches, and dozens of feet of wire. Before you buy, check to see if the kit manufacturer includes all these components.

If the manufacturer doesn't include the parts, you'll probably have to decide between a 12- and a 24-volt system—except that there *is* no choice: Use 12 volts.

There might be a few extenuating circumstances, for example, if you already own a stack of 24-volt avionics. Yes, there are some advantages to 24 volts. For example, the wiring can be lighter because the same power can be passed at half the current. And yes, many production aircraft now use 24 volts.

But the preponderance of evidence is solidly on the side of 12 volts. Components cost half as much. Battery chargers can be bought at many hardware and automotive

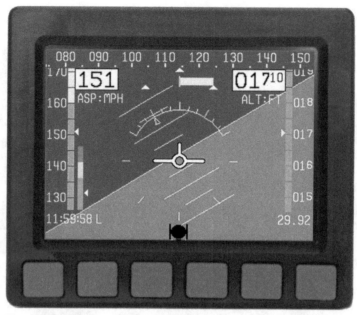

Fig. 5-20. *This Dynon EFIS-10A displays airspeed, altitude, pitch, roll, yaw, rate of climb, magnetic heading, G-level, vertical speed, and turn rate and doubles as an encoder for the altimeter. Total cost is less than a complete suite of separate instruments.* Dynon Avionics

stores for $20 or so. If the battery goes flat, you can jump-start using the battery in your car.

Forget 24—go 12.

If you're planning to fly at night, a few more items will be necessary: *position lights* (left and right wingtip and tail) and an *anticollision light.*

The position-light requirement can't be met just by having bulbs stick out at the various locations. The lights must be visible within certain angular limits. For example, both wingtip lights must be visible from in front of the aircraft. The individual lights must be visible through 120 degrees in the fore and aft directions. If you want the FAA inspector to approve your kitplane for night VFR, you'll have to prove compliance. The easiest way is by using approved position-light assemblies.

The same is true of the anticollision light—angular coverage requirements must be met. Either strobe lights or rotating beacons are acceptable. The usual practice is to install either a single strobe at the top of the horizontal stabilizer or one light in each wingtip.

In the cockpit, *post-mounted* instrument lights are the easiest to install, but the small spotlight-style lights are more flexible. Getting the light full and even is the biggest problem. Proper instrument lighting is another one of those arts where solutions generally are found only by trial and error.

The only time a landing light is legally required is when the aircraft is being operated for hire, which a homebuilt can't do anyway. So you don't have to put one in. If you do, keep in mind that hot filaments don't like vibration. Wing-mounted lights have a lower rate of bulb failure than cowling lights. But then, they don't provide even coverage either. Your choice.

AVIONICS

The primary reason to install electrical systems is to power the avionics stack. Again, avionics selection should be determined by your mission.

If your mission is to transit the airways with more electronics than James T. Kirk, boldly go ahead. Fifteen years ago, my friend Bret built a Questair Venture. He spent almost $60,000 for the kit and engine—and another *$50,000* on avionics— everything from an autopilot to a moving-map display.

But he had everything figured out. The alternator provided enough power; the panel had enough space. The weight of all that equipment caused other problems. He had enough payload capacity to carry a toothbrush, but he had to buy toothpaste at his destination.

An exaggeration? Of course. But Bret knew what he was giving up. A lot of home-builders don't. They go hog wild at the avionics store and willy-nilly mount radios and gadgets in every nook and cranny. Then they wonder why their pride and joy climbs slowly and stalls viciously. Their spouses wonder why the credit-card bill is so high.

Avionics cost the homebuilder in three ways: pocketbook, panel, and payload. You want to minimize the first two and maximize the last.

Pocketbook is obvious. As a rough rule of thumb, count on spending an average of at least $1,000 for each gadget (comm radio, VOR receiver, GPS, and so on). Some, such as basic global positioning system (GPS) units, are cheaper; most aren't. These tiny little boxes eat a lot of money, especially if nothing but "top of the line" is good enough for you.

Panel space is somewhat ignored until too late. If the plane is tandem seating, there might not be much room for anything but a compact navcom and a transponder. Side-by-side seated airplanes have wide panels and should have sufficient room for all but the most extravagant layouts.

Avionics weight can add up in a hurry, chewing into useful load. For modern avionics, count on about five pounds for each box, including tray, cables, and anten-nas. Two nav-comms, a transponder with blind encoder, a GPS, and a control panel will weigh about 25 pounds. Used, older-model avionics might be 50 percent heavier.

Let's take a more specific look at the type of avionics you might install.

Communications transceivers

A NORDO (no-radio) airplane in these days is a real problem. Even when you oper-ate from uncontrolled fields, the world is full of ninnies who can't see an airplane until they hear it on the radio.

With a simple knock-around aircraft, just a hand-held communications trans-ceiver will do. There's nothing wrong with mounting an antenna on the aircraft, hooking up a power plug to the electrical system, and installing a small plate to hold the radio (Fig. 5-21). The radio itself will cost about $400, and you can pay as

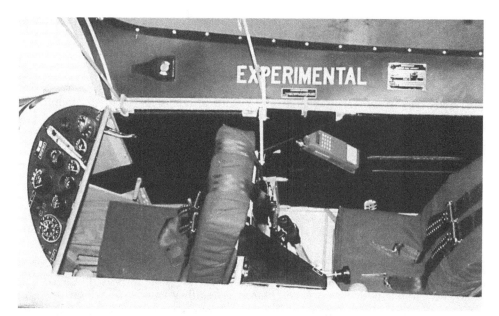

Fig. 5-21. *The only radio in this Sonerai is a hand-held nav-com held securely and conveniently in a bracket on the cockpit sidewall.*

little as $30 for a simple whip antenna or $100 + for a fancy blade type. External antennas work much better than the little "rubber duck" included with the radio. One nice hand-held feature is its built-in theft-proofing. You don't have to leave it in the airplane as a target for thieves, and you can walk around and listen in to the flying chatter at airshows.

But hand-helds aren't the perfect solution. They're low in transmitted power, and thus they are shorter-ranged and less likely to catch the listener's attention. The displays and controls aren't designed for easy in-flight use.

More conventional comm radios will cost around $800 on up. Nav-comms (that include VOR navigation as well as the communications transceiver) start only a little higher. If you don't plan on installing a GPS-based moving-map display, consider buying a nav-comm that incorporates a GPS receiver instead of a VOR.

Whether you buy two comm radios depends on the type of airspace you fly through, whether you're flying IFR, and whether you're used to dual radios. An audio control panel is necessary to allow a single microphone and speaker to be used with both radios. Control panels are light (three pounds or fewer) but cost up to $1,000. And like the radios themselves, they need to be installed and hooked up.

Transponders

Transponders aren't required everywhere, but it's tough to fly a distance without needing one. The transponder itself will cost about $1,500, and a blind encoder will cost another $200 to $400.

While one can find good deals on many used instruments, a used transponder may not be the right solution. Until recently, transponders incorporated a high-powered transmission tube that had a limited life. A used transponder can check out okay, but if there's only a hundred or so hours left on the tube, you'll be replacing the transponder soon. Since the tube is the single most expensive part on the unit, replacing it makes little sense. Modern solid-state transponders eliminate the tube.

Navigation

Combination nav-com prices start around $1,000. However, some nav units include the indicator (in the form of a digital display), and others don't. For these, the cost of the indicator will drive the package price upward.

Of course, the hottest navigation item is receivers for the satellite-based global positioning system (GPS). There are hand-held GPS receivers for any budget; mine cost less than $100 at the local camping-goods stores. It doesn't have any specific aviation-oriented features, but it does point a needle in the direction I need to fly.

GPS moving-map system with color displays are very popular. The manufacturers tend to control the prices fairly closely, so it's tough to find a real bargain.

Emergency locator transmitters (ELTs)

ELTs are fairly cheap and light—around $200 and two pounds. One nice feature is that installation consists of bolting it to the structure; little or no wiring is needed.

ELTs aren't required in single-seat aircraft, but $200 is cheaper than spending a week (or eternity) in a crashed airplane.

Intercoms

Most homebuilts are noisy. I recommend wearing a headset, so why not buy two and install an intercom? An intercom is light and costs less than $250. You could just use a portable intercom, of course. But you'll find that a panel-mounted unit gets rid of a lot of wires in the cockpit.

Tunes

Why ride bored on long trips? Install an automotive CD player or even an MP-3 player. Don't bother with speakers—converter boxes allow connection with the standard aviation headset. Don't hook it up directly—aviation headsets have 600-ohm impedances, and the stereo requires 8 ohms. The aviation environment is harsh, though. Pick out a cheap, rugged unit.

Antennas

One point to know about antennas: They're just wires of particular lengths. You don't get a longer one for more powerful radios; the wire length is set by the transmit and/or receive frequency. Because the precise length is critical, and commercial models are cheap, buy one rather than build. Since fiberglass and wood both freely pass radio wavers, people with airplanes made of these materials can install custom antennas internally, leaving the exterior uncluttered. Several companies, such as RST Engineering (*www.rst-eng.com*), offer kits.

However, check with the kitplane maker first. Several composite kitplanes contain graphite, which will block radio signals almost as well as metal.

Another factor regarding antennas is the establishment of a *ground plane.* Consider it as the base that the radio waves "push against" when the antenna transmits. For nonmetal airplanes, the ground plane usually consists of metal foil or mesh at the base of the antenna. Metal-airplane folks don't have to worry because their entire airframes handle the job.

Saving money on avionics

The prices listed earlier are approximate, based on typical advertisements. Careful shopping can find even greater savings. Avionics list prices are the maximum you should pay; most places sell for less.

And like everything else, you can save on avionics cost by buying used. Have an avionics shop test prospective purchases. Sometimes the equipment is still installed in an aircraft, and you can verify operation yourself. Solid-state units are hardy—if a unit does not fail in the first 10 hours of operation, it'll last forever. Used avionics are a good deal.

A couple of things to be wary of, though. First, the U.S. government is planning on phasing out the entire LORAN system in a couple of years. You might find a really good deal on a LORAN receiver—might I suggest that you pass it up?

Other government actions made some older communications transceivers illegal in the United States. Talk to a local avionics shop before committing to a used radio to ensure that it's still legal to operate.

An option for those handy with a soldering iron is a radio kit. RST Engineering sells kits for common avionics, with prices equal to those of used equipment. This will add to your construction time but is a cheap way to upgrade your panel.

Installation

The standard width for regular aircraft radios is 6.25 inches. Allow at least 12 inches behind the panel. The heights of the units vary but typically run around 1.5 to 2 inches.

The components are not solidly mounted to the panel; instead, a tray or rack allows the unit to be removed easily for service. Some trays have a set of electrical contacts that automatically connect the unit to power and antenna; others merely allow the unit to slide out once it's disconnected manually. Keep in mind that the trays might not be included with the purchase of used avionics.

Space-saver units are designed to fit into a standard 2.5- or 3.25-inch instrument hole and must be mounted from behind. These items do not have trays; instead, connectors attach directly to the radios. If one of these radios is removed for repair, the connector must be secured if you are going to fly the plane. Companies sometimes offer prewired harnesses for these radios. Speaking from personal experience, take them up on it!

Don't neglect cooling. The traditional mounting procedure is to stack the radios atop each other, which concentrates the heat. Solid-state electronic gear runs far cooler than the old tube radios but is also more sensitive to temperature variations.

One last point about installation: Carefully study the warranty when you buy new avionics. Some companies only give warranty coverage if factory-certified technicians perform the installation. Other companies give triple the warranty if certified techs do the job. Solid-state avionics are wonderful, but the little chips aren't as tolerant of mistakes as vacuum tubes. One little snap of static electricity, and they're gone.

There are certain functions, such as antenna tuning, that take specialized equipment. But handling the wiring harnesses isn't any different from handling any other electrical wiring. Discuss your situation with the avionics manufacturer. The manufacturer may allow you to install the wiring and only require an outside checkout before the boxes are connected.

You may wish to leave expensive or complex installations to a professional. Then, if it blows up, you don't have to pay for it.

A final option is a prebuilt avionics setup. Companies will build your entire panel for you, leaving you to install the panel in the airplane and hook up electrical power and antennas. I'll not mince words: This is expensive. But if you want a full-up panel and have the extra money, this is the way to go.

By now, you should have decided on your kitplane's equipment and other factors. You know which engine will be installed, what instruments to use, and the right avionics for your mission. It's time to get ready to build.

6

Preparations

THE KITPLANE AND its equipment have been selected. Delivery times vary from weeks to months. There is no need to sit around moping—there are plenty of things to do in the meantime.

SHIPPING

At the time the kit is ordered, you normally select the shipment mode. It usually boils down to having it shipped commercially or picking it up in person. In either case, getting the kit from the factory to the workshop is not trivial.

All kit prices are F.O.B. (*free on board*) the manufacturer, which means that shipping costs are extra. These can vary from a few hundred to a thousand dollars or so, depending on the size and weight of the kit and the distance you live from the manufacturer. You'd like to save the money, but as with fabricating a part rather than buying a completed one, the hassle is counterbalanced by the savings.

Let's take a look at some of the issues.

Crating charges

It obviously costs the kit company time and money to prepare your purchase for shipment (Fig. 6-1). The company passes that cost on to you in what are called "crating charges."

Crating charges run from $50 to $1,000 on up depending on the size of the package. Most of the time, you minimize the crating charges by buying the entire kit all at once rather than by subkits. For instance, Glasair charges $695 for crating the entire kit at once. If you buy the four subkits individually, the total crating charge comes to $1,250.

If you pick up the kit yourself, you may avoid some of the crating expenses. Some companies have a "will call" charge (basically to cover the cost of loading the kit onto your trailer or truck). This charge runs around $200 or so.

Fig. 6-1. *This GlaStar is being crated for shipment, a process that the buyer pays for one way or the other.*

Factory pickup

Materials kits and/or multiple subkits usually can be carried in the back of a pickup. If a pickup is too small, a rental truck or trailer probably will be necessary. Most kit companies have experience with buyers picking up their kit in a rental truck and should be able to tell you what size will be necessary. The various rental companies (U-Haul, Budget, Ryder, and so on) offer quotes either in person or via the Internet.

You can rent the truck near home, or fly to the factory's town, rent a truck, pick up the kit, and drive it home. A 26-foot rental truck costs about $40 per day and 80 cents a mile. Calculate the total expense for round-trip or one-way, considering time and distance, and choose lower cost or better convenience to your preference. Check with the manufacturer to find out if anyone else in your area has a kit on order. You might be able to carry both and share expenses.

Also, check on one-way rental. I ran a comparison with one of the local agencies. A two-way rental of a 26-foot truck ran $40 per day and 79 cents a mile. A 400-mile round trip, then, would cost $356 if I could complete it in a single day. The one-way rental cost just $110, and it included the mileage and allowed me to take two days.

Some composite and tube-and-fabric kits include components 15 feet or more long. The Glasair, for instance, comes in a single crate about 25 feet long. The longer the truck, the higher is the rental rate. In these cases, a trailer might be the best solution.

Fig. 6-2. *This Sonex kit is being loaded onto a rented trailer.*

Get the crate size and weight from the kit manufacturer, and then check around with your friends. Simple flat trailers are best, but the addition of a few boards might make a boat trailer suitable. If you can't borrow a trailer, you'll have to rent one; they go for about $20 to $40 a day (Fig. 6-2).

Carry lots of rope and bungee cords to secure the kit to the trailer. It wouldn't hurt to take along some boards and a saw in case extra support is needed. The crates generally aren't waterproof, so take some sheet plastic and a staple gun. Those blue plastic tarps are pretty cheap.

Depending on the kit, the trailer and crate will weigh 1,000 to 3,000 pounds. Small pickup trucks can handle lighter kits, but heavier loads and/or steep terrain might require a full-size pickup. You could rent a truck, but if you've got to rent both the truck and the trailer, you're probably better off having the kit shipped.

How much will you save by hauling it yourself? Assume that the factory is 500 miles away, you own an adequate truck, and a rental trailer costs $30 a day.

Fuel mileage is going to be moderately low on the way down and worse on the return. Speed is the same way—you'll hold the speed limit without much problem going to the factory, but the weight of the kit will reduce your average speed on the return leg.

Count on getting to the factory in a single day, and we'll assume that your fuel consumption is about 15 mpg. You'll need a hotel room overnight. Pick the kit up at the factory the next morning. The weight of the crate reduces speed enough to require a day and a half on the return leg. Mileage drops to 8 mpg. For simplicity's sake, we'll assume that gasoline costs $2.25 a gallon.

You drove 500 miles at 15 mpg and 500 at 8 mpg. Gas cost is about $215. Two nights in motels, nothing fancy, cost $50 a night. Meals and incidental costs probably come to another $50. Total rent on the trailer (three days) is $90.

For about $455, you've delivered the kit to your door. In the preceding case, a kit manufacturer quoted $385 for truck freight. It is best to run the numbers carefully before deciding which way to go.

There are other drawbacks to hauling it yourself. You'll probably have to take at least one day off from work. The trailer can sufferflat tires or even break free from the tow vehicle. Any damage to the kit on the way back comes out of your pocket. And how well are you going to sleep while your $30,000 kit lies unattended in the motel's parking lot?

There's usually a breakeven point between commercial shipment and factory pickup. Thus 100 miles is ideal for hauling it yourself, whereas 500 miles is "iffy"; across country, it might be better just to bite the bullet and write the check for commercial shipment.

Commercial shipping facts of life

Thus, in addition to the *terra incognita* of aircraft construction, you're probably going to have your first dealings with a commercial trucking firm. We see the trucks on the road, but few of us have ever interacted with the shipping companies. Let's go over a few points that might prepare you if you're having the kit shipped.

The cost of shipment, as mentioned at the beginning of this chapter, depends on the size of the package and the distance to be shipped. Figure 6-3 illustrates the cost of shipping a complete RV-7 kit to various locations in the United States. Another company reported that overseas surface shipment (i.e., by boat) costs around $3,000.

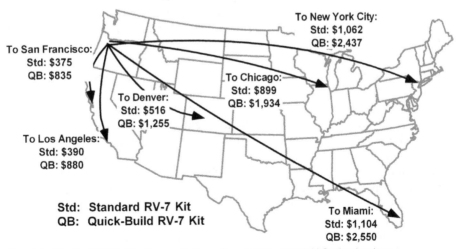

Fig. 6-3. *Typical RV-7 shipping costs from Oregon to various U.S. destinations.*

First, make sure that the kit company is absolutely clear on *where* the shipment is supposed to be delivered. If you're going to build the kit somewhere other than your home, make sure that that is the only address the kit company gives to the shipper.

The shipper's instructions probably will direct the driver to call you when the crate arrives in your home town. Don't think that you'll be able at that point to convince the driver to deliver the crate to an address other than that specified in the bill of lading. That's where they're *legally* required to deliver it. The only way to change it is to have the kit company generate a new bill, coordinate with the shipper, and so on. It's far easier to settle the issue before the kit is shipped.

The kit company may ship the package "freight collect." This means that you pay the shipper directly when the kit arrives. Determine the acceptable modes of payment prior to kit arrival.

There's another problem that may come up. When the kit company ships a package, it provides the weight to the shipper. The shipper doesn't always verify this weight; reportedly, shippers actually weigh only 10 percent of the packages they carry.

There have been cases of a kit manufacturer understating the weight of the shipment to give the buyer a price break on shipping costs. This works out nicely— unless, of course, the shipper discovers the "mistake." Since *you* are the addressee, you'll have to pay the higher freight costs before the company will deliver the kit.

It isn't as simple as that, though. Shipping companies have been known to use the false weight as an excuse to reclassify the shipment into a much more expensive rating class. The probability of such a situation arising is low, but it has happened. There's not much you can do to prepare yourself either. Good luck.

The unloading problem

Having the kit shipped can be far less stressful. But there's one aspect many people overlook: Truck drivers are paid to drive, not to be stevedores. They knock on the addressee's door, point out the package, and watch the crate get taken off the truck. And they work normal working hours, when you and I aren't home.

In plain English: They don't do the unloading. They won't slide it into the garage for you. And you're *still* going to have to take a day off from work.

Unloading is a serious problem. The crated kit might weigh 1,000 pounds and be 20 feet long and 5 feet square. The driver expects a forklift. How are you going to unload it?

One way uses a set of 10-foot-long 2 × 6 boards. Pick up about five of them. Have the manufacturer call when the kit is shipped, and have the manufacturer instruct the shipper to call on arrival (COA) before attempting delivery.

When the shipper tells you that the kit has arrived, set up a day and an approximate delivery time. Cajole a couple of friends into helping. When the truck arrives, have the driver back into the driveway. Arrange the boards into a ramp, and carefully slide the crate down to the pavement.

Or you can use the 2 × 6s to build three sturdy sawhorses. Have the driver back the truck into the driveway, and tie a rope from the crate to something sturdy inside the garage. Have the driver pull the truck forward a few feet until the slack disappears and the crate slides partially off the truck. Set a sawhorse right behind the rear bumper, directly under the crate.

Have the driver pull forward a couple more feet. Set a second sawhorse behind the bumper, again. As the truck goes forward again, the crate will be pulled past its center of gravity. The end closest to the garage will tip down; you and a friend should be ready to make sure that it lands atop the first sawhorse.

When the crate is almost all the way off the truck, set the last sawhorse behind the bumper. Inch the truck forward. As the last edge of the crate comes free, you and a friend should guide it safely atop the sawhorse. This method is illustrated in Fig. 6-4.

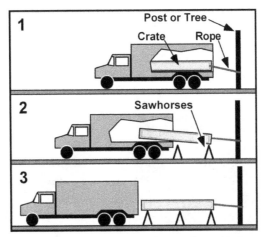

Fig. 6-4. *A typical method of unloading a kitplane crate from the back of a truck. The crate must be guided onto the sawhorses as it comes out.*

Another solution is to pick the crate up at the freight office. Borrow a trailer, and get a couple of friends to leave work a bit early. The freight company will already have the crate off the truck and may use its own forklift to load it onto your trailer. Because the shipping charges call for the crate to be delivered to your home address, the freight company will reimburse the difference.

If the shipping company doesn't have a local office, arrange to meet the trucker at a convenient loading ramp. Often, industrial districts have a public ramp for this purpose, which is literally a ramp that ends abruptly at truck-bed height. You can slide the kit off the truck, and you and your friends can carefully load it onto the trailer.

If you don't have a batch of strong-backed friends, pay more money and have a local moving company take it from the shipper's office to your house. At least that company will actually take the crate out of the truck and put it where you want it.

Inspection

No matter how the kit comes into your hands, inspect it immediately. If there is external damage to the crate, open it and inspect it *before* you accept delivery. Damages will have to be noted on the acceptance paperwork. Signing the shipment papers doesn't get the trucker off the hook; it merely certifies that there was no obvious external damage at the time of delivery.

Damages that weren't apparent on delivery should be reported to the shipper and the kit manufacturer. Any shortages aren't the shipping company's fault unless a hole was knocked in the crate and components might have fallen out.

Inventory the contents of the crate as soon as possible. The kit should include a complete parts list. Skip the bolts, washers, and other standard parts if you wish, but verify that everything else is there. Probably the easiest way is to check off each part as it comes out of the crate (Fig. 6-5). Some parts (Fig. 6-6) should be checked for fit right away, of course.

Retain your inventory list, even once you're sure that everything is there. Why? Eventually, during construction, you're going to have trouble finding a certain part. If you see that it's checked off your inventory list, then you know the part was there at some point.

Finally, don't throw *anything* away until you're sure that it isn't necessary. Too obvious? Perhaps. In some case, though, portions *of the crate itself* are used as jigs

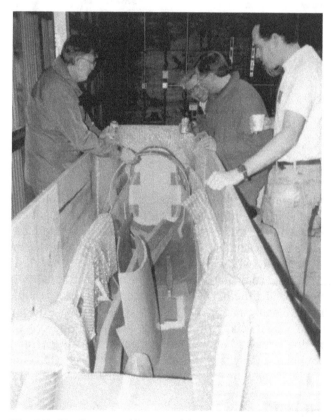

Fig. 6-5. *Hauling everything out of the crate at once is a good way to lose something. Check each item off the packing list as soon as it comes out.*

Fig. 6-6. *Go ahead, have your picture taken in the cockpit. Everybody does it.*

for component assembly. So don't burn the crate for firewood until you're sure that it isn't needed to hold the wings during assembly.

Other components

Anything not included with the kit usually can be shipped by a lower-cost means such as parcel post or UPS. Instruments, paint, tires, and other items take only a week or so to arrive, so don't order them too much in advance to save storage space.

Engines are too heavy for anything but truck freight. Lycomings come in a box about 3 feet square by 2 feet high, which means that you can carry the engine in a pickup truck if necessary. Rotaxes are much smaller and lighter.

Remember, there's little reason to buy the engine, instruments, or other components immediately. As mentioned in Chap. 5, the engine must be pickled if it won't be run for a while. Even when pickled, the engine crankshaft has to be turned over a few times every month, so why do it more months than you have to? Find out the typical delivery time, and order your engine when you're ready.

CASE STUDY: FROM FACTORY TO GARAGE

The fast composite Glasair III (Fig. 6-7) caught Bruce Bateman's eye back in 1990, and he bought the kit just before Oshkosh that year. The kit was shipped from the factory just north of Seattle about a month after he'd placed the order. It took four days for the shipment to reach Bateman's California home town. Shipment was via Consolidated Freightways freight collect; Bateman paid the shipper when the kit arrived.

He opted to use a local moving company to bring his kit from the shipper's freight yard to his home. "I could have had Consolidated deliver directly to my home,"

Fig. 6-7. *Bruce Bateman selected the powerfull Glasair III for his homebuilt project.* Stoddard Hamilton Aircraft.

Bruce Bateman.

Fig. 6-8. *Bateman's kit was carried to his house from the interstate carrier's freight yard by a local moving company on a lowboy trailer.*

he remembers. "But then I would have been responsible for offloading the crate—something beyond my capability since I don't have a forklift."

He arranged to accept delivery at the freight yard, where the shipper used its forklift to move the 1,600-pound crate onto the moving company's "lowboy" trailer. As can be seen in Fig. 6-8, a lowboy trailer carries the load only about a foot and half from the ground. This eases the unloading problem.

Fig. 6-9.
Ramps and shipping dollies were used to unload the crate. The movers then rolled it directly into Bateman's garage

Bruce Bateman.

At Bateman's home, the movers used a couple of heavy-duty dollies and some ramps (Fig. 6-9) to offload the crate from the side of the flatbed and move it into his garage.

It was a tight fit. The crate had to go diagonally in his 20 × 20-foot garage, and he had only inches to spare. Fortunately, he had measured and calculated the fit in advance. He installed four heavy-duty dolly wheels at the corners of the crate. In that way, when he worked on the kit, he could open the garage doors and roll the crate partway out of the garage for a little elbow room. He then removed, inventoried, and shelved the contents.

His total shipping expenses were $600.33 for transportation from Arlington to his home town and $440 for the local mover. Add in the crating cost, and Bateman's total cash outlay to take delivery of his kit was about $1,540.

WORKSHOP REQUIREMENTS

The workshop is like a budget: No matter what you have, more is needed. You'll need more tools, more space, more light, and more power. The following are some general requirements for the homebuilder's workshop. Specific requirements, including tools, will be covered in later sections.

Space required

As mentioned in Chap. 2, the kitplane manufacturer usually will indicate the space required to build the aircraft. Most specify a two-car garage.

Is that a garage big enough to hold two mid-1960s Cadillacs or sized for two Hyundais? Garage sizes have changed drastically in the last 20 years.

Suffice it to say that the more room you have, the better off you are. Sure, it's not absolutely vital. I've seen a Q2 built in a two-bedroom apartment. I watched an Osprey (a large amphibian) being built in a one-car garage. I've seen pictures of a Zenith being built in a canvas "temporary" automobile garage.

But it's much, *much* easier with sufficient room.

This is a moot point for most of us. We have a basement, a garage, or a spare bedroom, and that's what we have to build the airplane in. Most of us can't afford to build a workshop in the backyard. At best, we might rent a hangar. But the majority of kits are assembled at home, in whatever space is available.

If you have enough room for the single largest complete component (usually the fuselage, except for one-piece wings), you have a chance. Building in such limited quarters will be difficult. Moving components around will be a chore—imagine trying to arrange a hoist to lift an engine in place when you barely have enough room to turn around.

Some of the oldest jokes in homebuilding are about guys who have to disassemble the house to get the completed airplane out of the basement. Many of these stories are true. It can easily happen unless you think things through completely. A 35-inch-wide fuselage won't necessarily fit through a 36-inch door. Wing attach plates might project out; the hallway beyond the door might take a sharp turn. If aircraft egress room is questionable, slap together a full-sized plywood mockup and experiment.

Clever solutions can be found. The Osprey builder had a garage door 8 feet across, with a minimum fuselage width of 8-½ feet. He developed a slick little system to give him extra clearance (Fig. 6-10).

Fig. 6-10. *Kirk McCarty built his first homebuilt, an Osprey II, in a single-car garage. The fuselage was wider than the door, so he cut away part of the door pillar and used a jack to temporarily support the building. Once the fuselage was through the door, he replaced the missing section and removed the jack.*

If building room is especially tight, you'll eventually have to complete construction somewhere else. You'll probably end up paying rent. But the longer the move can be delayed, the better.

Power

There are two main questions regarding your shop's electric power: Is it adequate, and is it convenient?

If you plan on using any 220-volt power tools, of course you'll make sure that the shop is wired for it. Most industrial-grade floor tools require 220-volt *three-phase* power. While the right voltage is usually available (electric clothes dryers usually require 220 volts), three-phase power often isn't.

How many circuits are available for standard 120-volt tools? A 1/3-hp motor takes between 5 and 10 amps. Standard house wiring is 20 amps, so two small motors are all you'd want on any single circuit. Of course, you're not likely to use more than one tool at a time, unless someone else is helping. At least two 120-volt circuits should be available.

Modern garages and basements shouldn't have any problem, but be careful with older homes. Some 10-amp circuits are still out there; a moderate-sized table saw will blow the fuse. I have a friend who lives in one of these houses. His motto is, "A penny saved is a fuse." He's just kidding, but if you're stuck with low-current wiring, you might have to pay an electrician to upgrade the workshop area.

Convenience is usually of greater impact to the homebuilder. You'll be roaming all over the workspace with cord-type electric hand tools. Where are the outlets? In my garage, all the outlets are on one wall. Extension cords are a fact of life.

They're hard to escape in any case. Outlets are usually on the walls, whereas the aircraft sits in the middle of the floor. One or two 25-foot cords should suffice.

Outlet strips are useful accessories. These plug into an outlet and provide six or so additional outlets controlled by a single switch. Workbenches and tools limit access to the wall outlets themselves, and the outlet strips bring power closer to where it's needed.

Outlet strips sell for around $10 and can be found on sale for half that. Most come with their own circuit breaker, which, unfortunately, is usually rated at lower than 20 amps. Try not to use an outlet strip with several large tools running at the same time.

One feature to ignore is outlet strips that have *power protection* or are *surge-limiting*. This feature is intended to protect delicate electronics (such as computers) from power fluctuations. Power tools aren't sensitive; they *cause* surges.

Lighting

For shop lighting, nothing can beat ceiling-mounted fluorescents. The fixtures are cheap, $10 or less (without tubes), if you watch for sales. Shop carefully for tubes; prices vary from $1 to $3 depending on if they're marketed for home use (individually wrapped, found beside the rest of the bulbs at the grocer's) or shop use (no box, piled in a rack in a hardware store). I buy a box of 10 for $20.

The fixtures can be either screwed to the ceiling or hung from hooks. I hang mine; in fact, I've set up groups of hooks in various areas and move lights as needed. An electrician could add ceiling outlets, but I just tack up extension cords using cable holders.

Because of the shadows cast by ceiling lights, you'll need portable lighting as well. The standard shop trouble light works fine and acts as an extension cord, too. Bulbs specifically designed for trouble lights cost around $3 each. Regular light bulbs work, but their filaments are a bit sensitive to shaking and bumping. I'm a klutz— I use the special bulbs.

There are also portable fluorescent bulb holders. While these are more awkward to handle, the long tubes cast fewer shadows and have less glare.

Heat

If you live in a cold part of the country, you'll want some heat for your workshop. You don't want to spend 30 hours a week shivering in the garage and handling frosty aluminum. The only alternative is to stop work during the coldest months, but these are months that you'd rather work because you can't fly.

Just because you live in a warm climate doesn't excuse the need for occasional workshop heat. A composite aircraft requires approximately 70°F, as do wooden kitplanes. Tube-and-fabric models are fairly impervious to temperature until it comes time to apply the fabric and dope. And you'd like to rivet skins on aluminum kits while the skins are warm; otherwise, they may get a bit wrinkled in warm weather.

Comfort heating might be taken care of by a portable electric heater costing $40 or so. But building an airplane isn't a stationary enterprise. You'll be moving around too much to benefit much from a small heater.

Options for full heat abound. Attached garages often contain the house's furnace, and the ductwork is accessible. Install a few junctions, and the furnace will heat the garage as well. You won't get the temperature up too far because the thermostat is still in the house. But every little bit helps.

Large kerosene-burning space heaters sell for $200 and up. Wood stoves, multiple electric heaters, and waste-oil-burning furnaces might fit your particular situation. You might even buy a used house furnace.

However, pay careful attention to local codes. For example, kerosene heaters are banned in my town but legal in the surrounding county. Your options when using a basement as a workshop (inside the house) are more limited than might be allowed for a separate garage.

Finally, don't forget insulation. Adding batting between the joists can do wonders. Garage doors are a particular nuisance—they don't seal worth a hoot and are a major source of drafts. Try a curtain of plastic stapled to either side of the door with the foot weighted down by boards or sandbags. Also, lay some scrap carpet on the floor of the work area. Cold cement sucks heat from your feet and legs, and a little bit of insulation goes a long way. It also keeps dropped parts from rolling too far.

Worktable

One thing necessary for many homebuilts is a worktable big enough to build the major components on. Generally speaking, worktables are usually 3 to 4 feet wide, with the length dictated by the needs of the kit. They can be as short as 5 feet and as long as 16 feet. A height of about 3 feet seems the most comfortable. However, if your overhead space is restricted, you may want to make the worktable a little lower.

The worktable might have a solid top or might be an open trusswork to allow access to the components from underneath. *Both* modes may be necessary; for instance, the fuselage may be built inverted on an open truss and the wings built on a solid tabletop.

A worktable can be built from any material, but if you find one made of anything other than wood, it's probably because the builder found a bargain on a used steel or even aluminum table. The top must be wood because you'll probably be attaching temporary supports to it.

The bigger the workbench, the longer/heavier will be the parts being assembled on it, and the more likely it is that you yourself may have to crawl around on it. So the construction can vary from the heavy-duty table shown in Fig. 6-11 to the medium-duty one shown in Fig. 6-12. When you can, use screws instead of nails. Nails tend to loosen over time, and screws are easier to remove if you decide to scrap the table at the end of the project.

In any case, construction must be sturdy. Not only is the airplane going to come together on it, but you may end up crawling across it to access a hard-to-reach portion.

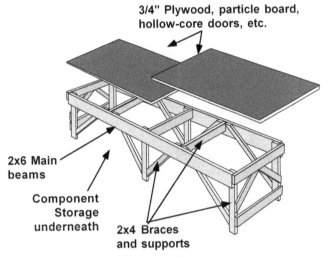

3/4" Plywood, particle board, hollow-core doors, etc.

2x6 Main beams

Component Storage underneath

2x4 Braces and supports

Fig. 6-11. *The traditional homebuilder's worktable, constructed from dimensioned lumber and plywood. Diagonal braces ensure rigidity and yet allow storage underneath.*

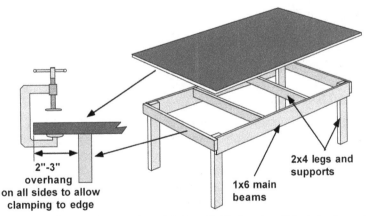

2"-3" overhang on all sides to allow clamping to edge

1x6 main beams

2x4 legs and supports

Fig. 6-12. *A simplified version of the worktable for aircraft that don't need full-length work surfaces. The legs can be 2 × 4s or 4 × 4s.*

The biggest requirement of the worktable is its precision. The table is the basis for accuracy when building the aircraft. You'll build your wing on the table, and if the table is warped, the wing will be warped, too.

The correct top material can cut down on warping. Plywood is most common but has a lower resistance to warping than particle board. But particle board tends to crumble. Hollow-core doors or drafting-table tops are great at resisting warpage, but they can't take the abuse as well. Some builders add a plywood overlay to such doors. Whatever you choose, get the thickest you can afford, ⅝-inch for an absolute minimum and ¾-inch or greater preferred.

The table must be precisely horizontal. This allows checking accurate placement of components by using a carpenter's level. This complicates workbench construction because your shop's floor probably isn't level. The workbench must be able to compensate.

When it comes time to adjust the table, three tools are important: a carpenter's level, a ruler, and a ball of string. Designate one corner of the table as a reference. Attach one end of the string to a block of wood, and stretch the string above one long edge of the table. Attach it to a same-sized block of wood in the far corner. Measure the distance between the string and the tabletop at points between ends. Shim up the top where necessary, eliminating any sag. This process is shown in Fig. 6-13.

Then check the level along the edge, and adjust or shim the table's legs to compensate. Use the string and ruler to eliminate sag across the short axis of the table. Make measurements every two feet or so along the table's length. Then use the level across the short axis. Don't touch the adjustments on the legs you've already set; change the other side's instead. Shim the table edge as necessary.

Repeat this procedure as many times as necessary to get the errors as small as possible. If I had to make a choice, I'd rather have a table that's a little off level rather than warped.

Check the table occasionally afterwards or if the table is moved. Again, the table is vital to airframe trueness; a mere degree of error could cause performance or handling problems.

Not all airplanes need a worktable. Some use special support systems such as those shown in Figs. 6-14 and 6-15. Templates are cut from plywood and attached to sawhorses. These templates match the shape of the fuselage or wing and hold it solidly in place.

Workbench

You can use the worktable for all the activities involved with building the aircraft, but you don't really want to. Parts with sharp edges can cut up the surface, or heavy components can throw the table out of alignment. It's a good idea to have a separate workbench for the rough, heavy work.

There aren't really any specific requirements for the workbench. Make it the same height as the worktable. Don't sweat alignment and warpage too much; however, you might use the bench as practice for the full-size worktable.

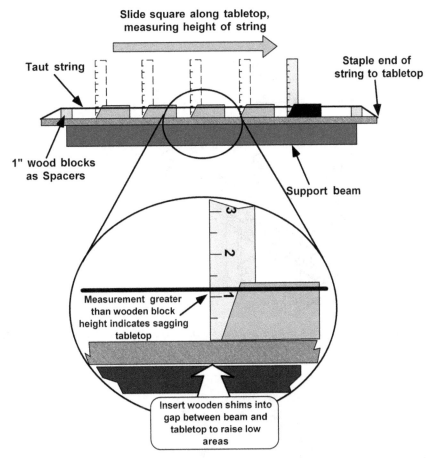

Fig. 6-13. *A taut string and a square can be used to eliminate sags from the top of the worktable. A laser level is a more high-tech solution.*

Or buy a commercial model. They come with drawers, electrical outlets, and other things that make building easier. I bought a used drafting table for $35; a used metal office desk would work just as well, with a sheet of plywood as a work surface.

If you're very limited in space, don't bother with a workbench. But fit one in, if possible.

Other items

Building an airplane generates a lot of dust. Shop vacuums are nice, but the little hand-held rechargeable models seem to work the best on tabletops. A simple push broom and dustpan suffice for the floor.

Don't plan on building the entire airplane standing on your feet. A lot of jobs involve assembling components at the workbench. Get a chair or a stool or two. I

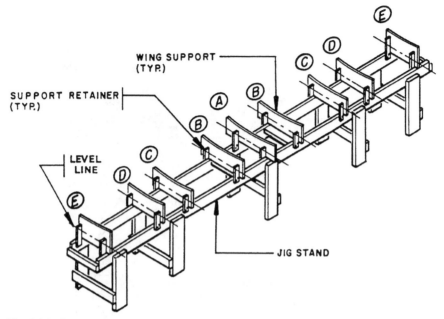

WING SUPPORT (TYP.)

SUPPORT RETAINER (TYP.)

LEVEL LINE

JIG STAND

Fig. 6-14. *To ensure a straight and true wing, the Glasair plans include the directions and templates to build a combination jig and worktable for wing construction.* Stoddard-Hamilton Aircraft.

Fig. 6-15. *Owing to its rounded edges, this Hummelbird fuselage rests on curved and padded plywood supports. Note the rollers that let it be moved easily.*

used to keep a lawn chair handy, and if I had to polish a lot of small parts, I set the chair in the sun and used my chest as a workbench.

One thing you'd like in the workshop is a telephone. You are going to get dirty, and there's no reason to be tracking dust into the house and getting greasy handprints on the phone. Probably the best overall solution is a cordless model. Otherwise, you can install a jack in the garage and plug in a cheapie model. The garage is the typical entry point for phone wiring nowadays. Phone service is supplied though a small four-conductor cable similar to the extension cable that is used from the wall jack to the phone.

First, go to Radio Shack and buy a four-position terminal strip, a modular phone jack, and 10 feet or so of four-conductor phone wire. When you get home, find the point where the phone line enters the house. This should be a small multiconductor wire traveling on its own. Find a good spot along the wire run inside the house to mount the terminal strip, and then carefully pull on the wire to gain as much slack as possible.

Cut the wire and strip back the outer shell to reveal the four conductors: red, black, green, and yellow. Strip a little insulation away from the wires on both sides. Mount the terminal strip on the wall or joist. Connect the red, black, green, and red wires on one cable to separate terminals on one side of the terminal strip. On the corresponding terminals on the other side, connect the four same-colored wires of the other cable. Also, strip back one end of the phone cable you bought, and connect the four wires to the same colors on one side of the terminal strip. Run the cable where you'd like to mount the phone, and install the modular jack per instructions.

There are other little comfort details you could attend to. As mentioned, an old piece of scrap carpet in the primary working area prevents dropped parts from rolling too far, as well as keeping your feet warmer. A radio makes the time pass pleasantly. Hang up a few pictures of the type of plane you're building both for inspiration and to answer the questions of visitors. You'll be spending a lot of time there; make it as homey as possible.

Two last points regarding the workshop: First, never turn down anything free. Second—and perhaps most important—be inventive. One local builder found an old rollaway typing table. He clamped a plywood top to it and had an instant portable workbench. A veterinarian gave me a broken safe, which I use as a filing cabinet for receipts and magazines. Grab up old carpet and blankets for padding, scrap lumber for braces, and glass jars to store parts and leftover liquids. The motors for many homebuilder's bench sanders have been liberated from junked appliances. Muffin pans work great for storing small parts—surely the wife would love to have a new set while the old set disappears into the garage.

It's one thing to specify what items are necessary to the workshop, but if you don't arrange them properly, every little task becomes a chore.

WORKSHOP ARRANGEMENTS

Let's count up the major components of the homebuilder's workshop. Worktables and workbenches. The aircraft's fuselage. Two wings. A heavy engine. Free-standing power tools, such as drill presses, a bandsaw, and a table saw. Racks of waiting parts and components. Cupboards of bolts and fasteners. Not to mention enough room to assemble the entire aircraft, an object 20 feet long with a wingspan approaching 30 feet.

Workshop space is not unlimited, hence the need for efficient layout.

Hand tool storage

Most of us keep our hand tools in a plastic or metal box. That works out fine most of the time. We can scoop up the box and carry it to the car, the broken shelf, or the unassembled bicycle.

But building an airplane means using the same tools continually in the same location. How much time do you want to waste digging through the mess in an old toolbox looking for the $^7/_{16}$-inch wrench?

Make it easier on yourself and set up a tool board, as shown in Fig. 6-16. The easiest way to make one is from white-painted pegboard. A variety of hooks and hangers is available that will hold your tools and keep them handy. The pegboard cannot be attached directly to the wall because the hooks stick through the board and hold on the backside. Attach your pegboard to a framework of 1-inch pine furring strip.

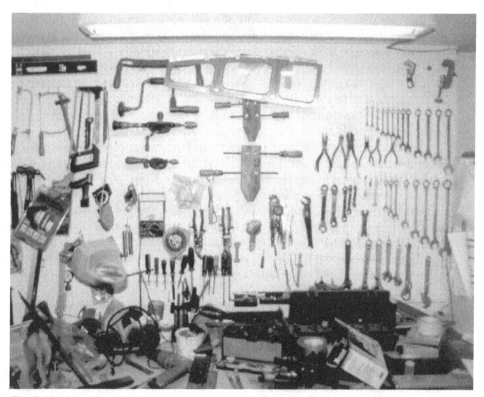

Fig. 6-16. *A tool board is the best way to store often-used tools.*

Arrange the tools you'll use the most at the most convenient locations. I like to draw the outline of the tool onto the pegboard to help keep track of where each one goes.

Tool boards are a marvelous way to keep track of tools. They work even better if you put each tool back after you use it. Every good homebuilder should develop this habit. Let me know if you figure out how—I certainly haven't picked up the knack.

The tool board can be seen from anywhere in the shop. If the tool is on the board, fetch it. If you can't find it, then start searching the worktable and workbench. At least you didn't have to dig through a toolbox first.

If you decide on storing some tools in a toolbox, pick up a rollaway unit with a multidrawered box. These units consist of a toolbox atop a wheeled cabinet. They normally are fairly expensive, around $150 for the most basic unit. Cheaply made bargain units can be found for half that, and for the amount of use they're likely to see, they'll do just fine.

Access

In kitchen design, the distance from the sink to the stove to the refrigerator and back to the sink should be no more than 10 feet; the same can be said about aircraft building. Minimize the distance between the worktable, workbench, tool board, and free-standing tools.

This rosy view of workshop design is discolored by real-world problems. The first is the worktable itself. The table is 3 feet wide and 12 feet long. Because you'll probably need to gain access to all sides, at least 2 feet of clearance must be maintained on all four sides. This makes the table workspace 7 feet wide and 16 feet long, or about the size of a typical car.

This not only restricts where the table can be placed, but its very size ensures that perfection can't be approached. There'll be plenty of times when the tool you want is on the opposite side of the table. So you'll have to walk all the way around and all the way back.

If possible, place your tools at the ends of the table rather than the sides. It seems closer when you don't have to turn so many corners.

The layout doesn't have to be fixed. My worktable occupied the center of the garage during fuselage construction. Once the gear was installed, the worktable was shoved off into one corner, ready for later use. (Don't forget to relevel and check the warpage whenever the table is moved.)

Storage of kit parts and components

An aircraft kit has hundreds or thousands of pieces that must be stored carefully until ready for use. Problems caused by bad storage can range from ruined material to chronically barked shins. While some general hints are given here, each type of material has specific handling requirements that must be followed. These are given in the appropriate construction chapters under "Material, Fasteners, and Safety."

The hardest pieces to store are large sheets of aluminum or plywood. They're awkward to handle, for one thing. For another, wood must be protected from rot and warping, and aluminum sheets are easily scratched or dented. At least easy access isn't a requirement, because you won't use a sheet every day. Some suggestions:

Ceiling racks. There's usually more headroom than you need, so ceiling racks move components and materials out of the way. Getting a sheet down requires special care, though.

Rolled. Aluminum sheets up to 0.032 inch or so can be stored easily in a roll. In fact, it's often delivered this way, so just leave it packed until it is needed. Fiberglass comes in rolls, which can be set up on racks hanging at the end of the table or off the wall, as shown in Fig. 6-17. Don't forget to cover the rolls in order to protect the fiberglass from dust and contamination.

Fig. 6-17. *For composite airplanes, the rolls of fiberglass cloth can be set up on racks similar to paper towels. However, the rolls must be protected against contamination when not in use.*

Wall racks. Ten sheets of 0.040-inch aluminum are only a half-inch thick. With a wall rack, they can be stored with minimal impact on working area. Wall racks for plywood must clamp the sheets tightly, or warping will result.

Off-the-wall racks. Flat sheets of aluminum can be stored in the darndest places. Why not under the bed or between sheets of scrap plywood? Curve a sheet around a corner. Tack sheets to walls. They'll snuggle down into ceiling trusses. Protect them from scratches with scrap plywood, leftover Christmas wrap, old blankets or mattress pads, or whatever.

Other components are smaller and easier to store. Racks for tubing and board lumber are easy to build but must support the load evenly. Underneath the worktable is a dandy place for tubing and aluminum extrusions and for larger kit components such as cowlings and canopies.

Smaller parts are even easier to store; however, access is more important. Nuts and bolts could be stashed anywhere, but you'll be needing them all the time. Plastic multidrawered storage cabinets (Fig. 6-18) are cheap and can be mounted anywhere or not permanently installed. A small easel-like mounting lets you bring all the hardware close to the job. Or mount your storage cabinet on the back of your roll-away tool box.

Fig. 6-18. *Low-cost plastic cabinets are ideal for small parts. The contents of each drawer should be marked either by hand with a felt-tip pen or by printing labels with a computer.*

Plastic storage cabinets aren't necessarily the best solution. The cheaper ones have awkwardly small drawers, and those with larger drawers are more expensive. The flat type, with the large top-hinged door, is better for access. A no- or low-cost alternative is to store the hardware in old cans, milk cartons, or what-have-you and build some cheap shelves to keep them on.

A cheap alternative is to mount some shelves and cut the bottoms off milk cartons to hold the parts (Fig. 6-19). This costs next to nothing, and the cartons are wide enough to make access easy.

Whichever method you choose, it's vital to clearly mark the exterior of each container, be it carton or drawer, with its contents. Some builders write the part info with a black marker (on masking tape if the surface won't take the marker). Others tape the actual labels to the outside of the container. Or use your computer and print off a set of self-adhesive labels.

Fig. 6-19. *Milk cartons and old boards make a cheap and adequate storage cabinet for small hardware.*

Storing finished components

Safe storage of finished components is slightly different. A ruined sheet of plywood merely drains your pocketbook; a ruined component wastes building time. Similar but more careful procedures must be used.

If allowed by the kitplane's plans, build the control surfaces first. They're easy to store on the walls or ceiling. You'd like to build the wings next because they can be stored similarly (Fig. 6-20).

Another good way to store wings is shown in Fig. 6-21. A board is bolted to the root fittings, and a pivot bolt slides down into the vertical part of a wheeled L-shaped trolley. This allows the wing to be turned horizontally for work (with the tip supported by a sawhorse) and then turned vertically and rolled into a storage area.

The fuselage is the hardest item to shunt aside temporarily. The wings are thin, but the fuselage is long and wide. One option, if your workshop has an open-truss ceiling, is to delay engine-mount installation and set the fuselage vertically.

Watch for mice. They love the little compartments of metal wings, and they enjoy chewing on wood and insulation. They'll happily build their nests in your composite

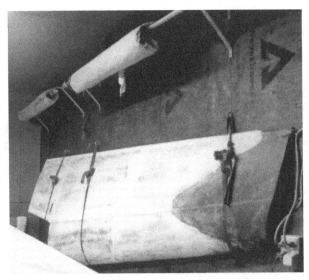

Fig. 6-20. *Control surfaces and wings can be suspended from walls or ceiling to get them out of the way.*

speedster and gleefully "piddle" on anything within reach. Ceiling and wall storage is a good first step toward prevention.

TOOLS

The kitplane plans should indicate what tools are required. But even an official tool list might not include everything. "Minimal tools" is a selling point; practically everybody has a power drill, so it looks better if that's the only drill the plans call for. However, a drill press will make your life far easier.

Where to get them

Two kinds of tools are actually needed for building a kitplane: *specialty tools* and *shop tools*. Specialty tools are those that are used mostly in specific professional applications. For example, most hardware stores sell tin snips, even aviation tin snips. But you'll have to go to a professional supplier to buy a hole duplicator or cable swage. These companies advertise in homebuilder magazines. The prices are uniformly high for the specialty tools and don't vary much from dealer to dealer. The best bet on these is to order the builder's kits available from various vendors.

Shop tools can be found at any hardware or discount store. Pliers, screwdrivers, electric drills, and the like can be bought almost anywhere. Prices vary drastically. Shop around.

There are several ways to lower tool costs. If the kit won't be delivered for awhile, keep an eye on local garage sales. Good deals in both hand and power tools often can be found. Similarly, some companies have a public outlet for selling used tools. While buying a moderate-quality new power tool is preferable to the clapped-out tools

Fig. 6-21. *Kirk McCarty built rollaway workstands for his Osprey II's wings. When rotated to the vertical position, a pin would lock the wing securely in place. The wings then could be stored in a small area.*

offered at these outlets, certain smaller items offer significant savings. One local outlet sells hole cutters in sizes up to 3 inches for $3. This is a third the cost of a new hole saw, and the used cutters are heavy-duty models instead of light hobbyist tools.

Another option is the discount tool merchants. Some are willing to bargain—mention that you saw the same tool at a lower price somewhere else, and they might offer a better deal. Or pick out several tools and make a single-price offer. I got a good deal on a drill press, vise, and some hand tools this way.

The quality of such tools varies widely, making careful evaluation necessary.

Evaluating quality

Price and quality generally are related. The more you spend, the higher the tool's quality is likely to be.

Admiral Gorshkov, the commander of the Soviet Navy during the 1970s, had a favorite saying: "'Better' is the enemy of 'good enough.'" In other words, there's no reason to buy higher quality than is necessary.

How good is good enough? For obvious reasons, you don't want to buy the cheapest. But you don't have to pay top price for tools of sufficient quality either. Stick with name brands where possible, but better prices and adequate quality are available in some off-brands as well.

Some signs of good-quality tools:

Polished surfaces. Cheap metal tools have an overall dull or matte finish. It takes extra effort to make 'em shine. Tool-mating surfaces (the area where the tool actually makes contact with the work, such as the inside of a socket) *must* be polished.

Rounded corners on non-tool-mating surfaces. Cheap tools are sharp everywhere, and you'll probably feel left-over ridges as well.

Fine machining on grip areas. The places where the hand holds the tool often have a crosshatch pattern in the metal to give a nonskid grip. The deeper, smoother, and finer the pattern is, the better it works.

Chamfered or radiused tool-mating surfaces. Sharp corners chip easily; the edges of the mating surfaces of good-quality tools should have a very slight curvature or be cut at a 45-degree angle to reduce chipping.

Straight edges. Look inside a cheap socket, and you actually can see a little reverse curve on the socket edge. This causes a sloppy fit and can damage parts.

Heft. Good tools are made of good steel; they're heavy and produce a satisfying ring when dropped.

Materials. Table-model tools are built of a variety of substances. Steel is better than cast iron, and cast iron is better than plastic. Steel has smoothly machined surfaces, cast iron has a rougher surface (especially in areas not in plain sight), and plastic is, well, plastic. This isn't a hard and fast rule; for example, there's nothing against a nonstructural motor cover or safety guard made from plastic.

There's nothing *against* buying the best-quality tools available. Some people use kitbuilding as an excuse to assemble a top-quality workshop.

Having cheap tools is no reason to rush out and replace the lot either. If it works, fine. But if the wrenches keep slipping or the Phillips screwdriver doesn't mesh in the slot very well anymore, it's time to consider replacement.

Sometimes you'll come across a do-everything tool, one that's supposed to do the work of several tools. While these often work out for odd jobs around the house, they usually can't hack intensive use.

I own a small screwdriver within which nest several smaller drivers. It's a handy little gadget; it's good-quality steel and brass, and it works quite well. However, the grip area is of poor quality. I've raised blisters after removing a single tight screw. The handles on the smaller screwdrivers are too small and awkward to use.

Despite this, I own about five of these. They're great for keeping in the desk or car's glove compartment, but I'd never consider building an airplane with one. Individual screwdrivers with large, comfortable grips are preferred.

More ways to reduce tool costs

Check out the plans, and find out when particular tools might be needed. For example, a good-quality Nicopress sleeve swage costs around $150, yet the swage generally isn't needed until the controls are being rigged. You might not need one for a year or more. In such a case, wait. Keep watching, and try to find one for less. The kit manufacturer might even start selling prefabricated cables. They'll cost more but will reduce building time.

Or you might find someone to lend you the tool. I'm all for "neither a borrower nor a lender be," but the specialty tools necessary are just that, special, and can't be used for much else. Someone who has completed his aircraft has little use for a hole matcher or bending brake. That person may not be willing to sell the tool (anticipating upgrades or maintenance) but might be quite willing to lend it for a few months or to let you come over and use it on occasion.

Check with your Experimental Aircraft Association (EAA) chapter. Some chapters own a variety of specialty tools for checkout to members.

The basic tool list

The following are the basic tools required for just about any homebuilt. These are the general-purpose tools, those that aren't construction-specific. Chapters 8 through 11 list the additional tools needed depending on the mode of construction.

You can get by with just a single example of most tools, but it gets fairly irritating when the only screwdriver gets misplaced, and work must halt until it's found. And *get* the right tools, and *use* the right tool. Remember the old saying, "When all you have is a hammer, everything looks like a nail." By having the right tools on hand, you won't be tempted to use a hacksaw as a file or whatever.

WRENCHES

Have at least one set of *combination* (open-end/box-end) wrenches in sizes from ¼ inch to at least ⅞ inch. The box end should be at a slight angle to the handle of the wrench. Acceptable-quality sets can be found for $40 or less.

An adjustable wrench (crescent wrench) is useful occasionally, but its use should be limited to those situations where a regular wrench doesn't fit or isn't available. Get one small one, and consider picking up a really big one eventually.

You'll eventually need a torque wrench. There are two types. The first has a dial indicator that shows the torque level. The second is the breakaway type, where the desired torque is selected, and the tool clicks when that level is reached. This type is preferred, of course, because you don't have to keep your eyes on the dial and have less chance of overshooting the desired amount of torque. These run around $60 or so.

SOCKET SETS

The minimum is a ⅜-inch-drive set (the little square hole in the socket is ⅜ inch across), with a ratchet wrench and sockets the same size range as the wrenches. The smaller sockets are probably ¼-inch drive, so an adapter should be part of the set.

An easier way would be to pick up a ¼-inch-drive set as well. Its ratchet wrench will be smaller and easier to swing in tight corners.

Eventually, you might need a ½-inch drive. As you might expect, this size is more expensive. Count on picking up the individual components as you need them rather than buying a full set.

SCREWDRIVERS

Screwdrivers come in two basic flavors, standard and Phillips. The standard type is the conventional flat blade, whereas the Phillips is the cross-shaped model. Standard screwdrivers come in combinations of length and blade size; get a variety. One of those items a builder always needs eventually is a short, wide-bladed standard screwdriver, so be sure to include one on your list.

Phillips screwdrivers come in three sizes. Get one of each to start; later, you'll probably need a couple extra in the two larger sizes (#2 and #3).

There are other types of screwdrivers, such as the Torx system. These still aren't very common in homebuilding.

Power screwdrivers have a mixed reputation. It's true that they make installing and removing screws a lot easier, but they do occasionally slip and scar the material.

PLIERS

Another tool with limited use on aircraft is pliers. Yet, when you need one, there's no substitute. Keep one around, and dust it occasionally.

Locking pliers (*vise grips*) are more useful. Pick up two or three; they're pretty cheap. I keep a sharp eye on the local hardware store's bargain bin and score one occasionally.

The main problem with most pliers is that they are designed to grip, and grip they do, with no regard for the damage they do to the material in the jaws. You don't want to use a pliers on an aircraft part—it is going to chew up the part or leave marks. Slipping a sacrificial piece of wood between the jaws and the part is an option.

A variation of the pliers is side cutters, or dykes. Get a medium-sized pair for cutting safety wire and electrical cable.

SNIPS

You'll need a pair of tin snips to cut thin aluminum. Aviation-style snips are comfortable and easy to use and aren't all that expensive. They come in straight-cut, left, and right varieties. If you're just buying one pair, get the straight. It'll cut curves, but not as tightly as either of the turning varieties. If you're cutting a lot of sheet aluminum with curves, by all means buy a left-cut and/or right-cut model.

Unless the money is really tight, though, buy offset snips. These are shaped more like a set of garden clippers; both handles are on the same side of the jaws. They keep your hand free of cuts and nicks from the freshly cut metal edge.

HAMMER

A standard claw hammer is required for no other reason than to nail together the worktable and workbench. A useful addition is a rubber- or plastic-tipped hammer. This can be used to tap tight-fitting parts into place without fear of marring the part or the surrounding structure.

AUTOMATIC CENTER PUNCH

The automatic center punch is used to place a small dimple on metal prior to drilling. It prevents the bit from walking.

The center punch works by a small amount of trickery. Place the point at the center of the desired hole. Push down with the palm of your hand, and the tool resists for a moment and then snaps downward.The punch can be set to make various dimple sizes and costs less than $10.

VISE

A bench vise is a necessity. Bench vises are sold by the width of the gripping area; a 4-inch model is probably about the smallest you should get. Bench vises are designed to be bolted to a tabletop. Clamp-base models can be moved easily, but they're more expensive, and larger sizes are rare. Forget the vacuum-clamp models. They just won't stick well to a wooden tabletop.

If portability is a requirement, get a standard bench model, place it on the corner of the table, and use C-clamps to hold it.

CLAMPS

Buy two types: C-clamp and spring clamp. The first is the conventional C-clamp. Get a number of them in assorted sizes. You'll never have enough, anyway. I'm just scrimping by with six, and I'd definitely need more if I were building almost any other type of aircraft.

The other is the spring clamp—the kind that looks like an enormous clothespin. Again, the number you'll need depends on the way your airplane is built. These are sized based on approximate opening width. Buy a few, each between 1 and 3 inches, and eventually get more depending on which you use the most.

Bar clamps are useful in cases where the clamping surfaces are widely separated. These are similar to C-clamps except that the top of the C is firmly attached to a bar that fits into the base.

SAWS

Numerous handsaws exist, but there's one you'll need for sure: a hacksaw. Most take 10- or 12-inch blades. Twelve-inch blades give the longest wear and fastest cut, but 10-inch blades are easier to use in tight quarters. Pick a hacksaw with long blade-mounting posts because the blade's less likely to pop off under pressure.

Remember, when mounting the blades, the teeth point away from the handle, toward the end of the hacksaw. The saw cuts on the forward stroke.

Buy other types of handsaws as the plans call for or on recommendation by another builder.

Hand-held power saws come in two basic flavors. The first is the jig or saber saw. This is useful for cutting a variety of material, from wood to aluminum and generally cost around $40 or so.

The second type is the circular saw. While not used for cutting aircraft materials, it makes workbench and worktable building a breeze.

If you can afford only a single floor-model tool, buy a bandsaw. A bandsaw drives a continuous-loop blade through a small slot in a steel worktable. It's the best all-around tool for any power cutting. While hand tools such as hacksaws or saber saws can be used, the bandsaw gives the easiest and most exact cuts. Saber saws drive their blade back and forth; this jiggles the piece being cut and requires a steadying hand, all while trying to steer an awkward saw along a precise path.

Bandsaws, on the other hand, run the blade in one direction: down through the worktable. The piece stays steady, and both hands can be used to guide the cut. Bandsaws can cut almost any material if the right blade and speed are selected. Bandsaw speed is stated in blade feet per minute (fpm). A speed of 1,000 fpm is needed for wood. Aluminum can be cut at the same speed or preferably slower, but steel requires no more than 150 fpm.

Bandsaw prices start at around $300. Blades cost around $15. One option to consider is the combination horizontal-vertical bandsaw. This can be pivoted downward for precise cutoffs of lumber or tubing. Such saws are designed for cutting steel tubing and have a good slow blade speed. They feature removable worktables for cutting in the vertical position like a conventional bandsaw. The worktables aren't all that solid, though, and their working height is close to the ground.

Table saws use circular saw blades in a fixed mounting. They're more important on plans-built airplanes than on kitplanes—the kitplane should include precut materials. Still, you *can* get blades that allow aluminum cutting on a table saw. They scream like banshees but make excellent long, straight cuts.

DRILLS

As an utter minimum, you'll need a hand-held power drill with at least a ¼-inch chuck (can grip and use bits ¼ inch in diameter and smaller). Reversibility is nice, but variable speed helps start holes in metal. Rechargeable drills give more flexibility, but the battery packs can be bulky.

An air-powered drill can make sense, especially if you're building an RV or other riveted metal airplane. While air lines are more awkward to move around than electrical cords, the drills themselves are small and handy.

Like a bandsaw, a drill press is vital. And drill presses are not that much more expensive than hand-held drills; small tabletop units can be found for about $100. Hand drills can't be held steady enough; the drill wavers slightly, and the hole gets oblong. Drill presses are well suited for drilling or cutting large-diameter holes, and even the cheapest models have at least a ½-inch chuck. Drill presses let you position the hole very carefully, too.

Fig. 6-22. *A floor drill press is better, but a bench model is adequate for many kitplanes. Note the grinder and bandsaw.*

Tabletop drill presses (Fig. 6-22) can't make very deep holes. Not only is the chuck-to-worktable space limited, but the very length of the drill bit also cuts down on the available room. Floor models are more flexible because they allow the work to be positioned properly no matter its thickness or the length of the bit.

A tabletop drill press worktable is clamped in place, whereas a floor model can be cranked up and down. The crank allows more accurate placement and, more important, repeated placement. After the tabletop worktable is swung away, it's hard to reposition it in exactly the same spot under the drill point.

The usual next step is to buy a nice set of drill bits, typically to ¼ undoubtedly need more of the ⅛-, ³⁄₁₆-, and ¼-inch sizes. So don't buy a huge set of bits. Get a set of eight, from ¹⁄₁₆ to ¼ inch in ¹⁄₁₆-inch increments. Replace them as necessary. In fact, buy extras of oft-used sizes.

Sizing drill bits in fractions of an inch is a concession to the home hobbyist. Machinists use a different system, assigning a number or letter to given sizes, as shown in Table 6-1. If the plans call for using a #30 bit, use that size. There are good reasons not to specify the nearly identical ⅛-inch bit. These numbered bits can be found at well-equipped hardware stores or professional outlets.

Reams are special drill bits. Conventional bits are designed to quickly open a hole of approximately a given diameter. Reams concentrate on exactly the correct size. They won't enlarge a ⅛-inch pilot hole to ¼-inch, but if you drill a hole to ⁷⁄₃₂

Table 6-1. Decimal equivalents of drill sizes

Most builders are familiar with drill bits available in fractional sizes such as ⅛ inch, ¹¹⁄₁₆ inch, and so on. However, the fractionals are just a subset of the bits available for precision drilling. Here are the decimal equivalents of drill bits from ¹⁄₁₆ inch to ¼ inch.

Size	Decimal Equivalent	Size	Decimal Equivalent	Size	Decimal Equivalent
¹⁄₁₆	0.0625	32	0.1160	³⁄₁₆	0.1875
52	0.0635	31	0.1200	12	0.1890
51	0.0670	⅛	0.1250	11	0.1910
50	0.0700	30	0.1285	10	0.1935
49	0.0730	29	0.1360	9	0.1960
48	0.0760	28	0.1405	8	0.1990
⁵⁄₆₄	0.0781	⁹⁄₆₄	0.1406	7	0.2010
47	0.0785	27	0.1440	¹³⁄₆₄	0.0231
46	0.0810	26	0.1470	6	0.0240
45	0.0820	25	0.1495	5	0.2055
44	0.0860	24	0.1520	4	0.2090
43	0.0890	23	0.1540	3	0.2130
42	0.0935	⁵⁄₃₂	0.1562	⁷⁄₃₂	0.2187
³⁄₃₂	0.0937	22	0.1570	2	0.2210
41	0.0960	21	0.1590	1	0.2280
40	0.0980	20	0.1610	A	0.2340
39	0.0995	19	0.1660	¹⁵⁄₆₄	0.2344
38	0.1015	18	10.1695	B	0.2380
37	0.1040	¹¹⁄₆₄	0.1719	C	0.2420
36	0.4065	17	0.1930	D	0.2460
⁷⁄₆₄	0.1094	16	0.1770	¼	0.2500
35	0.1100	15	0.1800	E	0.2500
34	0.1110	14	0.1820		
33	0.1130	13	0.1850		

inch and then use a ¼-inch ream, the hole will be perfectly circular with exactly the indicated diameter.

FILES

Files are cheap; stock up on an assortment. Files are described by their length, cross-sectional shape, cut, and grade. Length is obvious. The cross-sectional shape can vary from flat, to half-round, to round and other geometries. *Mill* and *flat* files are almost identical, but the mill file has one smooth edge. The smooth edge protects the opposite side when filing in tight places.

The *cut* describes whether the file's teeth are long and parallel or small and diamond-shaped: *single cut* and *double cut,* respectively; the double cut is more effective at removing metal.

Finally, the *grade* determines the true roughness of a file. There are four grades: *smooth, second cut, bastard cut,* and *coarse.* The coarse file removes a lot of metal quickly, whereas the smooth-cut file would be used for finish filing.

However, the grade doesn't tell the whole story. The grade is for comparison between files *of the same length.* A 12-inch second-cut file will be rougher than a 6-inch model of the same grade.

If you're facing a ragged edge of steel, the best file would be a 12-inch double-cut coarse or bastard file. For taking the last little notches out of a piece of aluminum, select a small single-cut smooth file.

Now you see the importance of picking up a variety of files. The *half-rounds* are the most useful, followed by the *flat* files.

You'll also need other aids to smoothing wood and metal, such as *sandpaper* and *emery cloth.*

BENCH GRINDER

This is another cheap power tool. Hand filing is slow, especially when working with steel. A bench grinder does the job much faster.

BENCHTOP BELT SANDER

Benchtop belt sanders usually include a belt that can be oriented both horizontally and vertically. Some have a small disk sander attached as well. These are tremendously useful for final smoothing and shaping after cutting aluminum or wood.

RULERS AND REFERENCES

A variety of rulers will be necessary, including devices that can measure angles and liveliness. A wide variety is easy to get because each is pretty cheap to buy.

A *12-foot tape measure.* You'll notice that the little metal tip appears loose. Don't try to tighten it. It slides inward to compensate for the width of the tip during inside measurements.

A metal ruler, 12 inches long, calibrated to one thirty-seconds of an inch. Buy a combination square, and you'll get both the ruler and a handy reference tool.

A *micrometer or vernier slide caliper.* This takes extremely accurate measurements depending on the skill of the operator. Study the instructions, and learn how to read the tool quickly and correctly. Slide calipers come in either decimal or fraction varieties (reads in 0.001-inch increments or 1/128 inch). I lean toward the decimal variety. Buy one that can take up to 1-inch measurements. Good-quality micrometers and vernier calipers are expensive, so plan on spending at least $20.

A carpenter's square. This is nothing more than a big L-shaped piece of steel with ruler markings on all edges. A square is used to help make good 90-degree angles. You'll need at least one. They are pretty cheap.

CLECOS AND CLECO PLIERS

Clecos are small spring-loaded temporary fasteners with prongs sticking out one end and a button on top of the other. Insert a cleco in the *cleco pliers,* squeeze the pliers, and the prongs get closer together. Insert the prongs into a hole, release pressure on the pliers, and the prongs spread to hold the cleco fast.

If you're trying to bolt two pieces of metal together, you can drill one hole through both pieces, insert a cleco, and then drill a second hole and insert another cleco. From that point on, each additional hole will maintain proper alignment. Figure 6-23 shows an example of their use, and later chapters describe their use further.

Fig. 6-23. *Clecos in action. Note the cleco pliers on the floor.*

Clecos are color-coded depending on fractional inch size. The code is

$3/32$ inch	Silver
$1/8$ inch	Copper
$5/32$ inch	Black
$3/16$ inch	Brass
$7/32$ inch	Silver
$1/4$ inch	Copper
$9/32$ inch	Black
$5/16$ inch	Brass

Note that the color code repeats after $3/16$ inch. You shouldn't have any trouble telling a $1/8$-inch copper-colored cleco from a $1/4$-inch copper-colored cleco.

Cleco pliers sell for about $8 and clecos for around 50 cents each. I've bought used ones for as little as 18 cents. It pays to find the lowest cleco prices, especially if you are building a metal airplane, because you might need hundreds.

To begin, just buy a few—two dozen each, for example, of the $1/8$- and $3/16$-inch types.

Power tool philosophy

Take a moment, now, to remember exactly *why* you're buying all these tools. *Finishing* the airplane is the target, here. There'll be a lot of obstacles to that goal. Don't add to the problem by excessive stinginess when it comes down to buying power tools.

Some kits say, "Can be built using ordinary hand tools." This is no doubt true—just like it's true to say, "Will be built a lot faster if you dig up $500 or so for a good bandsaw and drill press."

Sure, you *can* cut 0.090-inch aluminum with a saber saw. If there's not much cutting involved with your kit, you might be able to get away without a bandsaw. But until you have one, you cannot understand how much simpler and easier a bandsaw makes building.

I went that route. I started out with nothing more than a small benchtop drill press. I cut all my aluminum with a saber saw until I was faced with cutting a 2-foot-long piece of 3-inch-square $1/8$-inch-thick aluminum tube right down the middle to make two U-shaped channels. I broke down and bought a bandsaw for $350. It cut the tube in just five minutes. I haven't used my saber saw since.

Preceding sections have described a number of hand tools. Most are also available as power tools. There are even powered cleco pliers. Some of these tools are only available in the air-powered variety, but a compressor isn't a bad thing to have around the shop either.

Whether you'll want to go power or hand depends on how much of a particular operation is required. If your plans require just a little work with a set of snips, go ahead and buy the hand variety. But if you'll need to cut entire sheets of aluminum, go with a power snips instead. Power cleco pliers may seem a little silly—until you're faced with removing and installing 200 clecos on a wing panel.

So make your tool decisions wisely. Sometimes you just have to spend the money.

To minimize the "net" money spent, consider buying good-quality used tools rather than brand-new off-brand ones. The name-brand tools are long-lasting; building your airplane won't wear them too much. When you're up and flying, you probably can sell a good-quality bandsaw for only a bit less than you originally picked it up for. Your "net" cost, then, is practically nothing, although it still takes the same outlay at the start.

There are also several bargain power-tool outfits, such as Grizzly Industrial Tools (*www.grizzly.com*) and Harbor Freight Tools (*www.harborfreight.com*). These companies purchase tools in bulk from Asia. Quality tends to vary depending on the specific manufacturers the tools are obtained from, but I have several friends who have been quite happy with their purchases. Typically, the tools need some adjustment to work properly, and the instructions sometimes leave much to be desired. Shipping these tools is often expensive, but the companies occasionally distribute coupons for free shipping.

Safety items

No matter what type of airplane you're building, it involves working with fast-turning machinery and/or toxic chemicals. The first line of defense, therefore, is safety goggles. Using any sort of power tool puts you at risk for eye damage because they all fling small particles of material. You need two eyes to fly an airplane. Wouldn't you feel real silly spending three years building your dream plane and then losing your medical certificate because you didn't wear $3 worth of protection?

Don't count on being safe because you already wear normal eyeglasses. Dust gets blown around and will come in the sides. Get safety goggles instead. Keep them in a convenient place, and get used to donning them.

I did not wear safety glasses. One day I was drilling plastic, and a piece about $\frac{1}{16}$ inch across flew up and stuck me in one eye. A careful bit of tweezer work removed it, with no permanent damage. I wear goggles faithfully now, and so should you.

A set of ear plugs also isn't a bad idea, but the earmuff types are a lot better. Hearing loss is insidious, and riveting is deceptively hard on the ears. Buy several sets of hearing protectors so that assistants and visitors are well protected (Fig. 6-24).

One thing is for sure about working with homebuilts: The toxic chemical risk is not going to get any better. Many builders consider the biggest advantage of metal airplanes over composites is that metal aircraft construction avoids hazardous materials like epoxies.

But zinc-chromate primer, the traditional corrosion preventative, causes cancer. And most, if not all, aluminum on a homebuilt should be treated before assembly. There are somewhat safer alternatives, but they all require precautions.

In any case, there's no such thing as a safe paint. If you plan on painting the aircraft yourself, protection is vital. At the very minimum, wear a mask specifically designed for spray-painting use. Wear an old coat, gloves, and hat as well.

Chemical problems also come in the hyperallergic department. Back when VariEzes were under construction, a few builders would work contentedly for months without problems. Then, one day, they'd arrive in the workshop, and their skin would break out into an itchy rash.

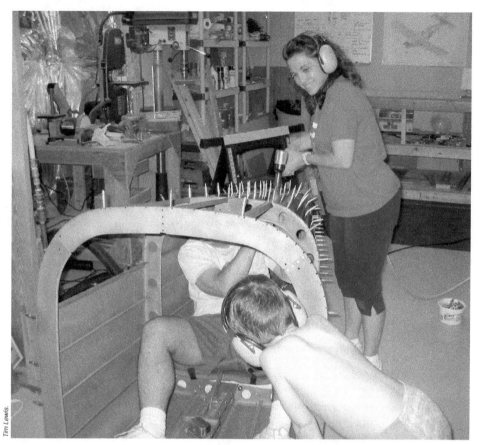

Fig. 6-24. *When building a metal airplane, hearing protectors should be available for both workers and visitors. Here, Kyle Lewis watches while his parents, Debbie and Tim Lewis, drive rivets on the family RV-6.*

Regular exposure to the epoxy had resulted in an allergic reaction. It's permanent; it never gets better, and it never goes away. Not everyone is susceptible. If it hits, though, forget completing the aircraft.

Modern epoxies reduce this problem dramatically. But it still can happen. Don't touch the chemicals, and don't apply epoxy with your bare hands. Buy rubber or plastic gloves, and don't work in bare sleeves. Incidental contact shouldn't cause a reaction, so don't worry about a few drops here and there. Several how-to articles show people applying epoxy or related mixtures barehanded, and that's bad, bad, bad.

Finally, toxic dust can be produced when you work with certain composite materials. A simple dust mask will suffice when you are cutting foam or sanding a completed structure. But more elaborate protection is needed when you are painting.

The best advice I can give is *read the label.* Don't assume that the company lawyers forced "Chicken Little" statements down the product manufacturer's throat. Follow the safety precautions.

While there are a number of smaller items you'll need, the tools just listed and summarized in Table 6-2 will suffice to start. Again, this is the *basic* list. More tools, and possibly more expensive tools, may be necessary depending on what type of aircraft you are builting. The individual construction chapters include an applicable list of additional tools.

OTHER SHOP ITEMS

There are a number of items that fall into the netherland between tools and aircraft parts. The best description for them might be *shop supplies.*

When you are building your airplane on a worktable, the parts occasionally must be held in place temporarily. Hence a supply of scrap 1 × 2 or 2 × 4 lumber is a good start. The 1 × 2 is commonly called a *furring strip;* the 2 × 4s are called *studs.* Set aside a convenient location to store these scraps. Cut them as necessary, and then reuse them. Similarly, you'll need nails to hold the pieces in place. Six- or eight-penny nails are about right.

Scrap plywood comes in handy. You can build forms for bending tubing or make templates out of it.

In fact, much of this scrap wood should be left over from building your worktable and workbench. As you build your airplane, you'll find that the scrap wood comes in handy; buy more as needed.

Hopefully, enough nuts, washers, and bolts came with your kit. Even so, pick up some extras. They always roll away into the dark corners or a different size must be installed for some reason. They aren't that expensive, and it beats stopping construction on a Saturday evening because you've run out of bolts. Buy them from an aircraft supply outlet, though, not from the corner lumber store. Chapter 7 explains the differences between aircraft-quality and commercial hardware.

One handy thing from the hardware store is threaded rod, sometimes called *readi-bolt.* It's like a very long (typically 3 feet) bolt without a head. Combined with a couple of blocks of wood or some scrap aluminum, it can be used to apply significant tension during the building process. Pick up a length of ³⁄₁₆-inch readi-bolt just to have around. *Do not* install it on the aircraft as a permanent part. It doesn't have much strength.

If you can find a low-cost source of cheap aluminum, pick up some 0.016- or 0.025-inch sheet for making patterns and templates. This doesn't have to be aircraft-grade aluminum because it is not going to be installed on the aircraft. You'll use it to make templates or to practice a particular technique on cheaper materials. Similarly, various sizes of aluminum angles are useful.

Hardware stores sell nonaircraft aluminum, but their prices are generally equal to what the homebuilder's catalogs charge. Check in the Yellow Pages under "Surplus," and call around. Often, the catalog companies offer a grab bag of aluminum parts.

If you do end up with some non-aircraft-grade aluminum, make sure that it doesn't end up incorporated into the aircraft. Spray it with orange paint, scratch it with a chisel, or do something else to make sure that it isn't mistaken for good aluminum.

Set up a scrap pile to dump your mistakes. There are two reasons for hanging onto scrap. First, there might be enough good material for another, smaller part.

Table 6-2. Basic tool list

Tool	Approx. Cost	Comment
Hand drill	$40	Cord-type preferred, air-powered good choice if building riveted airplane.
Bench drill press	100	
Floor drill press	300	Optional but preferred
Table bandsaw	150	
Floor bandsaw	300	Optional but preferred
Jig (saber) saw	40	Get variable-speed
Bench grinder	50	
Bench sander	120	
Combination wrench set	40	
Crescent wrench	15	
Torque wrench	60	Get "click" type
3/8-inch socket set	40	Get combination of 6- and 12-point sockets
Screwdriver set	10	
Pliers	8	
Snips	20	
Hammer	10	
Automatic center punch	10	
Files (set)	20	
C-clamps	10	Might need many
Spring clamps	5	Might need many
Bar clamps	30	
Tape measure	5	
Micrometer	65	
Carpenter's square	15	
Combination square	10	
Vise	50	
Cleco pliers	10	
Set of clecos	30	Might need lots more
Safety glasses	10	
Dust masks (dozen)	10	
Respirator w/filter	40	
Hearing protector	30	

Note: Actual price may vary depending on quality and local sales

I've often taken tubing from the scrap pile, cut off the ruined portions, and had sufficient material for a different item. Second, the aluminum scraps are recyclable. You can get a small return when your project is finished. Selected recycling centers won't take aluminum alloy, though, so check before taking it in.

You'll also need pens, pencils, and markers to draw lines; sharpeners for the pencils; and garbage cans, brooms, dustpans, and a dozen other small items that meet individual needs.

Paperwork preparations

You don't have to notify the Federal Aviation Administration (FAA) about your project until it's ready for inspection. However, at that time, the inspector will want proof that *you* built the aircraft.

The first item of proof is the builder's log. These are available on the market, or you can just buy a spiral-bound notebook. Make a log entry every time you work on the aircraft. Include the date, the number of hours worked, and the tasks performed. Especially note any variation from the plans. For example, my aircraft's plans call for the installation of a tail skid instead of a tailwheel. I designed and built a steerable tailwheel assembly and included drawings and descriptions in the log. Keeping a log on a computer is fine, but as with all valuable files, keep backups.

The second proof consists of pictures. Take photos of various stages of construction, including the changes. These don't have to be magazine-quality photos. Digital cameras are getting dirt cheap, and a cheap 1 or 2 megapixel camera is certainly sufficient to document your project.

One very high-tech way to take pictures of your building process is to set up a Web camera. These are so cheap that they're often just given away with a new computer. I know a man who rigged up a Web camera to take a photo every minute whenever the shop lights were on. At the end, he had a very rapid "movie" showing his airplane going together. When the inspector came to sign off on his airplane, the first thing the builder did was sit him in front of the computer and play the movie. When it was over, the inspector didn't even bother to look at the hard-copy building logs.

Finding help

One last thing to take care of before starting: Where are you going to find help when you need it? Here are some suggestions:

LOCAL CONTACTS

Your first step is to visit your local chapter of the EAA. This is the best place to find people building the same—or similar—types of aircraft. Call EAA headquarters at (414) 426-4800, and ask for chapter information.

Most chapters also include an EAA technical counselor. These are experienced builders and/or airframe and powerplant (A&P) mechanics who have volunteered to guide people through the building process. The counselors are available to answer questions over the phone and will come to your shop to inspect the plane in progress. Have them over often—the sooner problems are found, the easier correction will be. The counselor's suggestions have no legal bearing, but the counselors have vast experience, and their recommendations should be followed.

Your local EAA chapter may well have other members building the same aircraft as you. Even if it doesn't, though, invite the members to visit your shop occasionally. Those who are building other types of planes often have different insights into the ways problems can be solved (Fig. 6-25).

Fig. 6-25. *Builders of other types of kitplanes can offer new insights into solving construction problems. Here, RV-7 builder Dan Tracy (left) and Terry Dazey discuss the metalwork on Dazey's Murphy Rebel.*

Finally, back when you were shopping for the right kit, the manufacturer should have provided a list of other local customers.

ONLINE

Your kit manufacturer probably can point you to some online forums or e-mail discussion groups for your airplane. If not, you usually can find them by entering the aircraft name in an Internet search engine.

E-mail lists generally are found in Google Groups (*groups.yahoo.com*) and on the Matronics Web page (*www.matronics.com*). They are often excellent sources of information and advice.

One caution, though: The information on these sources isn't screened for technical accuracy. Sometimes folks just give dumb advice. So it's best not to act immediately on a suggestion—give it a day or so for other builders to chime in with objections and alternate approaches.

THE SPORTAIR WORKSHOPS

Probably the best way to gain the basic skills is through the sportair workshops offered by the EAA. These are offered at a variety of locations throughout the year and center on such topics as composites, RV assembly, sheet-metal work, welding, fabric covering, and so on. These are great to visit before you actually decide what to build because they can give you guidance on all type of construction.

All tools and materials are provided, and only about 25 percent of the two-day courses is spent in lectures, with most of the time being spent in hands-on workshop experience. Prices vary from $300 to $400 depending on the subject of the workshop. Discounts are given to EAA members.

For more information, visit *www.sportair.com* or call (800) 967-5746.

BUILDERS' ASSISTANCE CENTERS

Some kit manufacturers are setting up builders' assistance centers. These can vary from weekend courses to learn the basics of construction for their airplanes (Fig. 6-26) to sessions lasting several weeks. The later courses are specifically directed at

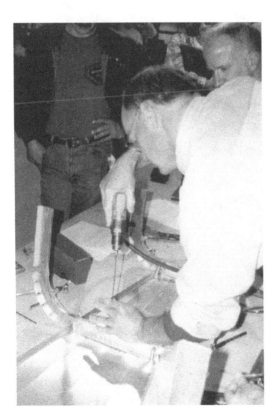

Fig. 6-26. *Builders' workshops are excellent places to learn workmanship.*

giving kit purchasers a head start, allowing them to perform critical tasks under the careful supervision of the factory experts.

These courses can cost from $100 to many thousands of dollars. If your goal is more oriented toward flying than the building process and you can afford it, a session at the builder center is a good idea.

PROFESSIONAL BUILDER ASSISTANCE

There are a number of companies that provide support for the builders of kit aircraft.

Careful! Remember that 51 percent of your aircraft must be constructed by an *amateur builder.* If it isn't, it can't be licensed in the experimental/amateur-built category.

If you get too much assistance from such a company, your FAA inspector may refuse to approve the aircraft. You might be able to license it in the experimental/exhibition category, but the restrictions are far, far tighter.

However, there are legitimate builder support centers. These provide facilities, tools, and on-site advice while *you* build your aircraft. These are legal and, in fact, encouraged by the EAA.

If you intend to avail yourself of for-hire help, coordinate your planning with the local FAA Flight Standards District Office that you plan to have certify your aircraft.

READY TO BUILD

You're all set now: You've picked out the kit and figured out engine and other options, your workshop is equipped, and your tools are ready. The kit lies on the worktable like an anesthetized patient awaiting the first touch of the surgeon's blade.

But it's not enough to know where the kit's appendix is located or, for that matter, how to rivet two pieces of aluminum together. The surgeon can't just hack away to the target, nor can the homebuilder be blind to the other requirements of aircraft manufacture.

There exists a body of standards that together are defined as *aircraft-quality workmanship.* Before the individual construction methods are detailed, let's look at some of the background details of aircraft construction.

7

Basics

THE INTRODUCTION OF KITPLANES changed the homebuilt industry forever. Not totally for the better, though.

In the early days of homebuilding, $20 would buy a 10-page plan set. The average person couldn't even attempt to start a project of this size on his own with such limited information. Prospective builders had to work closely with an experienced mechanic or builder to complete the project. The new builder would learn craftsmanship, not just how to build an airplane.

Nowadays, kits arrive with prefab parts. The pieces are cut to size, and holes are computer-drilled. All you add is the battery and the paint. Construction manuals are marvels of exactitude. Anyone who can keep a lawn mower running can build an airplane with little or no outside help.

But there's one crucial difference between keeping the Lawn-Boy spinning versus building an airplane: aircraft-quality workmanship. You can't drill just any hole when you are building an airplane, and you can't make parts out of just any material. In earlier days, the "old hands" taught the basic practices. Violations of accepted practice are subtle and may not be detected once the part is installed.

New builders typically worry about whether they'll pick up the skills necessary to construct the aircraft's structure. The tasks might be complicated, but they are rarely difficult. In most cases, kitplanes are designed to be tolerant.

Homebuilts are strong. Most in-flight structural failures are due to either serious deviations from the plans or overstress during aerobatics. If you take reasonable care, your airplane won't break up.

However, this is not the same thing as saying that the airplane is safe to fly. Poor workmanship can kill you just as fast, through jammed controls, warped surfaces, and parts that fail prematurely.

As pilots, we occasionally deride the Federal Aviation Administration's (FAA's) rulebook mentality and the minutiae contained in the Federal Aviation Regulations (FARs). However, there's one unfortunate truth: Nearly every regulation came about after someone flew an airplane into the ground. Overbearing as they might be on occasion, the air regulations are a good guide to safe flying.

So it is with aircraft workmanship. Every requirement, from self-locking nuts, to safety-wiring turnbuckles, to edge margin, grew out of an aircraft crash.

The ultimate source of workmanship standards is FAA Advisory Circular AC 43.13b, "Acceptable Methods, Techniques, and Practices." It's available through many sources. It's available online, although it's quoted in so many places that it's sometimes difficult to easily find a download site. A search for "AC 43.13 www.faa.gov" seems to point directly at the FAA's online repository.

This chapter reviews the basics of aircraft-quality workmanship. Read, learn, and remember.

THE NUTS AND BOLTS

Bolting two pieces together is simple. Run a bolt through matching holes, place a washer over the end, and then add a nut to hold the bolt in place. This is just about the same as working on a car. However, there are several rules to follow.

Probably about the biggest one is the requirement to use aircraft-grade hardware. Of course, anytime the word "aircraft" is associated with anything, you know the price is going up. Why not buy bolts at the local hardware store?

Over the years, the government has established a number of standards for the construction and testing of aircraft hardware. When your kitplane plans call for the installation of an AN4-32A bolt, it means something more than a mere manufacturer's part number. The number refers to a federal standard specification; the bolt you install must meet this spec.

The major specs are

- MS (military standard)
- NAS (National Aerospace Standard)
- AN (The granddaddy of them all; originally the Army-Navy Standard, it's been renamed Air Force–Navy instead.)

The specifications will list the minimum and, in some cases, maximum requirements for approval. These requirements include dimensions, tolerances, strengths, and finishes.

For example, bolt specifications will list minimum strengths in both *shear* and *tension* applications. The specification guarantees that the bolt can withstand strain up to a certain load. Shear loads are those that try to make opposite ends go in different directions. A tractor trying to pull a stump out of the ground applies a shear load; if the roots are too strongly set, the stump will break just above the ground. Tension loads are those that try to stretch the bolt. Figure 7-1 presents an example of each type of loading.

Most AN hardware is made from steel that is plated with cadmium. The cadmium gives the metal a goldish color and provides corrosion resistance. Bolts and washers also are available in aluminum. They're half as strong as the steel hardware, so do not use them unless the plans call for them.

Similar parts in the different systems meet different requirements, complicating substitution. An AN3 bolt and an NAS623-3 machine screw are similar in appearance; the machine screw's head is Phillips instead of hex. But if your plans specify the

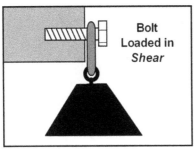
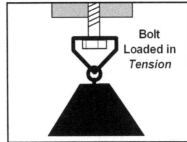

Fig. 7-1. *Shear loading versus tension.*

NAS bolt, don't substitute the AN variety. The National Aerospace Standard specifies a higher degree of heat treating. The NAS bolt is 33 percent stronger.

This doesn't mean that you should blindly substitute NAS bolts for AN bolts. There are good reasons to use AN bolts; cost and availability are big ones. A competent designer will make allowances for the type of hardware. In any case, the limiting load factor is usually set by the aircraft's structural design and material, not the hardware that bolts it together.

Because you're building an experimental aircraft, hardware-store bolts and nuts are perfectly legal. But the use of approved aircraft hardware gives you a guarantee of sorts. You know that an AN3 bolt can withstand more than 2,200 pounds of tension. At what point will the hardware store bolt break?

The person at the hardware store will give you a guarantee: "If they don't work out, just return them for a full refund." It's kind of like the old joke about parachute warranties: "If it doesn't open, we'll give you your money back." Hardware store bolts are made of cheap steel with limited heat treating. They're about a third as strong as approved hardware.

Many of the larger hardware stores *do* sell stronger bolts, which you'll hear referred to as "grade eight." The same problem applies, though: How do you *know* the bolts you are sold are indeed grade eight? It's best to buy your hardware from a quality aviation supplier.

Bolt basics

The standard bolts used in homebuilt aircraft are the AN3 through AN20 series. A bolt consists of two parts, the *head* and the *shank*. The head is the hexagonal portion that fits the wrench. The shank is the portion designed to slip into the hole. It's threaded at one end, with the unthreaded portion called the *grip*.

Whether a bolt is aircraft quality often can be determined by looking at the head. The usual marking is a four- or six-pointed star, possibly accompanied by a couple of letters. If the bolt has no markings on the head, it isn't an approved bolt. Typical head markings are shown in Fig. 7-2.

A typical bolt specification is *AN3-14A*:

- *AN* indicates the bolt meets the AN standards.
- *3* is the diameter of the bolt, in one-sixteenths of an inch.

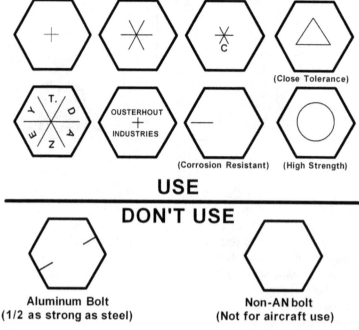

(Close Tolerance)

OUSTERHOUT
+
INDUSTRIES

(Corrosion Resistant)   (High Strength)

USE

DON'T USE

Aluminum Bolt
(1/2 as strong as steel)

Non-AN bolt
(Not for aircraft use)

Fig. 7-2. *Typical bolt-head markings.*

- *14* is the approximate shank length.
- "A" indicates that the thread area isn't drilled for a cotter pin.

The *diameter* is the width of the grip. There is a small amount of allowed tolerance, so don't be surprised if one bolt fits more tightly than another.

The *length* description is rather tricky. The first digit indicates the number of whole inches; the second digit is the number of ⅛ inches. So an AN3-14 bolt is about 1 and ⁴⁄₈ (½) inches long from the bottom of the head to the tip of the shank (Fig. 7-3).

Whether the threads should be drilled or not depends on how the nut will be secured. The undrilled type uses self-locking nuts. Most of your installations will use self-locking nuts, so buy undrilled bolts.

Castle nuts require drilled bolts in order to pass the cotter pin. Although special jigs are available to make it easier, avoid drilling the bolts yourself. Not only won't your work match the exact specifications of the standard, but the new hole also won't be cadmium-plated. It's okay only for nonstructural purposes.

Drilled-head bolts are available for special cases, for instance, if the bolt is installed into a threaded hole in a part. Safety wire is then passed through the hole to secure the bolt. Drilled-head bolts are identified by the addition of an *H* after the code for bolt diameter: AN3H-10A, for example. The AN76 series has a larger head with multiple holes.

As mentioned earlier, the AN3 through AN20 series of bolts have some small amount of shank-diameter variance. This amount of variance is acceptable in some applications but not in others. For example, the designer wants to minimize

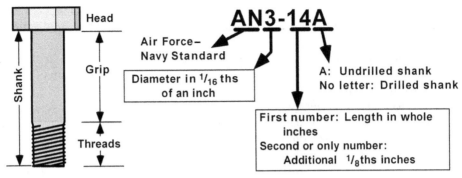

Fig. 7-3. *AN bolt specifications.*

slop in the wing attachments and so might specify a close-tolerance bolt. These are generally of the AN173 through AN186 series. They have a triangle marking on the head.

Occasionally, the designer specifies NAS bolts to take advantage of their higher strength. These high-strength bolts are marked by a small bowl-shaped depression in the bolt head.

The cadmium plating holds corrosion in check for most applications, but it isn't perfect. Over a long period, the moisture retained in wood can overcome the plating of a bolt installed through it. To make matters worse, the surface corrosion caused by contact with the wood tends to lock bolts in place, complicating removal and inspection.

Stainless steel bolts are available but are more expensive. Cheaper alternatives are to protect the bolt (by applying primer to the parts it holds together) or to varnish the hole to isolate the moisture from the bolt.

Sometimes you'll find bolts that have no grip length; the entire shank is threaded. These might be useful in lightly loaded applications but cannot be used in lieu of standard AN3 to AN20 bolts.

Nuts

Aircraft nuts must not come off accidentally. They undergo a lot of vibration that tends to loosen plain nuts. The consequences are obvious. There are three ways to secure them: *lock washers, self-locking nuts (stop nuts),* or *external safetying.*

AN-standard lock washers are used rarely. Weight is one reason—every bolt must have a lock washer in addition to the nut and a regular washer. Longer bolts are necessary, and installation is more complex.

Self-locking nuts are the most common method. The self-locking feature is implemented in one of two ways. First, the nut might incorporate an elastic fiber collar slightly smaller than the bolt. The pressure of the fiber on the threads then keeps the nut from turning. Second, the threaded portion of the nut can be manufactured with a slight distortion that adds sufficient friction to resist vibration.

The AN365 elastic stop is the most popular self-locking nut. Such nuts are cheap, less than a dime each in the ⅜-inch size. They can be reused as long as the fiber still holds the nut in place. If the nut can be turned by hand, junk it. The AN364 is a low-profile version, but it can only be used in shear applications. Use the AN365 unless your plans say otherwise.

Both nuts have their equivalents in other standards; the AN365 is paralleled by NAS1021N and MS20365 nuts, and the MS20364 and NAS1022N nuts can be used in lieu of the AN364.

Dash numbers after the main specification are used to differentiate sizes, for example, AN365-428. The dash number indicates the thread size. The system is rather awkward to memorize, but here's a shortcut: For AN4 and larger bolts, the first digit of the dash number is the same as the AN bolt diameter: AN4 takes AN365-428, AN5 takes AN365-524, AN6 takes AN365-624, and so forth. The shortcut doesn't work for AN3, so you'll just have to remember that it uses an AN365-1032 nut.

Elastic stop nuts are the most popular, but they have one drawback: The fiber insert is certified to only 250°F. This is adequate for most of the airframe, but such nuts can't be used forward of the firewall. An all-metal stop nut is necessary.

The most common of these is the AN363. By using threaded metal fingers, the hole diameter narrows slightly at one end and applies enough friction to prevent self-rotation. The AN363 uses the same dash-number scheme as the AN365 and is about 20 percent more expensive.

Self-locking nuts aren't the universal solution, though. If the bolt is intended to connect moving parts, the motion might overcome the nut's self-locking ability. An example would be a clevis attached to a control horn, hardly the place from which you'd want a nut to disappear.

Castle nuts are designed for these applications. They look like conventional nuts with notches cut out of one side like a castle's ramparts. The notches pass a cotter pin through the hole in a drilled bolt. The AN310 castle nut is for all uses, or the low-profile AN320 is used for shear-only applications. Just to add to the confusion, the dash numbers of the AN310/320 series actually correspond to the AN bolt size the nut goes with: AN310-3 for AN3 bolts and AN320-6 for AN6 bolts. AN310/320 nuts cost about twice that of the equivalent AN365 nuts.

For the belt-and-suspenders types, there's the MS17825 self-locking castle nut. This is a nice concept, but you can buy 10 AN365 nuts for one of these fancy varieties.

Washers

Washers have multiple functions. They act as shims, spread the compressive forces over a wider area, and protect the surface from rotation of the nut during installation.

The AN960 is aviation's standard washer. It comes in *regular* and *thin* (one-half the thickness of the regular washer). The light washer is used as a shim, especially when using castle nuts. Stock up on the regular variety, and get just a few thin ones.

The dash number of the AN960 washers give the size bolt they're designed for. And guess what? It's yet *another* system.

A typical washer spec is AN960-416. The three-digit dash number gives the diameter of the center hole in one-sixteenths of an inch, in this case, $^{4}/_{16}$ or $^{1}/_{4}$ inch. Light washers will have an *L* suffix, as in AN960-416L.

However, dash numbers of one or two digits indicate the bolt or screw size. An AN960-8 washer goes with a #8 screw, whereas an AN960-816 is for a ½-inch bolt. The right washer for AN3 bolts is the AN960-10.

As mentioned earlier, one function of a washer is to spread the load over a wider area. When the bolt goes through a softer material such as wood, you'd like to spread the load even more. Hence the AN970 flat washer, which has at least twice the diameter of equivalent-sized AN960 washer. There is no thin model of AN970. Just to make your day, the AN970 uses a different scheme for the dash number; the dash number on the AN970 is the same as the bolt it goes with. In other words, for an AN6 bolt, you could use either an AN960-616 or an AN970-6. But don't try an AN960-6 because it's too small. Figure 7-4 shows which nuts and washers go with which bolts.

		To Fit AN3 (3/16") Bolt	To Fit AN4 (1/4") Bolt	To Fit AN6 (3/8") Bolt
Elastic Stop Nut	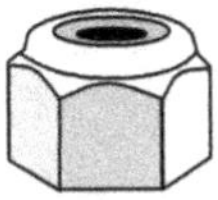	AN365-1032A AN364-1032A*	AN365-428A AN364-428A*	AN365-624A AN364-624A*
Metal Stop Nut	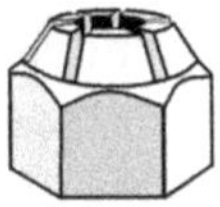	AN363-1032	AN363-428	AN363-624
Castle Nut	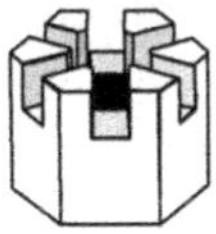	AN310-3 AN320-3*	AN310-4 AN320-4*	AN310-6 AN320-6*
Washer	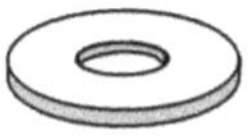	AN960-10 AN960-10L* AN970-3**	AN960-416 AN960-416L* AN970-4**	AN960-616 AN960-616L* AN970-6**
Cotter Pin	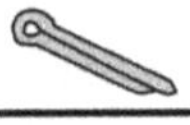	AN380-2-2	AN380-2-2	AN380-3-3

      * Thin unit... use only when called for
      ** Extra-wide washer

Fig. 7-4. *Aircraft hardware cross-reference.*

Bushings

Bolts are sometimes used as a pivot for a moving part. If the part moves over a broad angle or is highly loaded, the plans often will require installation of a bearing.

But if a part is lightly loaded, any bearing is overkill. In the short term, there'd be little wrong with just passing the bolt through a hole drilled in the part. But motion causes wear. Aviation bolts are hard and strong, much more durable than aluminum or wood. So the hole begins to enlarge and distort: *ovaling*. The pivot becomes sloppy, and eventually the part doesn't move right.

To avoid this problem, a *bushing* is installed between the bolt and the part. Bushings are tubular pieces of moderately hard metal with a hole drilled down the middle. They're pressed into a slightly undersized hole in the part, and the pivot bolt runs through the hole in the bushing.

Typically, bushings are aluminum, bronze, or a grade of steel that is softer than the bolt. The bushing wears at a slower rate than aluminum or wood. But its tubular shape makes it easy to replace: When the fit gets too sloppy, push out the old bushing and insert another.

There isn't really an AN standard for bushings—they aren't really under load. They just pass the loads to the components designed for them.

Other common hardware

Clevis pins are like bolts without threads, and they are used solely in shear applications. A typical application is in cable shackles and forks. The most common ones are the AN393 (³⁄₁₆-inch diameter) and the AN394 (¹⁄₄-inch diameter). A dash number divides the number of thirty-seconds of an inch between the head and the cotter-pin hole. In other words, an AN393-39 clevis pin in ³⁄₁₆-inch in diameter and ³⁹⁄₃₂ inch long, or 1 and ⁷⁄₃₂ inches.

Clevis pins and castle nuts are locked by the AN380/MS24665 *cotter pin*. The diameter and length are given by a bizarre little dash-number code. One big thing to remember about cotter pins is that they cannot be reused. Don't bend them straight and reinstall them.

Anchor nuts are lock nuts that can be attached solidly to the surrounding structure by rivets or screws. With the nut fixed in place, you don't have to hold it in place with a wrench whenever the component is removed or installed. If the plans call for an anchor nut, don't use the regular variety. You'll hate yourself if the part has to be removed later.

Typical installation

Let's assume that we're bolting a piece of ⅛-inch angle to a 1-inch-diameter steel tube. A good set of plans wouldn't leave us in suspense; they'd say, "Bolt the angle to the tube using an AN3-14A bolt, an AN365-1032 stop nut, and an AN960-10 washer."

More typically, they'd say, "Bolt the pieces together with an AN3 bolt," or ". . . with three-sixteenth bolts." Again, the best kits reduce your building time by telling you exactly what to do.

Let's operate on the assumption that the bolt diameter is the only thing specified. Selecting the right nut should be easy. We know the size of the bolt. If the bolt were meant to act as a pivot, we'd know to use a castle nut and cotter pin. Instead, the simpler self-locking nut will do. Because the assembly doesn't mount near the engine, a cheap elastic stop nut is acceptable. Should you use the AN364 shear-only style or the AN365 full-tension one? Without any guidance from the plans, better use the AN365.

Because neither of the items being joined is made of wood, we won't need the large-diameter AN970 washer. That leaves the AN960-10 washer; for right now, let's assume that we won't need the thin variety.

Which leaves the bolt. Because we're using a self-locking nut, whatever bolt we select should have the *A* (undrilled) suffix. The bolt's grip length must be approximately equal to the total thickness of the pieces being joined. Selecting the right bolt is awkward because the AN system refers to total length, not grip length. On AN3 bolts, the threaded area is about $7/16$ inch long. For AN4, the threads are $15/32$ inch, whereas AN5 and AN6 are about $1/2$ inch.

The threaded area develops the bolt's full strength in tension but is not designed for shear loading. So the threaded length of the shank should start just where the bolt emerges from the material. A little bit (at most, one full turn of threads) can still be inside. But you don't want too much coming out either. If the bolt is too long, the nut will reach the end of the threads before it applies pressure to the work. Up to three washers can be used to compensate for overlong bolts, but a shorter bolt would be lighter.

Sometimes, when all the bolts are too long, you might be tempted to lengthen the threaded area with a *die* (a tool used to cut threads). Don't. The AN specification calls for "rolled" threads, and your home threading makes sharp-edged ones. The bolt won't be as strong, and it'll be more susceptible to cracking. In addition, cutting new threads removes the cadmium plating, giving corrosion a place to start.

The absolute limit to bolt length is that no more than one turn of the threads should be left within the hole, and no more than three washers can be used to allow the bolt to apply its proper tension.

Select the correct grip length one of two ways: intellectually or practically. By adding the thicknesses of the material together, you derive the proper grip length. In other words, a $1/8$-inch thickness angle and a 1-inch-diameter steel tube need a bolt with a $1/8$-inch grip length. Look at Table 7-1, and you'll see that this means an AN3-14A bolt.

Or hold the pieces together, take a wire or piece of wood, and slide it into the hole to measure its depth. Then measure the length of the wire or wood or compare it with several bolts until the correct one is found.

Slide the bolt into the hole. It's strongly preferred that the head face up or forward so that if the nut comes off, gravity or the slipstream will hold the bolt in place. But if there's some reason you can't install the bolt that way, don't worry about it.

Slide the washer over the end, and then start the nut by hand. The fiber locking material inside the nut should make it impossible to turn by hand after the first turn or so. If the nut can be tightened all the way by hand, throw the nut away and use another.

When the nut gets hard to turn, take a combination wrench and socket wrench ($3/8$ inch for AN3 bolts) and tighten it the rest of the way. You'd prefer to use the socket on the nut and the combination wrench on the head. Otherwise, the cadmium gets scraped off the bottom of the bolt head, and the material below the head gets scraped. Turn the head if necessary; space constraints may dictate the situation.

How tight? AC 43-13 gives the numbers, but few homebuilders use torque wrenches for ordinary bolts. One builder told me, "Tighten it until the head starts to twist off, and then back off half a turn." It took me a couple of months to realize he was kidding.

Table 7-1. Approximate Grip Length

Dash Number	AN3	AN4	AN5	AN6	AN7	AN8
3	1/16	1/16	-	-	-	-
4	1/8	1/16	1/16	-	-	-
5	1/4	3/16	3/16	1/16	1/16	-
6	3/8	5/16	5/16	3/16	3/16	1/16
7	1/2	7/16	7/16	5/16	5/16	3/16
10	5/8	9/16	9/16	7/16	7/16	5/16
11	3/4	11/16	11/16	9/16	9/16	7/16
12	7/8	13/16	13/16	11/16	11/16	9/16
13	1	15/16	15/16	13/16	13/16	11/16
14	1 1/8	1 1/16	1 1/16	15/16	15/16	13/16
15	1 1/4	1 3/16	1 3/16	1 1/16	1 1/16	15/16
16	1 3/8	1 5/16	1 5/16	1 3/16	1 3/16	1 1/16
17	1 1/2	1 7/16	1 7/16	1 5/16	1 5/16	1 3/16
20	1 5/8	1 9/16	1 9/16	1 7/16	1 7/16	1 5/16
21	1 3/4	1 11/16	1 11/16	1 9/16	1 9/16	1 7/16
22	1 7/8	1 13/16	1 13/16	1 11/16	1 11/16	1 9/16
23	2	1 15/16	1 15/16	1 13/16	1 13/16	1 11/16
24	2 1/8	2 1/16	2 1/16	1 15/16	1 15/16	1 13/16
25	2 1/4	2 3/16	2 3/16	2 1/16	2 1/16	1 15/16
26	2 3/8	2 5/16	2 5/16	2 3/16	2 3/16	2 1/16
27	2 1/2	2 7/16	2 7/16	2 5/16	2 5/16	2 3/16
30	2 5/8	2 9/16	2 9/16	2 7/16	2 7/16	2 5/16

Anyway, massive amounts of pressure aren't necessary. The maximum torque of the AN3 bolt is 25 inch-pounds. If you're using a ratchet wrench with a 6-inch handle, that's only *4 pounds* (4 pounds × 6 inches = 24 inch-pounds) of pressure on the end of the handle. AN4 bolts specify a maximum of 11 pounds at the end of the wrench; AN5 gets all the way to 23 pounds. Less if your wrench is longer.

Tighten them up until they're snug, and don't apply insane amounts of torque. Use a torque wrench occasionally, and compare the readings with the figures given in FAA Advisory Circular AC 43.13.

Once the bolt is in place, look at the end of the nut. You should see at least one thread projecting past the fiber. If not, the bolt is too short and should be replaced with a longer one.

If there are more than three threads showing, the bolt might be too long. Try to rotate the bolt with the wrench without holding the other end. If it turns, the nut bottomed out before proper pressure could be applied. Remove the nut, and add another washer. Retighten, recheck, and repeat if necessary. If the bolt still isn't tight in the hole with three washers on it, replace it with a shorter bolt.

If you are installing a drilled bolt and using a castle nut, the nut must be in a position to allow the cotter pin to pass through the drilled hole in the bolt. The thinner washer (AN960-10L for AN3) can be used to ensure both proper torque and positioning. Bend both prongs of the cotter pin after insertion. Remember, never reuse a cotter pin.

Let's summarize the nuts-and-bolts of aircraft-quality workmanship, as shown in Fig. 7-5:

- Either a self-locking nut or a castle nut–cotter pin combination must be used.
- Do not use self-locking nuts on moving components.
- No more than three washers can be used per bolt.
- Little or no portion of the bolt's threads should remain inside the drilled hole.
- If possible, the bolt's head shouldn't be mounted down or backwards. This is a preference, not a hard-and-fast rule. There'll be instances when this isn't possible.

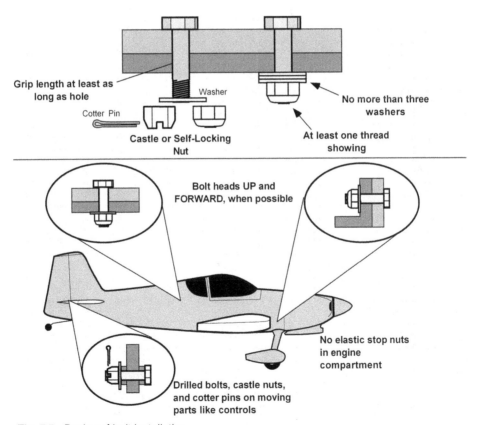

Fig. 7-5. *Basics of bolt installation.*

- Use aircraft standard hardware in all structural applications. Commercial-grade hardware is suited for nonstructural duties such as attaching placards, upholstery, and so on.

The preceding won't fit every situation that comes up while you are building your airplane but should give you a good start. Ask a local airframe and powerplant (A&P) mechanic or your FAA technical counselor for more information.

BASIC METAL CONSTRUCTION

Don't go skimming by this section just because you're building a composite airplane. You'll be making a lot of fixtures from aluminum and steel, and you should know the basic rules of metal construction. Let's start off by looking at the metals themselves.

Aluminum alloys

Pure aluminum is great stuff, but it must be alloyed with other metals for maximum strength. The type of alloy is given by a four-digit number, such as 2024 or 6061. The first digit identifies the major alloying elements. Alloy 6061, for example, consists of aluminum, magnesium, and silicon. The next digit indicates major modifications in the basic process, such as changes in percentages. The last two digits are essentially serial numbers for an exact alloy.

The alloy number is always followed by a dash and a *temper designation*, such as 6061-T3. Temper designations range from zero (i.e., 2024-0) to -T9 (i.e., 2024-T9), with the first digit after the *T* indicating the actual temper. Other numbers after this first digit are modifiers that don't really concern us. The temper indicates the degree of workability of the metal. T3 might crack if worked excessively, but you can fold a zero-temper sheet in half, and it probably won't break. However, it's only half as strong as T3 in the same alloy.

Alloy 2024 is the most common alloy. Other popular ones on homebuilts include 6061 and 7075. What's the difference? Strength, mostly. Of the common aircraft structural alloys, 7075 is the strongest, and 6061 is the weakest. Why not use 7075 exclusively? Well, because the extra strength comes at the expense of poor corrosion resistance and brittleness; 7075 is great for flat plates away from moisture but has a higher tendency to crack if flexed; 2024 is a good compromise that is not quite as strong as 7075 but less brittle.

It's still susceptible to corrosion, though. Pure aluminum won't corrode, so aircraft-grade 2024 and 7075 alloys are often clad with a thin layer of pure aluminum, hence "Alclad."

Most, if not all, applications using 2024 or 7075 will specify Alclad. The Alclad status, as well as the alloy and temper, will be printed on the sheet. The term "BARE" is a positive indication that the sheet is not clad. An example of corroded bare aluminum is shown in Fig. 7-6.

Most aluminum-monocoque aircraft use 2024-T3 throughout. Aluminum-tube aircraft (Murphy Renegade and CIRCA Nieuport) use 6061-T6 tubing. Alloy 7075 is fairly rare but often finds its way into special fittings.

Fig. 7-6.
Corroded unclad 2024 T-3 aluminum. While this piece can't be installed on an aircraft, it can be used for a template.

Steel

Steel has several advantages over aluminum. It's three times as strong and can be welded easily into complex shapes that are far more durable than a bolted or riveted aluminum structure.

Steel is quite a bit heavier, but sometimes it's worth it. Most of the lightest home-builts feature fuselages made from welded steel tubing.

Steel is a "magic metal" that we see too often to fully appreciate. The characteristics of a steel part can be tweaked by such processes as normalization, annealing, quenching, and tempering to produce exactly the required performance. Steel is not a modern wonder-metal governed by iron-clad (or steel-clad) patents. Quenching, for example, increases strength and hardness when the red-hot metal is suddenly plunged into a cooler liquid. A thousand years ago, swords made of Damascus steel were quenched in blood. Ick.

Shops specializing in more modern steel treatments are found in most cities. If you need some sort of special work done, you shouldn't have trouble finding a place to do it. But it's a rare kitplane that needs any steel work, especially custom treatment.

Like aluminum, steel alloys are identified by a four-digit code that specifies the alloying components. Unlike aluminum, one steel alloy predominates for aircraft: 4130, which incorporates chromium and molybdenum, called *chro-moly steel* for short. You'll hear it pronounced, "Crow-molly."

Alloy 4130 usually is sold normalized; that is, it has been heated to its critical temperature and allowed to cool at room temperature. It's required after any sort of heat treating (including welding) to equalize stresses and maximize the hardness.

Occasionally, you might find 4130 in the annealed condition. Don't use it on your aircraft without additional treatment. It's soft and easily formed but has no real strength. You can use it to build complex parts, but the parts must be heat treated afterwards. Look in the Yellow Pages under "Heat treatment, metal."

The steel alloy and condition are printed directly onto the metal. The condition is often abbreviated: 4130 N for normalized; 4130 A for annealed.

Other types of steels are *stainless* steels. These are used rarely for structural purposes. However, aircraft firewalls must be able to withstand a 1,200°F flame for five minutes, and stainless steel is one of the few materials that can take it. Alloy 304 is used commonly for firewalls.

The preceding discussion of aluminum and steel alloys and hardening is simplified; the matter actually is a bit more complex. For example, aluminum tempers aren't always single digits; further numbers indicate modifications. For example, a piece of aluminum angle might be labeled 6061-T652, indicating that the alloy was compressed to produce a permanent set as part of the heat-treating process. As far as the average homebuilder is concerned, only the first digit of the temper is of interest. Greater detail is given in FAA Advisory Circular AC 43.13.

Buying aluminum and steel

Hopefully, every speck of metal necessary came with the kit. In most cases, however, this is the exact quantity the aircraft requires. There's no obligation to supply extra to allow for occasional flubs. And you're going to make mistakes, especially at the beginning.

One kit manufacturer supplies templates for the most economical cutting of sheet aluminum for the aircraft. Unfortunately, several cuts are shared between two pieces. If you make one little mistake over a cut several feet long, both pieces are ruined.

It is best to develop an additional supply source. The kit manufacturer will sell you additional steel or aluminum, but if the part is made from sheet stock (raw sheet metal), you probably can get it faster and cheaper in other ways.

One way is by mail order through catalog companies such as Aircraft Spruce and Specialty and Wicks Aircraft Supply. Homebuilders are their primary customers, so you can buy small or large quantities.

For large orders where you can't wait, though, check the Yellow Pages under "Aluminum" or "Steel distributors." There's nothing specifically aircrafty about 2024 aluminum or 4130 steel. As long as a supplier can provide exactly the alloy and temper/condition required, there's nothing wrong with buying from public outlets. And you can pick it up the same day.

One glitch: These companies often have a minimum-order stipulation. Locally, several local outlets won't sell less than $100 worth of metal at a time. In other words, if you need just a 2-foot piece, you're out of luck. Also, their prices operate on a sliding scale—the more you buy, the less you pay per pound.

If you need just a single sheet or so, try to combine your order with that of another homebuilder so as to exceed the minimum and get the lowest possible price.

An option for builders in larger cities is surplus outlets. Companies that use aluminum generally buy it by the roll or sheet. Occasionally, odd pieces or the end of a roll is left over. Large companies find it more economical to sell the excess than to waste time trying to use every square inch. These companies often sell the scrap at public outlets at scrap metal prices. A sheet that costs $30 new might sell for around $10. It's not just theory; I've made several good buys in this way. I've got several sheets rolled up in my basement, ready to use.

One point to remember: Never buy ungraded and unmarked metal for your airplane. If you buy from a scrap dealer, make sure that the piece is marked with the original manufacturer's printing. Sure, you can tell whether it's aluminum or steel. But no one can determine the alloy or temper just by looking at it. Sheet metal has the alloy and temper/condition printed in repeating blocks, so only the smallest pieces will end up unlabeled. The printing often indicates the thickness, too.

The printing also indicates the *grain* of the metal. The grain is the long axis of the metal before it's cut and rolled. This becomes important when bending the material.

Cutting

Okay, you're working on your airplane, and you come across an instruction to cut a shape from sheet stock. The choice between a hand or power tool will depend on the thickness of the material and the physical size of the part. Selection criteria include speed, control, cleanness of cut, and ease of use.

Let's look at cutting aluminum first.

How about a pair of scissors? Don't laugh—on really thin aluminum (0.016 inch or less), they work quite well. Don't steal the wife's dressmaking pair—cutting aluminum quickly dulls the edge. Similarly, a paper cutter is fantastic for straight cuts in thin metal.

The most common tool for thin sheet aluminum is metal shears. The maximum thickness is limited mostly by your hand strength, about 0.040 inch is as thick as I care to try. Depending on the thickness, shears make a reasonably clean cut. But they can leave some pockmarks on the edges of thicker material. If you use the traditional "aircraft" sheers, long, straight cuts are somewhat of a bother because the two edges tend to get in the way of the handles and slice fingers. Offset snips alleviate this problem at a slightly higher purchase cost.

Another hand tool is the sheet-metal nibbler, which punches out sections of metal. Nibblers leave beautiful edges, and tight curves can be negotiated easily. Like most hand tools, they are limited in the thickness they'll cut. I overdrove a nibbler once, and the head broke off and whizzed past my ear. One major problem is slowness— the nibbler cuts, at most, an 1/8 or 1/4 inch per bite, and each bite has to be set up carefully. But nibblers work great for cutting instrument holes.

Hacksaws are pretty awkward for thin sheet but will cut away at plate as long as you're willing to saw. Blade selection is a compromise between speed and a clean edge. I prefer clean cuts over fast ones, so I get blades with 32 teeth per inch. But it still leaves a pretty rough edge.

If you won't really have that many cuts to do, you can opt for the humble saber or jig saw. This will handle aluminum to 0.125 inch and thicker, which is quite adequate for most homebuilts. Get fine-toothed blades for smooth cuts.

However, as mentioned in Chap. 6, a bandsaw is usually your best solution. It's really your *only* choice for thick aluminum. Set the blade speed to about 1,000 fpm (400 fpm is even better), and use a metal-cutting blade. Power sheers and even power nibblers are available, too.

Cutting steel is a bigger problem. Most hand tools just aren't suited for anything larger than the thinnest steel. Hand tools cut steel eventually, but your hands won't be worth much for a while afterwards. A bandsaw is a necessity for steel. Use a very slow blade speed; about 120 fpm.

Your first step is to mark the shape of the part on the sheet. The most accurate method is to paint an area with blue machinists' ink, and scribe the outline of the part into the dried ink. This works great when using a bandsaw, but a saber saw's shoe (the skid plates on either side of the blade) will leave long scratches on unprotected aluminum.

Don't scribe any lines over the part itself. Scratches disrupt the anticorrosion effect of Alclad and give cracks and corrosion a place to start.

An alternative would be to apply masking tape to the sheet and draw the pattern on the tape. If you are using a saber saw, stick the tape over a wide enough area so that the shoe won't contact unprotected metal. Contact paper, instead of tape, is another alternative. Or draw the shape on a piece of paper and glue it to the metal with rubber cement. The cement will peel off easily when you're done. Clean up tape residue with MEK or acetone.

If your plans include a full-scale template, scan it into your computer, print it on sticky-back paper, and stick it directly to the metal (Fig. 7-7). Or make a photocopy and glue it down with rubber cement. This cuts down on a lot of work because the templates generally include the locations of drilled holes, too. Again, if using a saber saw, include enough paper to keep the shoe from touching bare metal.

Be warned, though. Scanners and photocopiers often introduce size variations—a copy might be 5 percent larger or smaller than the original. Keep this in mind if precision cutting or drilling is going to be necessary. I like to scan in a short piece of ruler at the same time so I can quickly verify proper scale.

The easiest cutting will be with a bandsaw. Because the blade travels only downward, friction holds the piece flat on the table. Start and finish all cuts away from the outline of the part, if possible. When you've gained enough skill, you can try to shave the line closely. Otherwise, keep $\frac{1}{16}$ to $\frac{1}{8}$ inch clear of the line, and count on trimming the piece down with files or a bench sander. Otherwise, a minor jiggle could ruin the part. Don't try to cut corners too sharply—the blade might bind. Back off and come in at a tangent, if necessary.

Saber saws work differently. The blade goes back and forth, and if the metal isn't clamped or otherwise supported, it flaps. Since you generally are cutting out a full shape, this means that you have to stop often and reposition the metal so that the area you're currently working on is supported adequately.

This support is especially important when you are cutting thin metal. If you *must* use a saber saw on very thin aluminum, make a sandwich out of plywood with the metal in the middle. Use rubber cement, instead of catsup, to keep the aluminum from sliding around in the sandwich. Otherwise, use the same technique as

Fig. 7-7. *Templates printed on sticky-backed paper can be used as a guideline for cutting out and drilling holes in a part. The picture shows a pattern for a small subpanel, including the locations and sizes of the instrument holes.*

with the bandsaw. It's a lot of bother and mess, though. See if you can borrow a power sheers or get a little time on a bandsaw.

Kerf control

If you've never done precision cuts before, there's one thing you should be aware of: controlling the kerf. The *kerf* is the slot cut in the material by the saw blade. Since the blades are usually about ⅛ inch wide, using the saw cuts a ⅛-inch slot in the metal or wood.

Often, you'll be cutting out a pattern on the material. Basic nature is to drive the saw blade exactly down the center of the cut line for the part. But if you do that, the part ends up narrower than it should (Fig. 7-8). Half the kerf (¹⁄₁₆ inch) is taken off the edge of the part. If the part was supposed to be 3 inches wide, it's now slightly narrower.

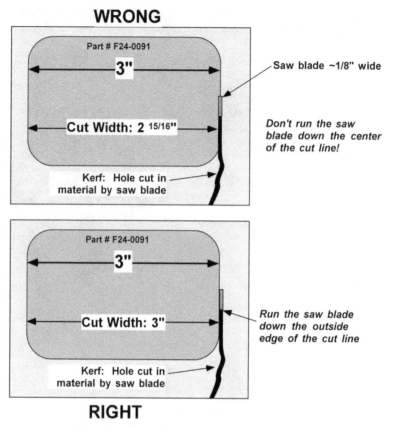

WRONG

Part # F24-0091

3"

Saw blade ~1/8" wide

Cut Width: 2 15/16"

Don't run the saw blade down the center of the cut line!

Kerf: Hole cut in material by saw blade

Part # F24-0091

3"

Cut Width: 3"

Run the saw blade down the outside edge of the cut line

Kerf: Hole cut in material by saw blade

RIGHT

Fig. 7-8. *Builders must execute "kerf control" to ensure that parts are the proper size.*

So keep the sawblade on the *outside* of the cut line. You're going to file or sand off a bit of material to make it smooth, so it doesn't matter if the part ends up a little wider.

Smoothing the blank

With a little experience, it takes very little time to cut the *blank* out from the sheet stock. It isn't a part yet. Feel the edges and realize that no matter what kind of tool you used, some burrs, ridges, and rough spots remain. It's vital to get the edges smooth with generous curves on inside corners. Premature failure of the part might result otherwise.

To illustrate the problem, take a cardboard rectangle about the size of a playing card. Cut a V-shaped notch in the side, and bend the card from top to bottom. The part will fail at the notch, right? Obviously, because there's less material there.

But take another piece and trim one edge, leaving a sharp-edged triangle in the middle, like a mountain on a flat plain. This piece fails at the sharp angles where the plains meet the slopes. Why?

Notches concentrate the stresses induced in the part. Inside corners, or the rough edges the saw leaves, cause unequal stress distribution. You'll hear the term "stress risers" to refer to these rough points.

By smoothing the edges and making inside corners with wide radii, you essentially confuse incipient cracks. They don't have an obvious place to start. Try the cardboard mountain again, but make the transition from plain to slope gradual. Quite a difference, huh?

To translate this example to an aircraft component, all inside curves should be gradual, and all marks left by the saw must be smoothed. Outside corners don't concentrate stress, but we'll knock the point off just for safety's sake. Besides, because those sharp corners don't contribute to the strength of the piece, they're just wasted weight (Fig. 7-9).

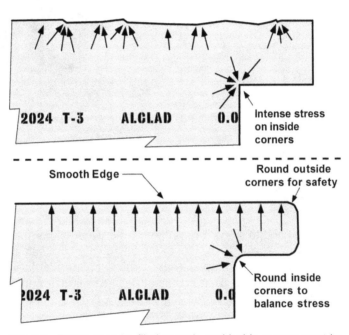

Fig. 7-9. *Edges must be filed smooth, and inside corners must be well rounded, or premature failure might occur.*

The easiest way to clean up the blank is to use the benchtop belt sander. If you don't have one, you'll have to use files.

Prepare to file the blank by clamping it in a vise between two pieces of wood. The wood keeps the jaws from scarring the metal. One edge of the blank should stick just above the top of the vise and just beyond the edge of the wood. This allows you to work on one edge and a corner.

Metal files cut in only one direction—from the handle toward the top. Don't saw back and forth with a file. Pulling the file back on the metal only dulls the file and makes machinists cringe.

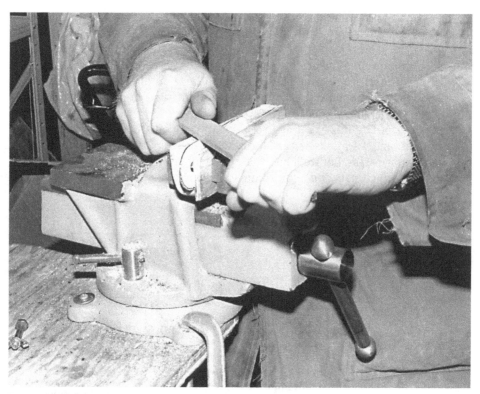

Fig. 7-10. *Smooth out metal edges by placing the file as shown and pulling it toward you. The file is at a 45-degree angle to the edge, and the handle end is farthest from the body.*

The best method to smooth the edges of the blank with a file is to draw-file, as shown in Fig. 7-10. Grasp the handle of the file in your left hand and the top in your right hand. Stand at the end of the vise, and place the file flat on the end of the blank's edge farthest away. Point the top of the file slightly toward you. Apply a little pressure and pull the file along the edge. When you reach the end, lift the file, set it back, and repeat. When the saw marks are almost faded, switch to a finer file and continue.

Smooth the curves with short strokes of the file. A good file can shape curves and corners with ease. Occasionally, tilt the file at 45 degrees to apply a slight bevel to the edge. Reposition the blank in the vise as necessary.

Use the fine file to remove the remaining tool marks. Remove the blank from the vise and polish the edges with small pieces of emery cloth. When finished, the edges have a shiny gleam with no tool marks and should be silky smooth to the touch.

Don't use steel wool to smooth the edges. The wool gets caught and leaves little bits of itself in crevices, which accelerates aluminum corrosion.

Those inside curves might require a half-round or rattail file. A bench grinder makes quick work of straight edges, or use a Moto-Tool with a grinder wheel for complex curves. Be careful not to eat away too much metal with power tools, though. In any case, finish up with the emery cloth.

Bending

Simple, straight-line bends are discussed here; Chapter 9 includes information on the more complex bends involved in making metal aircraft.

Before undertaking a bending job, take a look at the metal extrusions available. If you're just going to make a 90-degree bend, why not see if Wicks or Aircraft Spruce carries the correct size of angle aluminum? U-shaped fittings often can be made out of U-channel or square tubing with one side cut away. I was able to avoid making a thick aluminum U when I found the correct-size T section at a local surplus yard.

If a stock fitting isn't available, you'll have to make the bend(s) yourself. Follow three basic rules: Make the bends across the grain, observe the minimum bend radius, and allow for springback.

We all understand the concept of grain as it applies to wood. The grain in metal refers to the direction it was drawn through the rollers while being formed. The grain is the long axis of the metal before it was cut and rolled. The printing (alloy, hardness, and temper) is applied along this axis.

When the metal is bent, the bend line should travel vertically through the printing. As Fig. 7-11 shows, the labeling's relationship to the crease should be like the fold in this book, not like a matchbook cover. The piece will be 20 percent weaker if it is bent in the wrong orientation.

Every metal has a particular amount that it can be bent before it loses strength. Going back to our analogy, take a piece of cardboard (not corrugated) and carefully roll it into a wide cylinder. Release it, and you'll find that it's taken a set; it now includes a bit of a bend. But the cardboard is still whole, just as stiff as it was before.

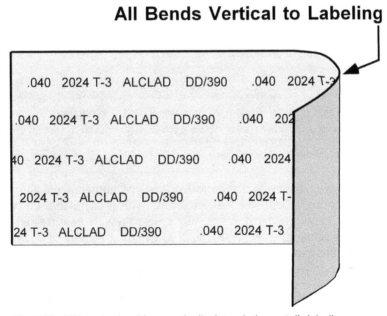

All Bends Vertical to Labeling

Fig. 7-11. *All bends should go vertically through the metal's labeling.*

Now grab both ends and bend the cardboard until it folds. It's not stiff anymore. All the strength has gone out of it.

This works the same way with metals. Each alloy, each temper, and each thickness has a minimum bend radius. Bent beyond that point, the metal might crack. The approved limits should be approached with caution, and it's better if more generous radii can be used.

Normalized 4130 steel must be bent with a minimum radius equal to three times the thickness of the material. That's pretty fair—a 0.080-inch-thick piece (and that's thick, for steel) must use a 0.25-inch radius. Steel that is 0.090 inch thick or less can be bent through a 180-degree angle as long as this minimum bend radius is followed. Thicker than 0.187 (³⁄₁₆) inch, the steel can be bent no farther than 90 degrees, and between 0.090 and 0.187 inch, 135 degrees is the limit.

Aluminum is slightly more complex. The minimum radius depends on many factors. Table 7-2 gives the values for commonly used alloys.

Table 7-2. Aluminum Bend Radius

Metal Thickness	2024 T-3		6061 T-6	
	Normal	Minimum*	Normal	Minimum*
0.016	0.050	0.025	0.016	0
0.020	0.064 (¹⁄₁₆**)	0.032	0.022	0.002
0.025	0.094 (³⁄₃₂)	0.044	0.031	0.006
0.032	0.128 (⅛)	0.064	0.48	0.016
0.040	0.168 (³⁄₁₆)	0.088	0.064 (¹⁄₁₆)	0.024
0.050	0.227 (¼)	0.128	0.089 (³⁄₃₂)	0.039
0.063	0.315 (⁵⁄₁₆)	0.189	0.126 (⅛)	0.063
0.071	0.362 (⅜)	0.220	0.150 (⁵⁄₃₂)	0.075
0.080	0.420 (⁷⁄₁₆)	0.257	0.178 (³⁄₁₆)	0.090
0.090	0.486 (½)	0.306	0.216 (¼)	0.108
0.125	0.750 (¾)	0.500	0.375 (⅜)	0.188

* *Absolute* minimum bend radius. Difficult to achieve without cracking or tearing the metal. Use the "Normal" value unless tighter radius strongly justified.

** Fractions shown are approximate.

One item of interest is that annealed steel and nontempered aluminum (2024-0, for instance) can be bent with very small radii. This is useful in some cases. However, such pieces cannot be used structurally unless they are heat-treated after bending. This makes sense in some cases, such as building dozens of complex aluminum ribs. Make them quite easily out of 2024-0, and then send them to be treated to T3. The treating process does include some warping, so you'll have to straighten them out a bit before installation.

The last factor to be concerned with is *springback*. Not to be confused with the South African deer, "springback" refers to the simple fact that metals don't stay where you bend them. Take a piece of aluminum and bend it exactly 90 degrees. When released, it'll flex back to some lesser angle. You actually have to bend the metal past your target angle.

How you set up for the bend depends on the material, its thickness, and the required radius. Thin, small pieces generally can be bent by hand when accuracy isn't an issue. The CIRCA Nieuport's aluminum tube fuselage uses pop rivets and 0.025-inch aluminum gussets to hold the longerons. One end of the gusset is riveted in place, then it is wrapped around the longeron to hold the transverse truss members. See Fig. 7-12.

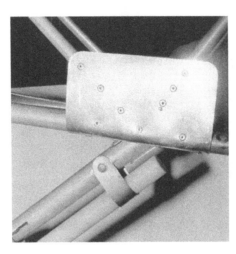

Fig. 7-12.
CIRCA Nieuports' fuselage truss is held in place by thin metal gussets wrapped around the longerons and pop-riveted in place.

But most applications require more accuracy and better leverage. One end of the sheet or plate must be held firmly while the other end is bent over past the proper position to allow for springback.

A hand tool called a *seamer* (Fig. 7-13) makes tight curves in sheet. The tool is like a pair of pliers with very wide jaws. Clamp it across the desired bend axis, and bend the metal.

The shop vise can be used for many pieces. Start by making a bending block out of scrap wood. The main feature of this block is a curved edge equal to or greater than the minimum bend radius you need. Whittle the wood with a jackknife, rasp, plane, or any other convenient tool. Sand the radius smooth when the approximate radius is reached.

If you're making a 90-degree bend, the radiused area will have to continue past the right angle. In other words, the top of the block will have to be either semicircularly in cross section or beveled. This is for springback allowance.

Clamp the metal in the vise between two pieces of wood, as you did for draw filling the edges, only make the bending block one of the pieces, with the curved radius at the position of the bend.

There are formulas to tell you exactly how much the metal is stretched and compressed during the bend and where exactly to place the bend line. Hopefully, the

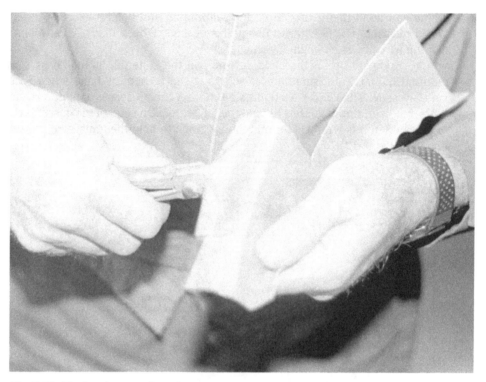

Fig. 7-13. *The hand seamer in action.*

kitplane's plans will tell you. Otherwise, don't cut your blank too close to the final size. Leave significant overlap, and trim it to fit when it's bent.

Now it's time to bend the free end of the metal over the bending block. A seamer works well here; otherwise, set a scrap of wood across the end of the metal and push on it with the palm of your hands. Don't just grab the end and push it toward the vise. This results in a wide curve instead of the desired radius. Think of it as pushing the metal sideways, not down. If you are bending thin metal, you can put the pushing block right on top of the vise. Otherwise, you'll have to get closer to the end.

Sometimes, the metal is too thick, or the bend is too awkward. It's acceptable to bash it with a plastic or rubber hammer to force it into place, but bend the free end evenly. Don't try to force the left corner down and leave the right corner standing. Force one corner slightly down, slide a bit over, bend that section down, and then slide farther over. Don't make any dents or crimps.

When you get the correct angle, remove the piece from the vise and trim the sides as necessary.

The preceding assumes that you don't have a bending brake. The brake automates the process; about all you have to do is clamp the metal in place and lower a lever. Brake prices vary from about $30 for light-duty ones to over $500. When I need to bend up some 0.025-inch aluminum angle, I built a light-duty one from scrap wood and hardware-store hinges.

Drilling

Now you have a part, not just a blank. But to get useful work out of it, you'll have to drill some holes so that it can be bolted or riveted in place.

Obviously, if a hole is drilled too close to the edge, the little bit of metal between the hole and the edge might tear away under stress. You must maintain a minimum *edge margin*, or distance, from the center of a hole to the nearest edge to maintain full strength.

The standard edge margin is twice the diameter of the hole. For example, the centerpoint of a ³⁄₁₆-inch (AN3) hole shouldn't be closer than ³⁄₈ inch (2 × ³⁄₁₆-inch) to any edge. This can be measured easily when only a single edge is concerned but is complicated when you must drill holes near several edges. By happy coincidence, the wide-diameter series of aircraft washers (AN970) has an outside diameter just slightly larger than the correct edge margin for the appropriate center hole. Align the washer so that its edges just come to the part's edges, and mark the center of the hole for drilling (Fig. 7-14).

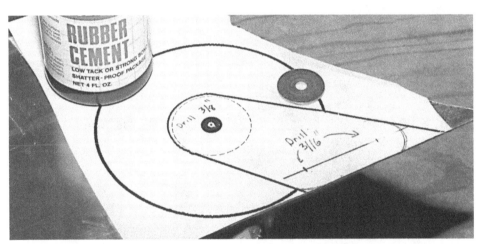

Fig. 7-14. *The AN970 series large-diameter flat washer is of slightly larger diameter than the required edge margin for its center hole. It can be used to quickly check for proper edge margin.*

In a similar vein, the minimum separation between hole centers is three times the diameter. Remember that the edge margin ensures full strength and applies to all holes: bolt holes, rivet holes, lightening holes. However, if the application isn't under high stress, you can shave the edge margin. A good example is lightening holes in plywood or metal ribs. They violate the edge-margin constraints, but there's very little force trying to crush them flat. Instrument holes also need not follow the edge-margin rules unless the panel is a structural member. The edge-margin parameters and a handy table are given in Fig. 7-15.

When you drill, you want the hole exactly where you place the bit. To keep the bit from skittering, use a punch to dimple the metal at the drill point. An automatic centerpunch is best—it can be worked with one hand, as shown in Fig. 7-16.

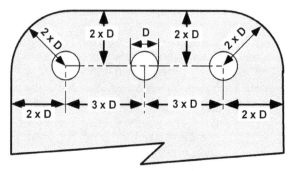

Diameter (D)	2 x D	3 x D
$3/32$	$3/16$	$9/32$
$1/8$	$1/4$	$3/8$
$3/16$ (AN3)	$3/8$	$9/16$
$1/4$ (AN4)	$1/2$	$3/4$
$5/16$ (AN5)	$5/8$	$15/16$
$3/8$ (AN6)	$3/4$	$1 \; 1/8$
$7/16$ (AN7)	$7/8$	$1 \; 5/16$
$1/2$ (AN8)	1	$1 \; 1/2$

Fig. 7-15. *Edge-margin basics.*

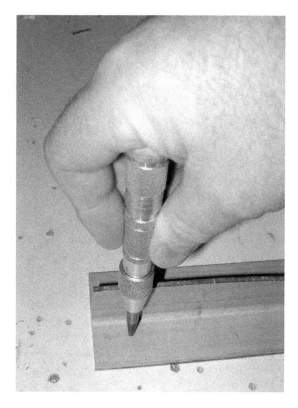

Fig. 7-16. *Use of an automatic center punch.*

But wait one minute. Where do you want that hole? Sure, the plans might show a precise location. But how precisely can you drill? If you drill two $3/16$-inch holes 2 inches apart on plate A, and plate B has the holes 2 $1/32$ inches apart, they won't bolt together. Factories with numerically controlled machines can drill holes exactly, as can some craftsmen. I certainly can't.

The only way to ensure that the parts will match is to drill them simultaneously. Drill all the holes on plate A with a $1/8$-inch bit. On plate B, drill only one hole. Cleco through this hole to the corresponding hole on plate A. As explained in Chap. 6, clecos are little spring-loaded temporary rivets that hold the parts in alignment during drilling. You could use small bolts, but clecos are much easier to install and remove.

Position both parts, and then clamp them in position. Hopefully, you've been able to move the pieces to the drill press; otherwise, use a hand drill. If you look through one of the holes in plate A, you'll see an undrilled section of plate B. Set the ⅛-inch bit in place, and spin the drill for a moment. This will mark the correct spot on plate B. Separate the pieces, drill ⅛ inch all the way through on plate B, and then clamp the pieces together using two clecos. You can drill the rest of the holes now because the clecos will maintain alignment. Insert a cleco as each hole is completed.

Sometimes you will want to drill a hole to match up with a hole underneath. In this case, an inexpensive hole matcher is a lifesaver. The hole matcher is two strips of metal attached at one end; at the other end, one strip has a post the size of the hole to be drilled. The other strip has a bushing with a hole in it that's the same diameter as the post. As shown in Fig. 7-17, slip the post into the existing hole, and slide the sheet to be drilled between the strips. Drill the sheet through the bushing, and the resulting hole will be lined up with the existing one. Insert a cleco, and go on to the next hole.

When all the holes are drilled, separate the parts. Drill holes to one size less than the final size (e.g., drill ¹¹⁄₆₄ inch for a ³⁄₁₆-inch hole). Then drill one hole in each part to the final size (a matching hole, of course), cleco them together, and final drill and cleco all the remaining holes.

Drilling directly to the final size is sloppy—the cut bits of metal rattle around and enlarge the hole. By first drilling a slightly undersized hole, only a slight amount of metal is actually cut during the final drilling. This ensures that the hole is exactly the desired size.

What size *did* you want that hole? It's not as simple as you might think. After all, there has to be some small amount of clearance to allow the bolt or rivet to pass.

Bolts are slightly undersized. For instance, an AN3 bolt, nominally ³⁄₁₆ inch (0.1875 inch), is actually 0.185 inch. But rivets are exactly the indicated size, so the

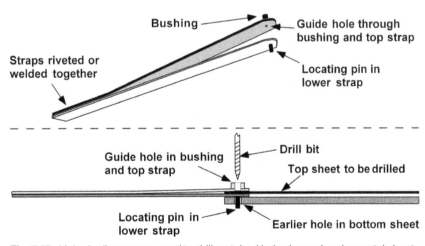

Fig. 7-17. *Hole duplicators are used to drill matched holes in overlapping metal sheets.*

hole must be a bit bigger. For ³⁄₃₂-inch rivets, use a #41 bit; ⅛ inch takes a #30 bit, ⁵⁄₃₂ inch takes a #21 bit, and a #12 bit is used for ³⁄₁₆-inch rivets.

The drilling process likely has left a ridge-shaped burr around each hole. You must eliminate these burrs for the same reason that you smoothed down the edges of the part—to eliminate stress concentrations. The best way is to take a deburring bit or countersink and twirl it by hand on each side of the hole. This cuts the ridge off and bevels the edge.

A major cause of burrs is dull drill bits. To sharpen a bit, hold it in your right hand with just the tip showing. With an underhand motion, loft it toward the nearest trash can. Then go buy another. Bits are cheap enough that you should maintain a stock of the smaller ones.

Some folks deburr a hole using a large-diameter drill bit. But the bit tends to chatter and leave small nicks even when it is turned by hand. While better than not deburring at all, this method isn't recommended.

Once the holes are deburred, the part is finished, although you might want to go over the edges with emery cloth to clean up last-minute dings. A small piece of metal has been magically transformed into an aircraft-quality part.

Depending on the final use of the part, it might require corrosion-proofing and painting. Guides for this process are included in Chap. 9.

AIRCRAFT CABLE AND ACCESSORIES

No matter what type of kitplane you're building, you'll have to work with aircraft cable at some point. Even composite homebuilts with cantilever wings and pushrod-actuated controls usually have a cable-controlled rudder. Many designs use cables for aileron and elevator as well, and biplanes and other ragwings still use them for external and internal bracing.

Aviation cable is rarely used for a trivial purpose. It moves control surfaces, keeps the wings attached to the airplane, and maintains the rigidity of structure. Cable failure still kills homebuilders. But today it's rarely the cable's fault; most often, it's just plain poor workmanship. Let's look at building aircraft-quality cable assemblies.

Cable construction

The basic unit of aircraft cable is a single wire formed of *galvanized* (zinc-coated) *steel* or *stainless steel*. The cable could consist of this single wire; if the wire were ⅛ inch thick, it'd be as strong as a ⅛-inch steel rod. And that's exactly what it *would* be, a solid, strong, and stiff steel rod. Unfortunately for those wanting to move control surfaces, a ⅛-inch rod won't go around a pulley. It has no flexibility. We could use a smaller-diameter wire, but we'd lose strength. But use a bundle of these tiny wires, and the resulting cable is almost as strong and far more pliant. You get both strength and flexibility.

Aircraft cable consists of a number of wires (usually 7 or 19) twisted around each other to form a strand. If more flexibility is needed for a given outside diameter, smaller wires are used in multiple strands, which, in turn, are also twisted together.

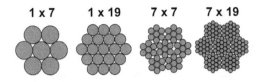

1 x 7 **1 x 19** **7 x 7** **7 x 19**

Fig. 7-18.
Cable construction is defined by the number of strands and the number of wires per strand.

Aircraft cable is specified by diameter, construction, and material. While all three factors affect the cable's strength, diameter has the greatest effect. One-sixteenth-inch cable can withstand about 500 pounds, and its strength approximately doubles for every 3/16-inch increase in diameter.

Construction affects strength to a lesser degree. A cable's construction is described by the number of strands and the number of wires per strand (Fig. 7-18). Hence, 1 × 19 cable has one strand consisting of 19 wires, whereas a 7 × 19 cable has seven strands, each of which has 19 wires. If both cables have the same outside diameter, the individual wires of the 1 × 19 cable are thicker, and some sizes can be 30 percent stronger than the same diameter in 7 × 19 cable. But the thick wires reduce flexibility.

Single-strand cable is best for bracing and other applications that don't pass through fairleads or pulleys. The so-called flexible cables (7 × 7 or 7 × 19) are best for control systems. The 7 × 19 construction is sometimes called *extraflexible*. For a given cable diameter, the wires of 7 × 19 cable are smaller and hence more easily damaged. Go with 7 × 19 construction if your design uses small pulleys. Otherwise, 7 × 7 cable gives reasonable flexibility and wears better.

The relationship between material and strength is interesting. For single-strand cable, galvanized or stainless steel cables are equal. However, galvanized is stronger in multistrand construction.

Here's a comparison of the wire strengths:

Diameter	Material	One Strand	Seven Strands
$\frac{1}{16}$ inch	Galvanized	500 lb	480 lb
	Stainless	500 lb	480 lb
$\frac{3}{32}$ inch	Galvanized	1,200 lb	920 lb
	Stainless	1,200 lb	920 lb
$\frac{1}{8}$ inch	Galvanized	2,100 lb	2,000 lb
	Stainless	2,100 lb	1,760 lb
$\frac{3}{16}$ inch	Galvanized	4,700 lb	4,200 lb
	Stainless	4,700 lb	3,700 lb

Cable selection

If you're building a kitplane, you shouldn't have to select the cable to use—it should be spelled out in the plans. But sometimes the plans just say, "Run a cable between the fitting and the rudder horn."

For certified aircraft, the FAA requires control cables to be at least ⅛ inch in diameter. This doesn't apply to kitplanes, of course. Some kitplane manufacturers voluntarily comply. But current practice tends toward ³⁄₃₂-inch cable for a number of reasons. Attaching a fitting to the smaller cable requires only one compression with a Nicopress tool, for example. In addition, the lighter cable is more flexible. It has sufficient strength for nonbearing applications, but the smaller size is less tolerant to damage.

Should you use galvanized or stainless cables? Galvanized cables cost half as much as stainless. But stainless is more resistant to corrosion, and its shiny appearance is more attractive. When the aircraft will be stored in salty or wet climes or the cable will be placed so that inspection is difficult, opt for stainless. Whenever galvanized cable is used, the annual inspection should include lubing with graphite grease or oil.

Cutting aircraft cable

Cutting galvanized or stainless cable is easy. Don't use a cutting torch—the heat will anneal and weaken the cable. It must be cut mechanically. Cable cutters are available for $15 and up, but a $3 cold chisel and a hammer work quite well. Wrap a piece of tape around the cable. Position the cable on a hard surface. An anvil works best or the flat top of a vise. Set the chisel on the taped cable, and rap it sharply with the hammer (Fig. 7-19). This might seem crude, but tools for cutting thick steel cable work on the same principle.

Cable terminations

Whichever type and size of cable are selected, we have to be able to attach the cable to the structure. A terminal must be added; either an appropriate fitting must be *swaged* directly onto the cable, or an eye must be formed to allow connection via other hardware. This termination must develop the same mechanical strength as the cable itself.

Swaged terminals are strong, fast, and attractive (Fig. 7-20). The terminals include a tubular section with the inside diameter equal to the cable diameter. The cable is inserted, and then the tubular section is compressed with a swaging tool. The steel of the fitting is forced into the crevices of the cable, ensuring a strong, permanent connection. It also produces an uncluttered appearance with no exposed cable ends.

Complete kits with extensive predrilled holes and jigging are apt to supply completed cable assemblies with swaged fittings. Other kits may include at least one end of the cable swaged, with the other end free to set the proper length on installation. The remaining kits require the builder to construct the entire assembly.

Unfortunately, swaging tools are expensive. They *start* at around $3,000. The usual practice is to have assemblies made for around $5 a fitting plus parts.

But this can add up in a hurry. A typical biplane might have 10 bracing wires per side, each with two fittings. That's $200 for both sides, plus cabane braces and controls. Not to mention redoing any mismeasured cables.

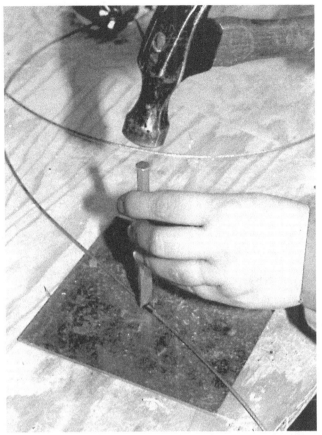

Fig. 7-19. *Aircraft cable can be cut with a cold chisel and a hammer. A solid surface is also required. Note the steel plate atop the workbench.*

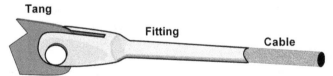

Fig. 7-20. *Swaged fittings are strong, attractive, and expensive.*

The alternative is to wrap the cable on itself and make a small loop or eye. Pass a shackle through the loop, and the cable can be bolted easily in place. This eye can be formed in three ways: *splicing, soldering,* and *Nicopressing.*

Unless you're a masochist or antiquer, don't even *think* about splicing. The cable's end is looped around a bushing or thimble and then spliced into itself. It's cheap—a marlinspike is the only tool required—but you'll need cast-iron fingertips and infinite patience. If you feel that you must try splicing, consult FAA Advisory Circular AC 43.13 and read your C. S. Forester. Don't say I didn't warn you.

Similarly, soldering is a lot of bother. After the cable is looped around a thimble, steel wire is wrapped around the cable and the free end. This wrapped section is then dipped in a molten-solder bath. It's too much effort, and the terminal ends up only 90 percent as strong as the cable.

Which leads us to the Nicopress system, where the loop is held by a swaged copper sleeve. It's ugly but perfectly serviceable. This has become the standard homebuilder's terminal—it's fast, easy, and costs only about 40 cents per end. It was developed by the National Telephone Supply Company.

The cost difference between Nicopressing and traditional swaging is due to the fact that copper is softer than steel. The copper Nicopress sleeves can be compressed easily by low-cost hand tools. The standard tool costs about $150. This bolt-cutter-like device can compress a sleeve in seconds. The miniswage applies pressure by tightening two bolts. It takes a lot longer, but the price is right, less than $20.

You *must* use a Nicopress tool to compress the sleeves. Don't use a vise or locking pliers. They don't work. One homebuilder used locking pliers on the flying wires of his Fly Baby. He died. Surely your life is worth $20.

Making a Nicopress termination

Let's assume that we're installing a cable terminal using the Nicopress system. To begin, pass a 1-inch-long piece of rubber hose over the end of the cable. Its inside diameter must be large enough to pass two cables. Automotive vacuum system hose is adequate and cheap.

Next, slide a Nicopress sleeve over the end of the cable. Make sure to use the correct sleeve; one intended for larger cable won't make good contact. Plain copper sleeves are used on galvanized cable, but zinc-plated ones must be used on stainless steel.

Bend the cable back on itself, and insert the end through the other hole in the sleeve and then through the rubber hose. If you're forming the eye inside one end of a turnbuckle, pass the cable through the turnbuckle before inserting it back into the sleeve and hose.

A loop has been formed, but it won't work on its own. Tension will flatten it out and concentrate the stress at one point, resulting in quick failure. A steel thimble placed inside the loop will equalize the forces. As with the sleeves, different-sized cables take different thimbles. The exception is $\frac{3}{32}$- and $\frac{1}{8}$-inch cable, which take the same size.

Before inserting the thimble, clip off the four points at the ends with a pair of sidecutters. This gives a tighter assembly. If you are using a turnbuckle, the ends of the thimble might have to be spread a bit to get it over the turnbuckle's end. If an AN115 shackle is going to be used, insert it now. The shackle *can* be added once the terminal is complete but not very easily.

Now the *terminal* (Fig. 7-21) is ready for compression. Make sure that the cable is lying flat in the thimble. Slide the rubber hose toward the loop, shoving the copper sleeve as close as possible to the thimble. Clamp the whole assembly in position using a hardware-store cable clamp around the hose.

Position the Nicopress tool around the major axis of the sleeve. The intent is to squeeze down the largest dimension to make the sleeve more circular, not flatter.

Fig. 7-21.
A sleeve ready for swaging. A short piece of rubber hose slid around the cable protects it from the clamp that temporarily holds the assembly together. Note the cut-off ends of the thimble.

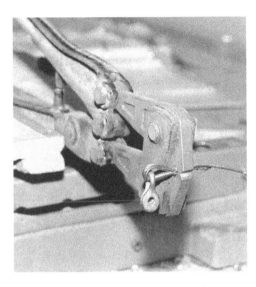

Fig. 7-22.
With the conventional Nicopress swage, it's often easier to clamp one handle. Note the shackle inserted through the thimble.

The standard tools have multiple grooves with a letter code to indicate cable size. The most common codes are

C ($\frac{1}{16}$ inch), G ($\frac{3}{32}$ inch), M ($\frac{1}{8}$ inch), and P ($\frac{5}{32}$ inch)

Incidentally, the same code is used when ordering sleeves. The miniswages are easier to figure out; they're just marked with cable sizes.

Center the tool between the sleeve ends, and begin compression. Use a slow, steady motion with the conventional tool (Fig. 7-22). If you are using a miniswage, tighten the bolts evenly and equally (Fig. 7-23).

When complete, release the pressure. Cables $\frac{3}{32}$-inch or smaller are ready at this point, but larger cables need at least two more compressions. Reposition the tool to the end nearest the thimble, and repeat the process. Finally, compress the sleeve at the opposite end.

Undo the cable clamp, and cut the hose free of the cable. Trim away the excess cable on the free end, leaving at least $\frac{1}{8}$ inch sticking out of the sleeve.

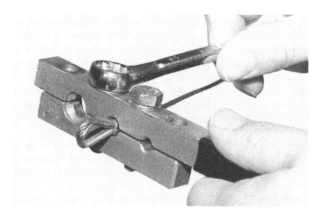

Fig. 7-23.
Tighten both bolts of the miniswage equally.

To develop full strength, the sleeve must be compressed to at least a certain thickness. Check the dimensions of the long axis with either a Nicopress gauge, caliper, or a micrometer. The maximum thickness after swaging should be

Max. Thickness	Cable Size
$\frac{1}{16}$ inch	0.1908 inch
$\frac{3}{32}$ inch	0.2674 inch
$\frac{1}{8}$ inch	0.3532 inch
$\frac{5}{32}$ inch	0.3965 inch

This completes the Nicopress process. But the cable actually has to get to something and connect to it. Let's look at cable-related hardware.

Tangs

A *tang* is a protrusion of metal intended for attachment of cable. In its simplest form, a tang is a short strap of aluminum with two holes drilled in it. One hole is used to bolt the tang to the structure requiring support, and the cable eye is formed through the other.

Tangs typically are more complex, if only to incorporate a bend between the two holes. A piece of angle or T-section can be used, although the pull imparted by the wire won't be lined up with the bolt holding the angle to the structure. This applies torque to the fitting and requires two bolts to hold it in place.

Tangs are easy to make, typically out of 0.080-inch 2024T3 aluminum or 0.050-inch 4130 steel. Because there is a lot riding on these fittings, make sure to follow all the metal-working rules, especially edge margin and bend radius.

In some cases, the AN42 through AN46 eye bolt can act as a ready-made tang. The thread sections are similar to regular AN bolts because they use the same dash numbers to indicate length. The diameter is equal to the last digit of the AN number plus one [an AN42-10 eyebolt is the same as 2 + 1 = 3, or an AN3 ($\frac{3}{16}$-inch) bolt].

Cable shackles

Tangs are great, but forming the cable eye through the tang's hole is a bother, especially if the cable ever has to be replaced. The AN115 cable shackle fixes this problem. It passes through the eye (although far easier prior to forming the eye) and allows the cable to be attached to the structure with a bolt or clevis pin (Fig. 7-24). When using a bolt, don't use a self-locking nut. Use a castle nut and cotter pin instead.

The dash number of the cable shackle indicates its rated strength in hundreds of pounds. An AN115-21 shackle, for instance, is rated at 2,100 pounds. This is the smallest shackle that can be used with $3/32$- or $1/8$-inch cable. The next-smallest size is AN115-8, and its 800-pound rating is sufficient only for $1/16$ inch.

One problem with shackles is the lack of throat depth. The side of the shackle might touch the edge of a control horn before the surface reaches full throw. If this is the case, replace the shackle with two flat straps of 0.050-inch steel or aluminum with holes drilled in either end, as shown in Fig. 7-25. Use aluminum for low-load

Fig. 7-24. *An AN115 cable shackle used to connect a rudder cable. Note the cotter pin in the bottom holding the clevis pin in place.*

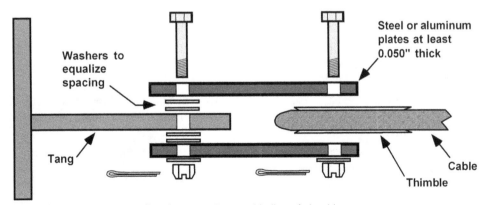

Fig. 7-25. *In some cases, flat plates can be used in lieu of shackles.*

application only, please. Sandwich the horn and cable eye between the plates, and bolt them through the hole and eye.

Because the horn is probably thinner than the eye, add washers on either side to keep the straps parallel. The plates shouldn't be allowed to clamp down on the thimble or cable. Use a longer-than-necessary bolt. Or slide a piece of tubing over the bolt to keep the plates apart.

Remember, when using bolts on moving structures such as control horns: Don't use self-locking nuts. Normal motion can loosen them. Use drilled bolts, castle nuts, and cotter pins. A clevis pin is a lighter and cheaper alternative.

Turnbuckles

Because cables apply tension, there must be a way to preload the cable to the proper amount. If the cable hangs slack, it can't do its job. The usual way is using turnbuckles.

They're not *entirely* necessary. It's possible to make the cables exactly the right size by building them in place. Run the cable between tangs, tighten it as much as possible, and then clamp it in place and compress the Nicopress sleeves. The cable probably would maintain the original tension, but is it the correct amount? What happens if the structure changes slightly? What happens if it's just a teensy bit off? How will you reattach the cable if it must be removed? These are all problems a turnbuckle will solve.

A turnbuckle is a brass barrel with removable fittings screwed into either end. The ends are threaded in opposite directions; when the barrel is rotated, both fittings move inward or outward simultaneously, hence increasing or decreasing the total length of the assembly. Up to 3 inches of adjustability are available.

Turnbuckle specifications cover complete assemblies and individual components (Fig. 7-26):

AN155 *barrel.* This is a tube-shaped piece of brass threaded at either end. The left-hand-threaded end is indicated by a ring scored around the brass.

AN161 *fork.* This has an end designed to be bolted or clevised directly to a tang, bellcrank, eyebolt, and the like.

AN165 *pin eye.* This is designed to fit between the tines of a fork and is secured with a bolt or clevis pin.

AN170 *cable eye.* Similar to the pin eye, the hole through this end unit is curved to match a cable and thimble.

The ends are steel and come in left-hand- and right-hand-threaded versions. A barrel is combined with two ends to form standard assemblies:

AN130. Barrel, cable eye, and fork; this is probably the most commonly used assembly and typically is used between the cable and a control bellcrank or tang.

AN135. Barrel, cable eye, and pin eye; this is a very common assembly used to attach cable to structure.

AN140. Barrel and two cable eyes.

AN150. Barrel and two forks.

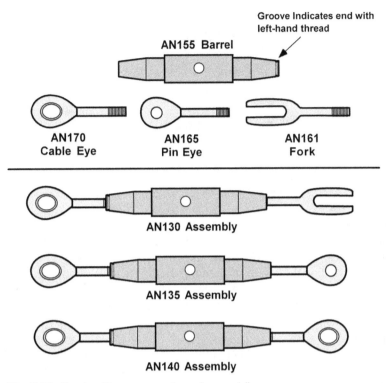

Fig. 7-26. *Turnbuckle components and assemblies.*

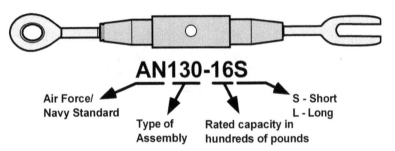

Fig. 7-27. *Turnbuckle specifications.*

Additional characteristics are given by a dash number. A typical specification is AN135-16S. The 16 indicates the strength in hundreds of pounds, and the *S* refers to the length category (Fig. 7-27).

The farther the ends are screwed into the barrel, the stronger is the turnbuckle. The strength ratings require that no more than three threads be exposed at either end of the barrel. If it is past this point, the cable must be replaced with a longer one (Fig. 7-28).

Turnbuckles come in two length categories: short and long, indicated by an *S* or *L* suffix (e.g., AN140-16L and AN135-22S). Short models are about 4 ½ inches in

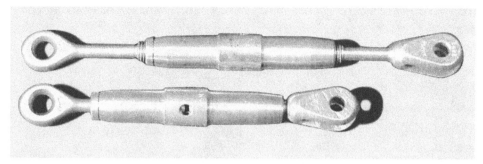

Fig. 7-28. *The minimum and maximum extension of a turnbuckle. No more than three threads can be showing at the end of the barrel.*

length, and the long ones are about 8 inches. The long version has a 1 ½-inch additional length adjustment or *take-up*. For example, the AN130-16S has a take-up of 1.125 inches, and the AN130-16L take-up is 2.875 inches.

The additional take-up of the long models is a *lot* of adjustment. But they can save you from making a whole new cable assembly if one ever comes out slightly too short. Cut the cable away from the short turnbuckle, and install a long one instead. Make sure that it has the same strength rating as the short one being replaced.

The barrel and the ends are both longer for the L models. The individual components use a similar dash-number scheme to indicate strength and length. So a typical barrel might be an AN155-32S. Turnbuckle end specifications also insert an *R* or *L* before the short or long identifier to indicate whether the end is left- or right-hand-threaded. Thus an AN161-16RS turnbuckle end is a right-hand-threaded short fork, good for 1,600 pounds.

Remember three main things when using turnbuckles. First, enough screw threads of the cable end must be contained within the barrel to allow the unit to meet fully rated strength. One cannot expect a single thread to withstand a 2,000-pound load. Therefore, make sure that no more than three threads are exposed at either end of the barrel.

Second, do not lubricate the threads. Turnbuckle rotation is necessary for installation, not while the aircraft is in operation. Lubrication would only make it easier for the turnbuckle to rotate on its own.

On this note, the third item is prevention of inadvertent slackening. Because tension tries to force turnbuckles to loosen, they must be safety wired to prevent rotation. The single-wrap method is fastest. Start with a new piece of safety wire about three times the length of the turnbuckle. Never reuse an old piece. It's tempting to just loosen the ends, adjust tension, and rewrap with the same wire. Don't because the changes reduce strength.

Push half the wire through the barrel's center hole. Bend each end in opposite directions (Fig. 7-29), and poke it through the cable eye or lay it the across the bottom of a fork. Wrap the ends of the wire around the shanks of the turnbuckle ends at least four times, and cut off the excess (Fig. 7-30).

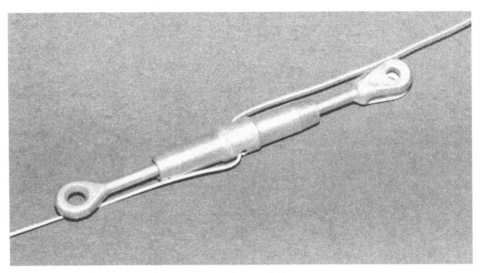

Fig. 7-29. *The initial step in safety wiring a turnbuckle.*

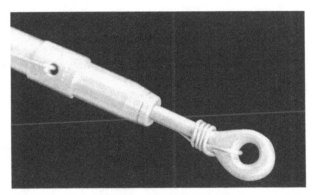

Fig. 7-30. *The safety wire must pass through the end of the fitting and wrap around the shank at least four times.*

The single-wrap method can be used on turnbuckles connected to ⅛-inch or smaller cables. Brass safety wire is acceptable on 1/16- and 3/32-inch cables, but stainless steel is required for ⅛-inch cable. Whichever type is used, it must be at least 0.040 inch in diameter. Cables 5/32 inch and larger can be single wrapped, but only with 0.057-inch stainless steel. This thicker wire is harder to bend.

Double wrapping is similar but uses two pieces of wire with the ends wrapped in opposite directions around the shanks. Don't wrap one wire directly atop another; instead, move down the shank a bit. Double wrapping lets you use 0.040-inch stainless steel on cables 5/32 inch or larger.

Double wrapping is actually the FAA's preferred method in all cases, but single wrapping is acceptable if the preceding rules are followed. To avoid safety wire entirely, buy the more modern MS212XX series turnbuckles. These use a quickly installed safety

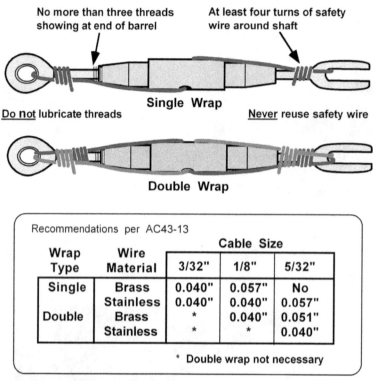

No more than three threads showing at end of barrel

At least four turns of safety wire around shaft

Single Wrap

Do not lubricate threads

Never reuse safety wire

Double Wrap

Recommendations per AC43-13		Cable Size		
Wrap Type	Wire Material	3/32"	1/8"	5/32"
Single	Brass	0.040"	0.057"	No
	Stainless	0.040"	0.040"	0.057"
Double	Brass	*	0.040"	0.051"
	Stainless	*	*	0.040"

* Double wrap not necessary

Fig. 7-31. *A summary of turnbuckle safety wiring.*

clip instead. Like all things fast and neat, however, they cost more than the ordinary item. A summary of turnbuckle use is given in Fig. 7-31.

Fairleads and pulleys

Bracing wires have it easy—they go in a straight line. But control cables must change directions. It would be nice if an aileron cable could go straight from the stick to the aileron without changing directions, but it isn't likely. And you can't just run the cable around a corner because the sawing motion would cut through the material of the corner, unless the friction frayed and snapped the cable first.

This is where *fairleads* and *pulleys* come in. They control and route aircraft cable.

Fairleads are simple objects with no moving parts. They prevent cable sag or change a cable run's direction slightly. They're sacrificial in nature; they wear away before the cable does. They still have to be strong and fairly hard; after all, if they wear through too fast or without warning, the cable could go slack or bind.

The traditional material is *phenolic,* made by a fiberglass-like process that substitutes other substances for glass cloth. It's commonly called *Micarta,* a name trademarked by Westinghouse. It cuts easily and doesn't splinter. Other materials used for fairleads include nylon and delrin (an advanced plastic). Anything softer than the steel cable could be used; some wooden aircraft plans just have the builder drill small holes through the structure.

Whatever material is installed, friction and wear are a part of its operation. Cable damage isn't eliminated, just slowed. Normally, fairleads shouldn't change the cable's direction more than 5 degrees; 15 degrees is the maximum, but you'll pay the price in excessive wear and stiff controls.

How do you measure the angle? Sometimes you can eyeball it with a protractor, but often the area isn't that accessible. Take a piece of thin, stiff wire, the stuff used for model airplane pushrods and the like. Bend it to 5 degrees with a pair of pliers. Run one leg of the resulting V shape alongside the cable, placing the bend point inside the fairlead. Tape the leg to the cable.

Now move the other end of the wire to match the cable on its side of the fairlead. If the V flattens to match the cable, the fairlead installation is acceptable. If it must bend sharper, the cable angle exceeds 5 degrees.

If it failed the 5-degree test, rebend the wire to 15 degrees and try again. If the cable bend is still greater, replan the fairlead location. If it's between 5 and 15 degrees, the installation is acceptable, but expect greater wear and control friction.

If such a sharp angle change is contemplated, make sure that the area is easily accessible for preflight inspections. Also, design the fairleads to allow removal of the complete cable. Don't just drill a $\frac{3}{16}$-inch hole in a piece of delrin and run an $\frac{1}{8}$-inch cable through it. The cable must be replaced someday, and such a fairlead would require cutting the cable.

Why does this make a difference? In the first place, it's easier to make up a new cable the same length as an existing one. Cutting the cable will make it a little harder to reconstruct the actual length. Second, the cable run will not be as easy to get to when the aircraft is complete. Rather than diving headfirst under the instrument panel to Nicopress a rudder-cable sleeve, it's simpler to build the assembly outside and install a completed cable. The fairlead design should include either a slot running to an outer edge or a wide enough hole to allow passage of a turnbuckle end.

If the angle is too much for a fairlead, a pulley (Fig. 7-31) will be necessary. Pulleys generally are also made of phenolic with a ball-bearing center. The usual specification is by diameter and *bore* (the bolt hole in the center). Flight control pulleys are typically AN210 or MS20220 models. Pulleys with a diameter of 2 inches or less shouldn't be used to change the cable's direction more than 15 degrees. This limitation is for primary flight controls; this angle can be exceeded for secondary applications.

The biggest problem in pulley use is intolerance to misalignment. The pulley's edges should parallel the cable within 2 degrees. Beyond that, excess cable or pulley wear might occur. This can be a problem if the pulley is located close to the part moved by the cable because the angle might change throughout the control range. Fairleads have an advantage in this regard because alignment is unimportant. With a fairlead, all that matters is total angle.

One danger with pulleys is the possibility of the cable slipping out. Always include something to prevent the cable from jumping the pulley. Traditionally, a short strip of aluminum or steel, bent in a U shape, is slipped over the end of the pulley and held in place by the pulley's pivot bolt. Or the pulley brackets can be long enough to allow a bolt or a cotter pin to be used, as seen in Fig. 7-32.

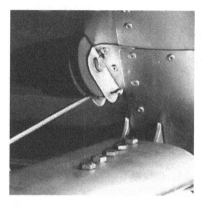

Fig. 7-32.
All pulleys must include a keeper to prevent the cable from jumping off. Note the large cotter pin in the pulley bracket.

Putting it all together

Let's run through a sample cable installation. One end of a 1 × 19 eighth-inch stainless steel cable is going to attach to a tang and the other end to an eyebolt. An AN115-21 shackle will be used at the eyebolt. Because this is a straight bracing application, no pulleys or fairleads are necessary.

One end of the turnbuckle must be attached to the cable, whereas the other will be bolted to the tang. This calls for an AN13 assembly, with a cable eye at one end and a fork at the other. The 2,100-pound strength of the cable requires an AN130-22. Because the cable is only being used for bracing, we won't need the extra take-up of the long-series turnbuckle. Hence we'll use an AN130-22S.

Take your reel of cable and install an eye (using an AN100-4 thimble) at the end. Slip the AN115-21 shackle over the thimble before forming the eye. Because this is ⅛-inch cable, each sleeve requires three compressions, the first in the middle, the second on the end nearest the eye, and the last on the other end. Because this is stainless steel cable, use a 28-3-M zinc-plated Nicopress sleeve.

Measure the distance to be spanned by the assembly. Cut off that length of cable beyond the newly installed eye. Because the turnbuckle requires approximately 4 inches of the distance, this leaves enough extra cable. Set the turnbuckle to a little longer than the midpoint of its range.

Slide a piece of shrink-wrap tubing over the end of the cable, followed by a short hunk of rubber hose and a plated Nicopress sleeve. Clip the points off another thimble, and force its ends apart slightly to get it through the turnbuckle's cable eye. Run the cable through the eye and over the thimble, and secure it using a hardware-store cable clamp, as discussed earlier in this chapter.

Using a clevis pin or a short AN3 bolt, temporarily install the shackle in the eyebolt and the fork end of the turnbuckle in the tang. Pull on the free end of the cable to tighten it as far as possible, and then snug down the cable clamp. We can't develop very much tension yet; that's why the turnbuckle is set a little longer than the midpoint. It gives us a little extra to pick up the slack.

If the cable is used in conjunction with another cable (an X bracing or elevator cables, for instance), don't compress the sleeve yet. Similarly install the other cable, too. Adjust the cable lengths so that they have equal adjustment ranges.

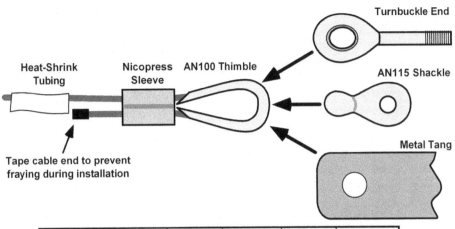

Cable Dia.	Material	Strength*	Thimble	Sleeve	Shackle
1/16"	Galv.	480	AN100-3	18-1-C	AN115-8
	Stainless	480	"	28-1-C	"
3/32"	Galv.	1000	AN100-4	18-2-G	AN115-21
	Stainless	920	"	28-2-G	"
1/8"	Galv.	2000	AN100-4	18-3-M	AN115-21
	Stainless	1760	"	28-3-M	"
5/32"	Galv.	2800	AN100-5	18-4-P	AN115-32
	Stainless	2400	"	28-4-P	"

* In pounds, for 7x19 Cable (7x7 for 1/16")

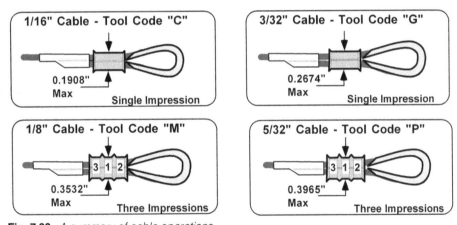

Fig. 7-33. *A summary of cable operations.*

When satisfied, make the three compressions on each Nicopress sleeve. The steps to this point are summarized in Fig. 7-33.

Once the ends of the cable are attached, turn the barrel to tighten the cable. How tight? If the plans call for a particular value, you'll need a tension gauge or tensiometer. Prices range from $25 to more than $1,000. The values are given based on the ambient temperature because the wires change length at a different rate than the airframe. For instance, the plans might specify a setting of 25 to 35 pounds at temperatures between 40 and 60°F and 20 to 30 pounds above 60°F.

If a recommended setting isn't available, tighten the wires until they give a satisfying twang. Watch that the structure doesn't bend if the cable is tightened too far.

Then safety-wire the turnbuckles, and the installation is completed. To summarize:

1. Use plain Nicopress sleeves or copper-plated sleeves on stainless steel.
2. Cut the points off the tips of the thimble before use.
3. Use one compression on ³⁄₃₂-inch cable and smaller; larger cables need at least three compressions.
4. Ensure that the free end of the cable sticks out of the end of the sleeve at least ⅛ inch after compression.
5. Don't lubricate the turnbuckle threads.
6. When in use, no more than three threads can show beyond the turnbuckle barrel.
7. Safety-wire all turnbuckles.
8. Install cable keepers on all pulleys.

A tool warning

Earlier in this chapter we looked at a small, inexpensive hand swage for Nicopress sleeves. I bought mine though a homebuilt supplier; while slower than the bolt-cutter type, it makes perfectly acceptable compressions. It's called a Swage-It.Unfortunately, there is a look-alike tool on the market that *does not* compress the Nicopress sleeves properly.

Normally, when installing ³⁄₃₂-inch sleeves, only one compression is necessary. The standard tool compresses the entire length of the sleeve. The substandard tool does not—it only presses a short length of the sleeve.

I didn't buy this tool from the local hardware store. It came from an aviation parts seller. If you have an opportunity to buy a low-cost swage, take a ³⁄₃₂-inch Nicopress sleeve with you. If the compression surface in the ³⁄₃₂-inch slot is much shorter than the sleeve, don't buy the tool.

BUILD TO LAST

When are you building your airplane for? Rather odd question, huh? Let me put it another way. When you imagine flying your airplane, what flight are you thinking of?

Are you thinking of the first flight?

Are you thinking of the fortieth hour, after which you can start taking friends and family for rides?

Are you thinking of your first trip to Oshkosh?

These are all nice goals—all nice things to look forward to. But these are all short-term goals. What about the fifth anniversary after completion? What about the *tenth*?

It takes a lot of work to get to the first flight of a homebuilt. But during the long building process, don't forget that you want the plane to last *beyond* the first flight. You'll feel pretty stupid if you have to rebuild it just a couple of years later.

Upcoming chapters discuss the protection required for specific construction materials. This section presents some general advice on steps to take during construction to make the plane last.

Opposites detract

It's funny to think about, really. Of the four major materials used in the construction of homebuilt airplanes—composites, steel, aluminum, and wood—the last three have adverse reactions when placed in direct contact with each other.

Take steel and aluminum. Stick them together, add a little moisture, and introduce some pollutants or salt from the ocean. The result? Dissimilar-metals corrosion. The moisture and contaminants form a weak electrolyte, and the differing electrical potentials of the materials cause an infinitesimal current flow. The current flow causes *galvanic corrosion*, which eats away at the aluminum.

Anywhere steel and aluminum are going to be in contact, one or both surfaces should be primed so that the metals do not come in direct contact. *Don't* just bolt a steel control horn to an aluminum aileron and spray primer over the entire assembly. Treat the horn and the aileron *separately*, and *then* bolt them together.

AN bolts are cadmium-plated. Cadmium is pretty much neutral to both steel and aluminum. However, make sure that you use a washer under the nut. Otherwise, the nut will grind off its cadmium plating when you wrench it tight and probably scrape through the primer on the aluminum as well. You'll end up with raw steel in contact with raw aluminum and a repair job in five years or so.

What about wood? Spruce is typically 12 to 15 percent water. So steel or aluminum in direct contact with wood corrodes. Before attaching them, prime the metal and varnish the wood.

Varnish the bolt holes as well. While it would take a long time for corrosion to threaten the strength of a bolt, the first stages will jam the bolt in the hole and make it tough to remove. This is one case where the higher cost of stainless-steel bolts makes some sense.

A lotta shaking going on

Unless you're building a sailplane, your airplane is going to vibrate. Vibration can do a lot of bad things to metal—metal fatigue is one of the most famous. Over a long time, flexing weakens metal. This is one of the advantages of wooden or composite airplanes.

Structurally, you shouldn't have to worry. A good designer will ensure that nothing vibrates enough to cause fatigue. Don't forget, though, that there's a lot of stuff *you* install that might suffer from this problem.

Run a metal fuel, oil, or brake line too far without supports, and vibration eventually may cause a fracture. When stiff parts such as tubes run a long distance, be sure to provide adequate bracing along the way.

Never use a metal pipe between two components with differing vibration levels. A classic example is the fuel line from the gascolator to the carburetor. The engine mounts allow the engine to shake a bit. Obviously, you need a flexible hose that can ride with the difference in motion between the ends.

Finally, don't forget that different parts react differently to the same vibration input. If they touch, the harder one is going to gradually cut into the softer one.

A friend bought a 20-year-old Thorp T-18 Tiger. This airplane has a fuel tank made from aluminum mounted behind the instrument panel. The airplane's throttle is mounted at the bottom center of the panel. The throttle cable is like most—a flexible steel sleeve around a stiff wire. The sleeve on my friend's plane had been led under the fuel tank. It only touched it slightly at the aft end.

Over twenty years, through, vibration had turned the sleeve into a saw. It had almost cut through the bottom of the fuel tank.

You could have the same trouble with electrical wires. They may rest gently on the edge of a bulkhead. The insulation will protect them for a long time, but inexorably, the bulkhead will cut its way through.

There are two ways to handle this problem. The easiest is to interpose a pad of some sort between the two pieces. This pad can be either soft or hard. A soft one, of course, will have to be replaced periodically, but it's the best choice when protecting two soft items or when one of them is a moving part such as a cable.

Probably your best bet, though, is a standoff. A standoff solidly holds one component away from the other, preventing direct contact. Standoffs often include their own padding. A good example is a cable clamp.

All right—let's take our gas tank and throttle-cable example and see how we might add protection. A soft pad? Well, you might slide a plastic or rubber hose over the cable as you install it. Still, it'll have to be inspected periodically and replaced as necessary.

A hard pad should be attached securely to the softer of the two items, the aluminum fuel tank in this instance. A piece of steel could be bonded or riveted to the tank. But if the bonding ever releases, you'll be back at square young. In the case of a fuel tank, you may not want to drill rivet holes.

You'd have to drill holes to install a standoff, too. But you may be able to install it on the instrument panel itself, and completely eliminate putting holes in your gas tank.

Maintainability

Your repairman certificate only *authorizes* you to work on the aircraft. It doesn't make it easier. That part is up to you—and should be something you keep in mind all through construction.

Each annual inspection will require a visual inspection of every moving part. Most of the time, lubricant must be applied. Your plane should include inspection or access panels to perform these critical functions. If it doesn't, consider adding some. These panels should be big enough to allow you to reach in and replace the accessed parts as well. Being able to see a problem doesn't mean that you can get your hands in to fix it.

For instance, most Fly Babies have the belly area completely covered with fabric. A while back, when I was doing some fabric work on the tail of my own Fly Baby, I cut out a 3-foot-long section of belly fabric and installed a large removable inspection panel (Fig. 7-34). This is a bit more complex and a bit heavier that normal, but it greatly eases working on the controls, brakes, and center-section structure.

Fig. 7-34. *The large removable access panel in the belly of the author's Fly Baby greatly simplified maintenance.*

Actually, a fabric-covered airplane gives the builder the best opportunity to add inspection panels. The fabric skin is nonstructural; you're limited mostly by aesthetics. Any holes to be added to metal, composite, or wood structures should be coordinated with the kitplane manufacturer.

Some parts, notably certain types of engine hoses, must be replaced on a regular basis. Keep this in mind during your firewall-forward layout.

While you're thinking of the engine compartment, consider the cowling. The initial test-flying phase will involve uncovering the engine a *lot*. Don't make this task any harder than it has to be. Make the top half of the cowling, at least, quickly removable like the one shown in Fig. 7-35.

Finally, just about any part can go bad eventually and must be unbolted and replaced. When building the plane, it's usually easy to reach over and get a nut hand-started and then tighten it with a wrench. This isn't necessarily true once the airplane is completed. It doesn't make as much difference with structure, but components

Fig. 7-35. *The cowling of this Straus Subaru-powered Merlin can be opened quickly by removing the wire through piano hinges located on either side.*

such as electrical regulators and brake cylinders will need replacement during the life of the aircraft.

Use anchor nuts in these instances. An anchor nut is, essentially, a regular self-locking nut with a pair of mounting lugs attached. They stay in place when the bolt is removed; they don't have to be held with a wrench when the bolt is turned. The lugs are secured by either $3/32$- or $1/8$-inch rivets, either solid ones or pop rivets. Homebuilt supply companies sell a wide variety. They cost a bit more than regular nuts, but the difference for the simplest varieties is only a dime each or so.

A final workmanship point

When are you building for? In truth, you are building for every single hour your airplane is flying. But when is the danger the worst?

Well, the first flight is an obvious danger point. If something is real bad, it'll probably cut loose then. However, there's a troubling trend in the homebuilt aircraft accident statistics. As Fig. 7-36 illustrates, almost 6 percent of homebuilt accidents occur on the first flight. Then the likelihood drops off—only to peak again in the 50- to 70-hour period.

What's happening here?

Part of the reason is the increased occurrence of accidents due to weather—once past the 40-hour point, folks start flying cross-country trips and start getting caught by low ceilings and storms.

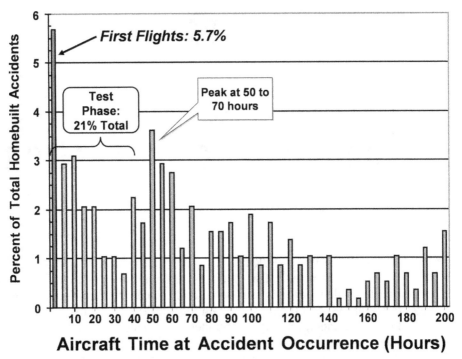

Fig. 7-36. *Homebuilts with between 50 and 70 total flight hours suffer a disproportionate number of accidents. The first annual condition inspection is vital to detect problems before they cause accidents.*

Most of the cause, though, is mechanical. Basically, it appears that many problems won't pop up until the plane has a few hours under its belt. Vibration and stress both play a part in the rise in the accident rate.

The lesson? Don't relax your vigilance just because the plane has passed its test period without a problem. Continue to inspect the airplane carefully, and be ready to detect the wear and tear of the initial flying period. At the first annual inspection, bring in another pair of experienced eyes to look at your plane.

This chapter has presented some of the basic rules for working with common aircraft hardware. For further information, see FAA Aircraft Advisory AC 43.13-1A or Stanley J. Dzik's *Aircraft Hardware Standards Manual and Engineering Reference*. Both books are available through the EAA.

It's time to look at specific construction materials. The next four chapters cover composite, metal-monocoque, and steel and aluminum tube structures and wood construction. The wood chapter also discusses fabric covering.

Specifically, each chapter covers the materials, the fasteners, the tools, basic procedures, prevention of degradation (i.e., through corrosion, rot, or ultraviolet radiation), and correction of typical errors.

If you've already decided on a construction method, at least skim the other chapters. As mentioned several times earlier, no kitplane uses only one method.

8

Composite construction

WHEN SOMEONE SAYS "AIRCRAFT KIT," our first mental image is usually of a speedy composite homebuilt. For good reason. Composite kits are the closest thing to the plastic scale-model aircraft we've built since childhood—add glue to the edges of the pieces, and then clamp them together (Fig. 8-1). Instead of a 2-ounce tube of styrene cement, the kitplane builder needs gallons of epoxy.

While composite kitplanes build fast, reduced construction time is just a by-product of the main goal—velocity. While the manufacturer might market the

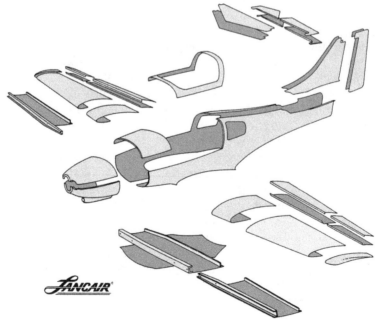

Fig. 8-1. *In schematic form, the Lancair kits are a little different from those for dime-store plastic models. The main fuselage consists of a left and a right half, with a bulkhead or two added for rigidity.* Lancair International, Inc.

plane as a fun flyer, it won't give up one knot in cruise: no open cockpits or bare, uncowled engines. The ads might mention short-field capability, but it is always followed with ". . . and can still cruise at over. . . ."

Fast airplanes need slippery, low-drag shapes. Composites are the easiest ways to turn theoretical curves into actual hardware. Designers turn to them whenever their goal is the fastest airplane for the given horsepower (Fig. 8-2). A Van's RV-7 gets good performance with a Lycoming O-360. But on the same engine, the fixed-gear Glasair IIFT cruises 15 percent faster.

Fig. 8-2. *Designers generally select composites when they want the best performance for a given engine. The Twister from Silence Aircraft cruises at 150 mph with an 80-hp Jabiru.*

But you pay for the extra performance. Literally. The Glasair kit costs twice as much. Once the engine and other nonincluded items are factored in, the Glasair costs about 30 percent more.

Some people believe that composite kitplanes are faster because the materials are lighter than aluminum. It doesn't really work out that way. Generally speaking, traditional composite aircraft weigh a bit more than the equivalent metal home-builts, while those made with more expensive graphite or Kevlar weigh less. For instance, a Glasair IIFT has an empty weight of 1,250 pounds, while an RV-6 weighs 1,100 pounds.

No matter the construction mode, the weight and performance depend on the skill of the designer and the parts manufacturing process.

In Chap. 2 we discussed the two different methods for making composite parts: molded and moldless. Let's take a look at how a molded part is made.

The part begins as a *plug* (Fig. 8-3), which is a full-sized replica made from wood, metal, or other convenient material. A *form* is molded around the plug. When the form is cut open and the plug removed, a negative impression of the part remains. This mold then can be used to duplicate the plug.

Fig. 8-3. *GlaStar fuselage plugs. These were used to form the molds from which kit parts are made.*

Surely you've seen detective movies where the murderer leaves a footprint in the mud. By pouring plaster of paris into the print, the inspector can exactly reproduce the shape of the miscreant's foot.

In kitplane terminology, the foot is the plug, and the print in the ground is the form or mold. Instead of plaster of paris, the inspector could use fiberglass cloth and resin to make a high-tech molding of the evildoer's bunions.

While fiberglass is lighter than plaster of paris, it's still too heavy for structural use on aircraft. Multiple layers of fiberglass are needed to get the strength sufficiently high. Weight then skyrockets.

This is where the composite materials come in. Instead of making a solid fiberglass part, a piece of lightweight stiff foam is inserted between layers of glass. This core makes the layup thicker without adding much weight. The complementary characteristics of the foam and the fiberglass/resin layup result in stronger and lighter parts.

As mentioned earlier, composite kit production starts with the mold. First, it's covered with a *release agent,* which is a material the resin won't stick to. Otherwise, the newly made part would end up permanently bonded to the mold.

From this point on, parts can be made in either the *wet layup* or *prepreg* method. Wet layup is the most common. A layer of fiberglass cloth is laid over the mold and soaked with resin. Another layer is added, and the process is repeated for a given number of laminations. The foam then is set in place, and further laminations added.

Then the layup is covered with a plastic bag, and vacuum is applied. The vacuum pulls excess resin through the cloth and away from the layup. When the resin's cured, the part is removed from the mold.

The prepreg method is used by several companies; Lancair (Fig. 8-4) was the pioneer in the homebuilt industry. Rather than making sequential fiberglass layups, Lancair uses sheets of fiberglass preimpregnated with the optimal amount of resin (referred to as "prepreg"). Premature curing is prevented by freezing the sheets until needed. The prepreg sheets are not as limp as dry fiberglass but are still formed easily. They are laid in the mold, the core material of Divinycell/Nomex honeycomb is added, and the specified number of additional prepreg sheets completes the composite. The assembly is vacuum bagged as in the wet-layup method but is cured at high temperature inside large ovens.

Which is better? Only a composite expert without an axe to grind could tell you, and I'm not one of them. Both sides claim advantages. Listen to the manufacturers, and make your own decisions. Either method seems to make acceptable aircraft.

Lancair International, Inc.

Fig. 8-4. *The Lancair series is the best-known example of a kitplane made from prepreg composites.*

Similarly, there are two different systems for joining parts: epoxy systems and vinylester systems. We'll discuss the differences later in this chapter. Structurally, either is fine if applied according to instructions.

This chapter covers the basics of working with composite kitplanes. Details of fiberglass work applicable to every other type of construction are also included:

- Preparing composite kitplane parts
- Joining composite components
- Preparing fiberglass components for finishing

Most composite airplanes still require some riveting and aluminum work, which is covered in Chap. 9.

ADDITIONAL TOOLS

In addition to the tools specified in Chap. 6, you'll need a number of other low-cost tools. Prime among them is a gram scale reading 0 to 500 grams and a hot-melt glue gun. The glue gun can be picked up at any hardware store. The scale might be a little harder to locate, but an ordinary diet scale with metric weights works if you can find one.

Builders using epoxies instead of vinylesters might consider buying an epoxy pump (Fig. 8-5) instead of the scale. The pump will dispense the correct ratio of resin and hardener with no bother. Make sure that you buy a pump with the same ratio as the epoxy used on your kitplane. The pumps cost around $150.

Fig. 8-5. *An epoxy pump is probably the easiest way to dispense the appropriate proportions of resin and hardener. Note that the pump is contained in a box to keep the epoxy warm.*

If you're building a kit that uses vinylester resin instead of epoxy, you'll need one nasty little item: a hypodermic syringe, without needle, of 15-cc capacity.

This might require a little delicacy. Walking into a drugstore and asking to browse through the syringes is likely to gain a few raised eyebrows and a possible visit from the local gendarmerie. And *then* you can try explain away the gram scale.

All kidding aside, you shouldn't have much trouble buying the one or two syringes you'll need. A bigger problem might be the syringe's capacity. A 5-cc model is pretty large, and ordinary drugstores might not sell them. More common are 1-cc models (used for insulin), and 3-cc models are about as large as most drugstores go. Check at farm or veterinarian supply stores for the larger models. Whichever you buy, it must be graduated at one-tenth cubic centimeter increments.

Ordinary scissors can be used to cut the fiberglass cloth, but spend a few bucks more for industrial model scissors. They'll cut better and last longer. Many builders also recommend a rotary cutter that looks remarkably like a household pizza cutter. It's different, though. And it costs around $20.

A Dremel Moto-Tool or equivalent comes in handy for all the trimming and shaping necessary with composites. These aren't very expensive, and most hardware stores carry a variety of cutters and shapers.

One thing composite construction does require is more disposable supplies. Find a local source for unwaxed paper cups and tongue depressors. Pick up a bunch of 1-inch disposable paint brushes. Buy an assortment of sandpaper; buy several packs—you'll need them.

Ditto with paint brushes. You'll use them to apply the epoxy. Small rollers come in handy, too.

It's not really in the tool department, but you'll be needing one more item if you're building a composite aircraft: a glass-cutting table. This table will be used to dispense the fiberglass cloth and provide a hard, clean surface on which to cut it. For the cutting surface, Lancair suggests using a piece of 1/8-inch-thick high-density polyethylene, sometimes called "tileboard," which can be found at home-supply stores. Figure 8-6 shows a typical layout.

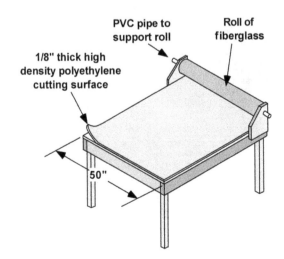

PVC pipe to support roll

Roll of fiberglass

1/8" thick high density polyethylene cutting surface

50"

Fig. 8-6.
A combination dispensing/cutting table for fiberglass cloth is a vital part of composite kitbuilding.
Lancair International, Inc.

MATERIALS, FASTENERS, AND SAFETY

When working with composites, you enter a high-tech world of aircraft homebuilding.

Materials

CLOTH

The most common material you'll be working with is fiberglass cloth. Literally, the cloth is formed from fibers spun from molten glass. It's then rather loosely woven to allow stretching and shrinking to fit compound curves without cutting or tearing. In Britain, it's called "glassfibre," which is probably a more correct term.

Fiberglass cloth comes in two varieties: *E glass* and *S glass*. The difference is in the formulation of the glass fibers themselves. S glass is better but more expensive. You'll also find cloth made with Kevlar and graphite. These advanced materials are usually stronger but are always more expensive.

There are four basic weaves of cloth: *mat, unidirectional, bidirectional,* and *triaxial*. Mat isn't really a weave at all. It's just random fiberglass fibers pressed together. Mixed with resin, it's used as a filler. Mat is definitely nonstructural.

Unidirectional cloth, or "uni" for short (Fig. 8-7), has most of the fibers in one direction. It's like a lot of long, strong ropes laid side by side. They're kept together by fill yarns spaced an inch or so apart. Uni is very, very strong in the direction the fibers run. Properly laid up, it's about 25 percent stronger than an equivalent cross section of 4130 steel. However, it has no crosswise strength because the fill yarns are only meant to hold the cloth together during handling. Uni is used in places like spar caps, where the designer knows the load will be applied in only one direction.

Bidirectional cloth (called "bid") has the same number of fibers going in two directions oriented 90 degrees to each other. It isn't as strong as uni but can take the strain in both directions. However, rarely is it as strong in both directions; one direction is usually weaker. The weaker direction is called the "fill," the stronger is

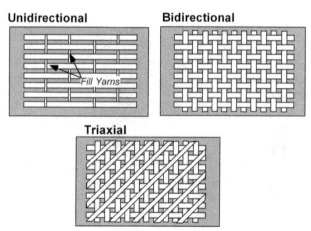

Fig. 8-7. *Unidirectional (uni), bidirectional (bid), and triaxial (triax) fiberglass cloth.*

the "warp." Unidirectional cloth, as an example, has a very high warp strength and a limited fill.

The strength and pliability of bid are determined mostly by the weave, count, and weight. The weave determines the stability and pliability of the cloth. For example, the *plain* weave is an ordinary crosshatch pattern. It's not especially pliable, but the pattern minimizes the tendency for the individual yarns to slip out of place under load.

The count is the number of yarns per inch in each direction. A typical cloth might have 24 × 22 count. The weight is given by the number of ounces per square yard. Coupled with the count, it's an indication of the size of the individual yarns. As with steel cable, the thicker the individual strands, the stronger and less pliable is the material.

Bid cloth is also referred to as boat/tooling or industrial cloth. Boat/tooling cloth is generally of plain weave; industrial cloth's weave is more pliable. Otherwise, pay attention to the weight. Lightweight boat or tooling cloth, for instance, comes in various counts and weights.

Triaxial cloth (called "triax") is sort of a combination of uni and bid. It has the two yarns of bid and a third yarn running crossways. It's a heavy-duty cloth that (of course) weighs more and is more expensive.

While this is useful background information, your fiberglass should come with the kit. A problem arises if you run out. There is no AN standard fiberglass—at least, none that is marketed to homebuilders. As such, it is ordered by the manufacturer's part number. One weaver could sell a cloth by the number P/N 8810, and another could sell a weaker, different cloth under the same number.

In other words, if you must buy additional cloth, *buy it from the same fabric manufacturer.* Otherwise, it might be dangerously inadequate. Your safest course is to order replacement cloth from the kitplane manufacturer. It might cost more, but you're more likely to get the right stuff.

Cloth should be stored in a clean, dry place. Any dirty or wet cloth should be discarded.

FOAM

Various types of foam are used in composite aircraft. As mentioned earlier, manufacturers use it to make lighter, stronger kit components.

Whether you'll have to work with it depends on the kit itself. Glasair builders make foam wing ribs. Other kits use styrofoam to help form the trailing edges of the wings and other complex shapes.

The advantage of foam is its light weight and easy workability. Most foams can be cut with ordinary knives or a heated wire and shaped with a variety of inexpensive tools. Foam parts can be made easily and will be very light, but they can't take any sort of load.

The usual practice is to make a component from foam and then cover it with fiberglass. Sometimes the foam is removed after the fiberglass cures (as in fuel tanks), but otherwise it's left in place as part of a builder-constructed fiberglass sandwich.

Styrofoam is the original homebuilder's foam. Light blue in color, it has a gritty, rather rough surface owing to its large cell structure. Don't confuse true styrofoam with the material in picnic coolers; coolers use expanded polystyrene with a smooth surface. A hot-wire cheese cutter works quite dandy with styrofoam. The technique is discussed later in this chapter.

Styrofoam dissolves in petroleum products and reacts to the vinylester resins used with some kitplanes. In these cases, designers usually select urethane foam. It weighs the same as styrofoam but is smoother owing to its smaller cell structure. It's colored green or tan. Never carve it with a hot wire; it emits a hazardous gas when heated.

A relative is urethane polyester foam. Its density can be up to 10 times more than styrofoam or urethane foam, which allows it to withstand higher compression loads. Polyvinyl chloride (PVC) is another high-density foam.

If you run out of foam, pick up an identical replacement. Don't go merely by sight. Never store foam where it can be exposed to direct sunlight. It deteriorates rapidly.

OTHER MATERIALS

Peel ply is a Dacron tape that won't adhere when used in a layup. It's used to prepare a laminate surface for later glassing or to reduce finish sanding.

Microspheres were very big 20 years ago but have been replaced by glass bubbles (microballoons) or Q-cell. Both are used the same way, to thicken epoxy or catalyzed vinylester to fill low spots or bond awkward shapes together. Cabrosil is similar.

Flox is flocked cotton fiber mixed with epoxy or catalyzed vinylester to thicken the material, similar to glass bubbles. The flocked cotton acts as a binder; flox produces a strong and durable surface when cured.

Use of these materials is covered later in this chapter.

Fasteners

Of course, the primary fastener will be the epoxy or vinylester resin used to impregnate the fiberglass cloth and bond aircraft components together.

EPOXY

Epoxies are two-part systems consisting of resin and the hardener. In theory, they're very similar to the epoxy glues you buy at the hardware store. But aircraft epoxies are optimized for fiberglass layups and large-area structural bonding.

Mixing ratios run between 2:1 and 4:1 resin to hardener depending on brand. Each product has different characteristics, such as viscosity (thickness) and curing time. Curing times vary from a few minutes to several hours. This is the time until the mixture hardens; actual curing for full strength might take far longer.

The resin-to-hardener ratio for epoxies is fixed; variance of more than 5 percent or so might cause problems. Accuracy is important.

VINYLESTERS

Vinylesters are multipart systems including resin, promoter, accelerator, and catalyst. To maximize shelf life, the resin is shipped in the unpromoted state; the user is expected to promote the vinylester resin a gallon at a time as needed. A typical ratio is 5 cc of promoter for each gallon of resin.

To use the vinylester resin, a small amount of MEKP catalyst is added. Remember, epoxies are optimized for a given mixing ratio of hardener to resin. Ten percent too much hardener can greatly affect the strength of the bond.

Not so with vinylesters. The catalyst doesn't add to the bond strength; it merely activates the resin. Catalyst-resin ratios can be matched to the environment and the task at hand. By adjusting the ratio, gellation times (time until the mixture sets to the point that it can't be worked) can be cut in half or more.

Figure 8-8 illustrates the ratio-temperature-time relationships for a typical vinylester resin. Use only the ratios approved by the kit maker; too little or too much catalyst might result in an understrength bond.

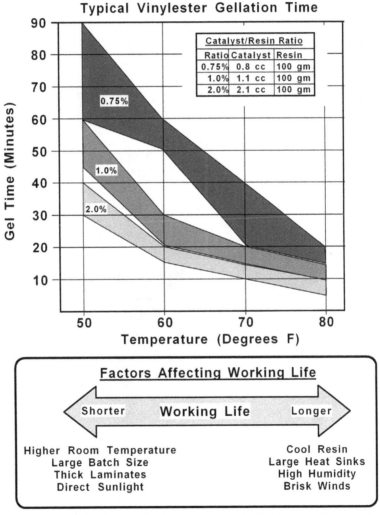

Fig. 8-8. *While epoxy has a fixed ratio of components, vinylester proportions can be selected (within certain bounds) for various cure rates.*

When necessary, accelerator can compensate for lower-than-optimal working temperatures. Above 80°F, it isn't needed. Three cubic centimeters of accelerator per gallon of resin usually suffice for temperatures between 65 and 80°F. Check the specific instructions included with your chemicals.

Store the materials in the proverbial cool, dry place. Vinylesters often are shipped in slightly permeable plastic buckets. Not only does this reduce the shelf life, but it also makes your shop stink of styrene. If you won't be using the material for awhile, transfer the resin to airtight metal or glass containers. Keep in mind, though, that it does have a limited shelf life.

Safety

Composite construction is like mixing a cocktail with the devil's chemistry set. You have to know exactly what you're doing. Composite aircraft are built using a wide variety of chemical substances, some hazardous and some supposedly benign. Proper safety precautions are vital.

Allergic reactions aren't as common as they used to be. But their relative rarity doesn't make them any more fun. Most of the reactions happened during the Long-EZ days. A builder would happily work along for six months or so, and then, just entering the shop caused his skin to break out in a rash. Some builders developed breathing problems.

Even those who protected their skin weren't safe; repeated incidental contact would still bring about severe sensitivity. And the effects were permanent.

A drive for safer epoxies arose. Safe-T-Poxy was developed, and it eliminated most of the sensitivity problems. A further advance was the use of vinylester systems, for which there are no recorded cases of allergic reaction.

So, technically, if you're using these modern epoxies or vinylesters, you can work with bare hands, right?

The safe levels of exposure to dangerous materials are always being revised. I don't think any limit was ever revised *upward*. We always seem to discover that the dangers are greater than previously believed. What's considered "safe" now might be defined as hazardous as long-term trends become apparent.

Remember that reactions to these materials traditionally have come as a result of long-term exposure. Just because you don't have a reaction during the first week doesn't mean you're *not* going to.

So don't apply epoxies or other materials with your bare hands. Wear rubber gloves or approved skin barrier cream. And provide adequate ventilation.

Other problems? Be advised that directly mixing vinylester promoter and catalyst produces a lot of heat and possibly an explosion or fire. Keep them well separated. A small reaction also can occur when catalyst is added to vinylester resin. Wear safety goggles—splatters can cause serious eye injury.

Some filler material, such as cotton fiber and Q-cell, are very fine powders that can mess up your lungs if inhaled. Wear a respirator or a dust mask. Sanding the cured material stirs up the same problem as well as adding fine particles of fiberglass and resin to the air. Wear a mask while sanding, too. Wear long sleeve shirts as well.

One unexpected danger comes from disposing of excess mixed epoxies or catalyzed vinylester resins. Because they produce considerable heat, don't throw them into the trash until they've cooled.

A good way to dispose of partially cured resin is to pour it onto plastic sheets. The plastic should be on bare ground, concrete, or other nonflammable material. When the mixture cools, fold up the plastic and discard.

COMPOSITE KIT PARTS

One way composite kitplanes differ is in the inclusion of a large number of identifiable components. A tube-and-fabric kit might deliver a welded fuselage, but everything else is a jumble of generic parts. An aluminum airplane kit will have some metal sheets bent to vague shapes.

But you can look at a composite kit and see the swoopy shape of the fuselage (Fig. 8-9). The wings are obvious, even with the cutouts for the ailerons and flaps.

The primary operation for building the structure is gluing kit parts together and reinforcing the joints with fiberglass cloth. Examination of kit parts would be worthwhile.

Coatings and joggles

The inside and outside of the parts might differ in color. Cured fiberglass parts are a golden tan. The insides of the pieces of your kit are probably this color.

Stoddard-Hamilton Aircraft.

Fig. 8-9. *Molded composite aircraft are almost always supplied as complete kits and come with a high number of identifiable parts.*

Outside, there are three possibilities: the same tan as the inside, a dark gray, or a shiny white.

As mentioned at the beginning of this chapter, composite kit parts are laid up in a mold. The curing process generates heat. Any substance applied to the mold before the first fiberglass layup gets bonded to the outside of the finished part.

Want a hard-shell exterior finish? Apply a special paint to the mold, make the part, and it comes out gleaming and protected.

This is how a *gel-coat* finish is applied. In the fiberglass boat industry, it cuts down on the total production time and labor. The hull comes out of the mold and doesn't have to be painted. Bond the top deck in place, and trim, and the boat is ready for delivery. It works the same way with airplanes, too.

Some kit manufacturers deliver the exterior gel-coated, but most don't. If the company specifies that reinforcing cloth must be bonded to the exterior, the advantage of gel-coating is severely reduced. The surface and edges of the reinforcing fiberglass must be smoothed and blended in. The color won't match the gel-coat, so it'll have to be painted. Matching the gel-coat's exact color and sheen is impossible; hence the whole airplane has to be painted.

And if the whole airplane is going to be painted anyway, why add the gel-coat at all? It just adds weight. Glasair components originally were delivered gel-coated. The company saved 35 pounds per airframe when it stopped using the process.

Then again, the company went *back* to gel-coating for the Glasair's stablemate, the GlaStar. The composite fuselage shell of the new design doesn't require reinforcement on the exterior.

Instead of gel-coating, many companies apply a gray primer using the same procedure. The primer often contains an ultraviolet block to slow deterioration. As a third alternative, the parts are delivered in the natural honey tan of the resin and glass.

The manufacturer adds other useful features during the molding process. The most important feature is the *joggle,* which consists of complementary indentations applied to the two edges that will be bonded together (Fig. 8-10).

Consider trying to glue two playing cards end to end. It's a weak joint at best because the thin edges present little surface for the glue to work with. The edges could be overlapped slightly. This produces a strong bond, but one card is higher. The joint is obvious and hard to hide and on an aircraft would have to be carefully positioned to minimize the aerodynamic effect.

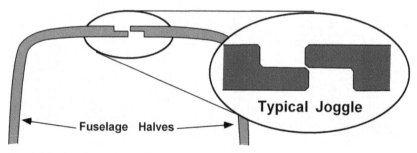

Fig. 8-10. *Joggles help produce strong and correctly aligned components.*

However, if the edge of one card were bent down in a Z shape, the pieces could be overlapped and still leave a smooth seam. This is a *joggle*. The tiny seam remaining can be filled and smoothed to completely hide the interface, or layers of fiberglass can be added for additional strength. The joggle also aids in the proper alignment of the two parts. Some kits use double joggles for additional strength and precision.

Storage

You won't be working on many of your parts for a while, so a moment or two to consider storage issues might be well spent. Those parts are expensive.

The main thing is to store them so that nothing—not even gravity—is trying to distort them. Don't lean your cowling up against the wall, for instance. Two years later, you may find that it has gradually taken a bit of a lean itself. Flat would be better; flat with something supporting the middle (so that it won't sag) would be best.

Recall from Chap. 2 our discussion of the temperature sensitivity of some composite materials. Keep the parts out of direct sunlight, and don't store them in an area subject to a lot of heat.

Preparation techniques

Composite parts require different degrees of preparation prior to glassing or bonding depending on the manufacturer's process and their location on the aircraft. At the beginning of this chapter I mentioned the release agent that prevents the part from bonding to the mold. Sometimes, traces of this chemical remain attached to the exteriors of the parts. These traces must be removed. Usually, all that's necessary is to scrub the area with water and let it dry.

The manufacturer's mold produces nice smooth parts, but such a surface isn't the best for glassing—you want a rough surface for the resins to grab. Therefore, any time you glass the exterior of the aircraft, the surface must be roughed up first with sandpaper. If the manufacturer primes the exterior, this coating must be removed as well. Otherwise, the resins will just bond to the paint, which flakes away from the structure at the first chance it gets.

After sanding, wipe the part down with acetone. Be careful if there are bare foam parts nearby—the acetone will dissolve them.

Little preparation normally is required for the insides of molded kit parts. Follow the kitplane manufacturer's directions.

A last note on preparation: When sanding, roughen the surface and then *stop*. Don't sand into the underlying fiberglass. Any damage to the cloth weakens the structure.

Cutting

Kit parts can be cut with normal tools, such as drills, hole saws, and saber saws. However, high-tech materials such as Kevlar will dull the tools quickly.

Excessive pressure causes heat to build, and heat is the enemy of composites. Let the tool cut at its own pace. If possible, back up the part so that you aren't pushing on the part itself.

Fine-shape the parts with the ordinary tools: files, sandpaper, and the like.

Bonding

Bonding kit parts together is very simple. The joggles are coated with the appropriate bonding agent (epoxy or catalyzed vinylester resin), and the parts are pressed together. Enough bonding agent must be used so that the glue extrudes along the entire length of the seam. If it doesn't, there might be some starved areas. Wipe the excess away before it cures.

The bonding agents can be pretty thin, and they have a tendency to flow out of the joints before curing. Often the resin is mixed with glass bubbles or flox to make a thicker, more viscous material. Later sections discuss the use of these materials.

The pieces must be held rigidly in place until the bonding agent cures. One way is to predrill small holes through the joggle, join the parts, and inset clecos. Hole spacing usually isn't very tight—between 6 inches to a foot seems typical for large pieces (Fig. 8-11). The holes can be filled once the clecos are removed.

Other methods include wrapping duct or wide masking tape around the part, using clamps (atop boards to spread out the force), and weighting down small pieces. No matter the method, take care to equalize the pressure because gaps might form if one area is held together too tightly.

Don't use clamps unless the kitplane manufacturer recommends them. If they are recommended, the manufacturer probably will specify exactly *how* the pressure should be applied. Follow the instructions—composite parts are not designed to withstand *localized* pressure. You can easily crush the underlying foam or honeycomb with even a small clamp.

Whether you're done at this point depends on the required strength. Most joints need reinforcement through additional fiberglass layups.

LAYUP PREPARATIONS

The basic operation is to cut a piece of fiberglass cloth to the appropriate size, lay it in place, and saturate it with resin. As in many endeavors, the preparations greatly affect the quality of the final product.

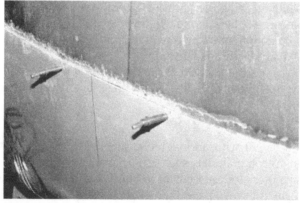

Fig. 8-11. *Clecos temporarily hold this composite fuselage together until the vinylester cures.*

The shop

General rules were given in Chap. 6. The need for a warm environment cannot be overemphasized. This is so not only because the resins cure slower when cold but also because they don't flow as well either. Your goal is to fully saturate the cloth without using excessive resin. The cloth works like a wick in normal temperatures, eagerly absorbing the bonding agent. But cold resin doesn't want to flow and doesn't want to soak into the fiberglass. You end up using more resin and end up with a heavy aircraft.

Everything associated with the glass work should be warm: resin, cloth, workbench, tools, and the like. Any cold item sucks the heat from the resin. If the workshop must be heated, turn the heater on an hour or so before starting.

Many composite builders keep their epoxy and/or their epoxy pump in a box heated by a 25-watt light bulb. This is a good, simple, and cheap idea. However, if you aren't going to be working for several days, turn the bulb off because some of the volatile chemicals in the epoxy could evaporate and result in bad bonds.

Whereas cold is the primary enemy of fiberglass work, dirt and dust vie for second. They'll contaminate the cloth and resin and deteriorate bond quality. Sanding residue is a primary culprit. Keep the roll of cloth covered until needed, and clean the shop regularly. Those little hand-held vacuums are neat, but beware—they exhaust the air sideways, which can stir up more dust if they are not handled carefully.

Composite aircraft builders require helpers more often because they are working with large structures and materials that have limited working times. Usual practice has one person mixing the epoxy while two others apply cloth or bond parts. Make sure that you have rubber gloves, respirators, and other protective equipment for your assistants, and carefully brief them on the dangers and procedures.

Preparing epoxy or vinylester resins

While you don't actually mix up your epoxy or vinylester until everything else is ready, subsequent information in this chapter will be easier to digest if you understand the process. To reduce repetition, I'll use the term "bonding agent" or "resin mixture" to refer to either epoxy resin with hardener added or catalyzed vinylester resin. Similarly, the term "mixing" also will apply to the act of adding the catalyst to vinylester resin. Where differences exist, I'll refer specifically to either epoxy or vinylester.

When dealing with the resin mixture, the primary concern is *pot life*, which is the time until the bonding agent becomes too thick to work with. Many factors affect pot life, which decreases with low humidity or in brisk winds. But temperature is the major variable. The warmer the materials, the faster the material cures.

Both systems have another common characteristic: *exothermic reactions*. They both give off heat while curing. The faster they cure, the more heat that is liberated.

The paradox should be obvious. The mixtures cure faster when ambient temperature is high. But curing raises their temperature, which, in turn, increases the reaction rate, which causes higher temperatures, and so forth. Atomic physicists call this a "chain reaction," and with radioactive materials, it makes large craters.

Chemicals are somewhat less energetic. The mixture just gets hot, and pot life is severely reduced. This overreaction is called an "exotherm." High temperatures are a danger to composite aircraft. Too-hot resin can melt foam and weaken the kit parts.

This is not a problem in a fiberglass layup—the bonding agent will be well spread out, and the heat will be dissipated. But a mixing cup concentrates the heat and supports the chain reaction.

The solution is to limit the batch size to less than 200 grams (about 7 ounces). The mixture still gets warm, so don't set it on anything that might be damaged by heat, especially completed fiberglass parts.

Sure, it would be nice to catalyze a whole bucket of resin and bond an entire bulkhead in place at one go. But you would quickly end up with a batch of rock-hard goop in a nearly glowing bucket. Instead, small batches will let you work on the next section even if one part has begun to cure.

Differences exist in the preparation of epoxies and vinylesters.

EPOXY PREPARATION

The epoxy manufacturer specifies the proper ratio of resin to hardener. It is important to closely approximate the specified ratio. Few, if any, aircraft epoxies use a 1:1 ratio, so you'll need some method to measure out the components.

As mentioned earlier, a common tool is an *epoxy pump*. It includes hoppers for the resin and hardener and dispenses a user-adjustable amount of each element with each pump of the handle. Check your adjustment on a regular basis. Alternatively, use a scale to weigh out the proper ratio of materials. A third method uses industrial-sized syringes to draw out and dispense the materials. Buy separate syringes for each component, or you'll be left with epoxied syringes and contaminated materials.

Mix the resin and hardener in an unwaxed paper cup using a wooden tongue depressor (or similar tool) to stir (Fig. 8-12). The cups can be reused as long as the epoxy inside has either cured completely or hasn't become too thick.

Mix for a minute or two. One of the goals of the layup process is to eliminate all air between the laminations. Your first step toward that goal is to keep from adding air into the resin mixture. Don't swirl the stick around like you're cleaning a brush. The more violent the action, the more air there is that ends up in the mixture. The resin and hardener must be blended thoroughly, but keep the air bubbles out. Scrape the sides and bottom of the container with the stick to ensure even mixing.

The curing time depends on the temperature and the epoxy brand. As mentioned earlier, the mixture will get warm. When it becomes noticeably thicker, mix up a new batch.

VINYLESTER PREPARATION

Vinylester mixing is similar to epoxy, with certain differences. To begin with, make sure that the promoter had been added to the resin. Do a gallon at a time, and slap a label on the jug so that you know that it's been promoted. Add the accelerator at the same time, in the amount appropriate to the temperature.

Fig. 8-12. *Proper mixing of the components is a must. Note the long sleeves and gloves for protection.*

Because the resin-to-catalyst ratio is so large, a pump makes no sense. Instead, use a scale for the resin and a syringe for the catalyst. A cubic centimeter is about equal to a gram of material, so for a 1 percent resin-to-catalyst ratio, add 1 cc of catalyst for every 100 grams of resin.

A kitplane manufacturer suggested hot-gluing the catalyst bottle and a paper cup to a piece of wood and keeping the syringe in the cup. That keeps the syringe and bottle together, and the cup prevents the syringe from leaking catalyst wherever it's set.

Mix the resin and catalyst the same way as epoxy; however, don't use a cup with partially cured resin already in it. It will accelerate the curing of the new mix.

Vinylester resins cure differently from epoxies. While epoxies thicken gradually, vinylesters reach a *gel point*. When gelling starts, the resin will develop a pastelike consistency for a couple of minutes and then solidify.

At some point during the curing process, epoxies and vinylesters reach the *green cure state*. In this condition, the resin mixture is partially cured and is somewhat rubbery. Loose edges of fiberglass cloth can be trimmed away easily with a razor blade. If you wait too long, the resin mixture takes on a solid cure that takes a lot of work to trim and shape.

Vinylesters reach the green cure state about a half-hour after gellation, whereas epoxies might take several hours depending on temperature. Epoxy and vinylester systems take 24 hours or so to cure fully.

Shaping foam

The foams used on kit aircraft can be cut and shaped easily using ordinary hand tools. Bandsaws and saber saws work dandy, but the stuff cuts easily with hand-saws or even penknives. When making a flat piece such as a rib, glue a pattern to the foam and cut it out.

Final shaping is done with standard abrasive shop tools. Files work fine, as does sandpaper. For the best use, wrap a piece of sandpaper around a chunk of scrap 2 × 4. You can buy a commercial sanding block, but the homemade variety works just fine.

A good foam-working tool is the Shurform, made by Stanley. It has a rough, self-cleaning cutting surface and a pair of handles like a wood plane.

If the piece of foam isn't big enough, glue two pieces together using a thick mix of resin and glass bubbles (covered in detail later in this chapter). However, don't get the mix too close to a cutting line. The cured mixture is difficult to cut, and any attempt usually damages the nearby foam. Similarly, don't use so much mixture that it oozes out the seam. This makes it tough to shape the surface of the piece. The joint should be 1/16 inch thick or less, and the mixture should stay at least ¼ inch down from the top surface. You'll fill the low area prior to covering with glass.

Hot wire cutting

With styrofoam, and *only* styrofoam, you can use the hot wire method. An electric current is passed through a fine wire, which heats the wire past the melting point of the styrofoam—like a large cheese cutter—and the hot wire treats the finest sty-rofoam like so much cheddar.

Hot wire cutting was in its heyday back with the Long-EZ and similar aircraft. The EZ's wings were composed of multiple sections of foam cut out by hot wire separately, bonded together, and fiberglassed. Most current composite kitplanes don't use this procedure for primary structure, but some require the builder to make canards, trailing edges, and other parts from hot-wired foam.

Commercial hot wire systems are available for cutting long shapes that include custom cutting frames and variable transformers and cost about $100 to $200.

For your typical kitplane, though, a simple homemade system works well enough. You'll need a spool of stainless steel (0.040- or 0.032-inch) safety wire, a couple pieces of dowel or tubing about 1 foot long, some nails, a 2 × 4 board about 3 feet long, and a battery charger (Fig. 8-13).

Form a giant U shape with the board on the bottom and the tubes/dowels as the uprights. Bore holes into the 2 × 4 to slide the tubes into place. Hammer a nail into the board near the base of each. Drill a small hole through the free end of the tubes. Then string a piece of safety wire from one nail, through the hole in the closest tube, through the hole in the other tube, and back down to the other nail. Tighten the wire by clamping a vise-grips to the base of one of the tubes and rotating it.

That's it. Connect the clips from the battery charger to the two nails. Turn the battery charger on. Battery chargers have built-in current limiters, so the meter should be showing a high output without actually pegging the needle. You probably won't

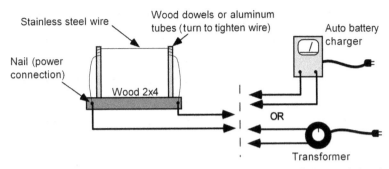

Fig. 8-13. *Small parts can be hot wire cut using a couple of pieces of dowel or tube, some stainless steel safety wire, a board, and an ordinary battery charger.*

see any change in the wire. It'll slacken up a bit with heat, so give the vise-grips another turn.

Take a piece of scrap styrofoam, and touch it to the wire. You'll hear a hiss as it sinks into the foam. If the styrofoam is visibly melting or the wire cuts a slot larger than its own diameter, the wire is too hot. This leaves pits and melted spots in the finished surfaces. The voltage must be reduced by adding more safety wire between the charger and the tool or by using an electric train transformer or other voltage control. Variable resistors are available from homebuilders' catalogs.

If the wire is slow cutting, the voltage (and hence the wire's temperature) might be too low. Reduce the wire length between the charger and the actual cutting wire. One way would be to clip the charger leads directly to the wire at the end of the tubes. Otherwise, you'll need a stronger power supply.

When the temperature is right, the hiss of the cutting wire should be faint. When the wire is withdrawn, fine streamers of blue plastic should follow. The surfaces of the cut areas should be smooth, with no pits and a lot of little blue plastic streamers.

When you're cutting a shape, two templates are tacked to the ends of a piece of foam. Again, you can slurry two pieces of foam together to get the right length, but don't get the slurry any closer than 1/2 inch from the prospective cut line. Turn on the cutter, and sink the wire into the foam until it contacts the templates on both ends (Fig. 8-14). Then slowly move the wire around the outline, taking care that each end of the wire is at the same position on the template at the same time.

Let the wire cut at its own pace; if you pull too hard, the wire will lag excessively and bow *inside* the foam, distorting the final shape. Cutting is a heck of a lot easier with two people, but coordinate so that you're both at the same point on the template.

When done, smooth the foam with files or a sanding block.

Installation

The first step to actually installing parts is determining where they go. The plans might specify a distance from a particular reference point, such as "12.5 inches outboard of wing root." Or there might even be a scored or raised line on the inside of the fuselage or wing to mark the spot.

Fig. 8-14. *The cutting wire is guided around templates tacked to both sides of the piece of foam.*

Wipe the area with acetone, and buff it a bit with 80-grit sandpaper. Exterior surfaces will take a bit more buffing to eliminate the smooth surface left by the mold. Make sure that all primer and mold release agent has been removed from the bonding areas.

Parts are bonded together with the standard resin mixture. Sometimes, though, the resin is too thin. It can tend to flow out of joints and produce a substandard bond.

To counteract this problem, various materials are added to the resin mixture to thicken it up:

Flox. Flocked cotton fiber and epoxy are used to bond parts together and form fillets. It's a white powder about the consistency of lumpy flour. It's mixed in with epoxy at a 2:1 fiber-to-epoxy ratio (by volume) to produce a tan putty-like substance. It can be mixed by feel; if the flox doesn't stand up on its own, add cotton fiber; if the mixture is whitish, add epoxy.

Mill fiber. The vinylester world's equivalent to flox is mill fiber. Instead of flocked cotton, very short fiberglass strands are added to the catalyzed resin. Again, you're looking for a putty-like mixture. Like flox, mill fiber mixtures form strong bonds that are ideal for structural applications.

Micro, glass bubbles, microballoons, Q-cells. These four common names refer to the same basic material—tiny glass bubbles added to the resin mixture for use as a nonstructural glue or filler. The bubbles look like powdered sugar. The term "micro" actually refers to now-obsolete microspheres, but the term is handy and remains with us. The ratio of bubbles to resin varies with the application. Nicknames describe the three basic ratios.

The first is "slurry" (or "wet mix"): glass bubbles and epoxy–catalyzed resin in about a 1:1 ratio. It's tan and a bit runny, looking like a cheap chocolate milkshake. This thin mixture fills the surface pores of styrofoam to seal it prior to applying fiberglass.

A thicker mixture, about the consistency of peanut butter, is called "thick micro" (or "thick mix"). It bonds nonporous parts or acts as a filler (Fig. 8-15). About three times as much balloons as resin mixture is used. The final product looks like peanut butter—creamy, please, with no lumps.

Fig. 8-15. *Thick micro looks and spreads like peanut butter.*

Finally, there's "dry micro," at about a 5:1 ratio. It's very thick and is used to fill the fiberglass weave before painting and to bond blocks of foam together.

All mixture ratios are by volume, not weight. You can learn to recognize the proper consistency and don't have to measure the quantities. The bubbles are always added to mixed epoxy or catalyzed resin, never just to the plain resin. And speaking of "nevers": Don't allow micro between layers of fiberglass. It weakens the bond.

Other materials are used depending on the kit. "Cabocil" is a glass powder similar to Q-cells but much finer. "Chopped strand mat" (usually simply called "mat") is similar to mill fiber but still in a clothlike form.

Kits use different mixtures in different areas. It makes a difference because mill fibers produce a stronger bond. Check the plans carefully.

Unless you're sealing styrofoam, you'll always use a thick mixture. It's solid enough that you usually don't have to clamp or jig the parts together while the mix cures. Generally, light components don't shift in thick micro because, like peanut butter, it "sticks to the roof" of your airplane. Keep watch, and be prepared to add some temporary external bracing if things don't stay in place.

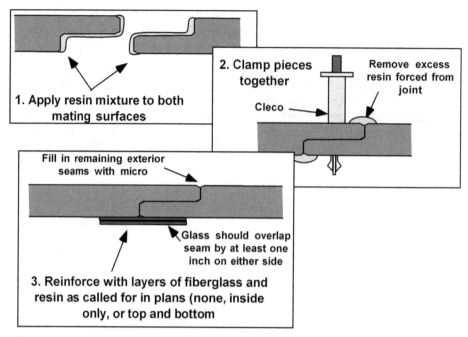

Fig. 8-16. *The joggled areas are bonded together with thick micro and held with clecos or other clamps until cured.*

There are two types of bonding operations. The first involves flat joins, such as two joggled parts or two overlapping flat components. In these cases, there's plenty of bonding surface available; all the builder has to do is apply a bead of thick micro (or whatever mix the manufacturer specifies) along each surface and press the parts together. Hold them together with rubber bands, bungee cords, tape, or whatever, as mentioned earlier. The tried-and-true cleco is often used, as shown in Fig. 8-16.

Some kits call for the "poor man's cleco," the pop rivet. While definitely cheap, pop rivets must be removed by drilling. They have their places, but if the plans say "use a cleco or a pop rivet," a cleco is usually the better choice. Because the pop rivet does provide a more solid attachment, it's better if the parts are heavy or if the parts will be handled repeatedly.

Make sure that the parts are solidly mated and that the joggles lie flat against each other. Wipe off the excess mix that oozes from the seams. If the plans call for it, add the specified number of layups to one or both sides.

The other type of bonding operation attaches two pieces at a sharp angle, one approaching 90 degrees. Examples include wing ribs and fuselage bulkheads. These pieces don't have a premade joggle to hold them in the proper position and don't produce a nice, flat, easy-to-glass joint.

The method used to join the parts varies with the kit. Most have you apply a layer of thick mix or flox to both parts and press them together. Others temporarily attach the parts with 5-minute epoxy or a hot glue gun and depend on the fiberglass layup for strength.

In either case, mix or flox must be used to form a fillet between the parts. A fillet eliminates the sharp inside corner formed by the two components. While fiberglass cloth can be formed to fit almost any shape, it has limitations. It won't tuck into a 90-degree corner. The thick mix or flox is used to make a fillet to smoothly transition between the two surfaces.

If your kit specifies a thick mixture, use a lot of it. When the parts are pressed together, the mix oozes out. Slide the curved end of a mixing stick along the joint to make a concave surface (Fig. 8-17). You'd like to see about a ⅜-inch radius. A fingertip is just about right, but make sure that you're wearing gloves or blocking cream. When the fillet is formed, scrape off the excess mix and return it to the mixing cup.

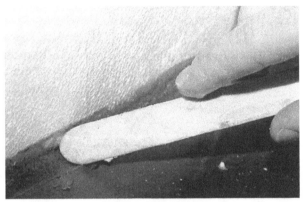

Fig. 8-17. *The fiberglass cloth cannot be applied to a sharp corner; a fillet must be added using thick micro. The end of the mixing stick is perfect for making the radius. Excess micro is scraped off and returned to the container.*

If the kit attaches the parts temporarily using 5-minute epoxy or hot glue, apply thick mix to the joint once the temporary glue has set. You can use the mixing stick or draw a nice bead with a metal cake-decorating tool (the resins will dissolve a plastic one). Shape the mix with finger or mixing stick into the desired radius. A summary of this process is shown in Fig. 8-18.

Once the fillet has set, it's time to add the fiberglass.

LAYUP OPERATIONS

When it's time to make a layup, you have one overriding goal: Saturate the fiberglass cloth with the right amount of resin. If you use too little, air bubbles remain in the laminate. Not only do the bubbles reduce the strength, but they also may allow water to seep between the layers of fiberglass cloth. If this water freezes, it will cause delamination (separation of the layups) and destroy the strength of the bond.

Too much resin adds weight for no gain in strength. Too much resin in a wet layup actually might cause the cloth to float, adding air bubbles as well as moving it out of position.

The following typical layup operation illustrates the step-by-step process.

Attachment Methods

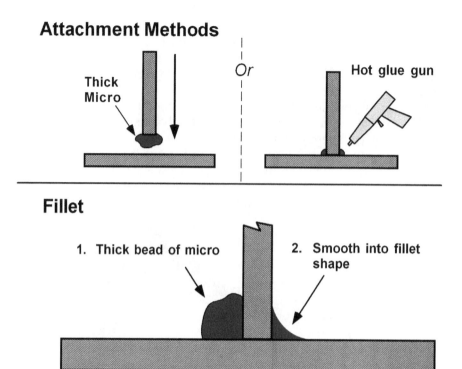

Thick Micro

Or

Hot glue gun

Fillet

1. **Thick bead of micro**

2. **Smooth into fillet shape**

Glassing

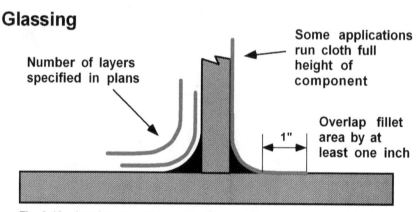

Number of layers specified in plans

Some applications run cloth full height of component

1"

Overlap fillet area by at least one inch

Fig. 8-18. *Attachment methods vary. Some foam ribs are bonded in place using thick mix (as shown on the left), but the prelaminated ribs from other designs are first attached with hot glue, and a fillet is added later.*

Foam preparation

In the rough-and-tumble environment of shop operations, foam parts sometimes get dinged slightly. The cloth tries to assume the shape of the dent and will leave an ugly depression. If it can't fill the dent, an air gap is left as a starting point for delamination. Ordinary dents are easy to fix with a thick mix of micro (Fig. 8-19).

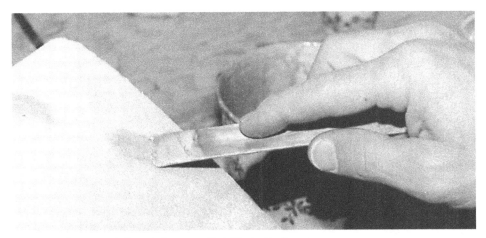

Fig. 8-19. *Dents and dings in foam are first filled with dry micro.*

Fill the holes, and scrape off the excess. Let it cure, and sand it flush with the rest of the surface.

Styrofoam's open cell structure soaks up a lot of resin, which adds a lot of unneeded weight and might produce a poor fiberglass bond. The surface must be sealed first. This is done with a *slurry,* which is made from resin mixture and glass bubbles in about a 1:1 ratio (by volume). Slurry should have a consistency of a cheap milkshake, pouring easily but still thicker than water (Fig. 8-20).

Pour the slurry onto the foam, and spread it around with a squeegee. The slurry is thin enough to fill the cells. Work it around, pausing every few seconds or so to scrape the squeegee across the lip of the mixing cup to remove the excess. Continue until the surface appears sealed.

Cutting the cloth

Usual practice is to cut out an oversized piece of cloth and trim it to fit. Lines can be marked with a felt-tip pen, but the remaining ink should not be contained within the bond area.

When you are using bid, the plans should specify the *bias* to which the cloth should be cut. Bias is defined as the angle to the weave. The usual practice is to cut at a 45-degree bias. In that way, the edges of the cut are slantways across a number of threads. A zero-degree bias, on the other hand, makes the cut along the same direction the threads run. The cloth edge unravels a bit easier.

The direction makes a difference strengthwise. The cloth is twice as strong when cut on a 45-degree bias. Follow the kit's directions when it comes time to cut, though.

Figure 8-21 shows examples of 45- and 0-degree bias cuts. The bias orientation doesn't have to be exact; the effect of a slight error is negligible. Eyeballing it is usually good enough. Uni, of course, is always cut directly across the weave.

If the edges of the piece are supposed to be cut away after the glass is laid in place, lay strips of ¾-inch masking tape along the borders. Mark the line on the tape, cut, and the remaining tape will hold the edges of the cloth together. Do not

Fig. 8-20. *To keep the styrofoam from absorbing too much resin mixture (resulting in excess weight and poor adhesion), its pores are first sealed with slurry.*

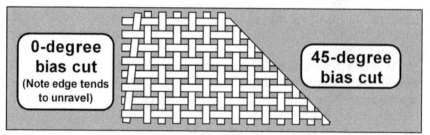

0-degree bias cut (Note edge tends to unravel)

45-degree bias cut

Fig. 8-21. *Bias and 0-degree cuts on bid. Note that the yarns tend to unravel on 0-degree cuts.*

use this method if the cloth's edges are to be glassed down because the tape *cannot* be neatly removed.

Two final points: First, don't scrimp. You must totally cover the area specified. Every fraction of an inch contributes to the total strength; if a cloth piece is too short, then the part is too weak.

Second, a pair of good, sharp scissors is important—they're cutting *glass,* not nylon or Dacron. A cheapie model dulls quickly, and a dull tool makes ragged edges. Don't fight it. Buy quality scissors, and sharpen them frequently.

Shaping and trimming

Take a look at a square of bid. The cloth has a loose weave; notice the small gaps between threads and the threads that aren't held firmly in place. Pull on the ends, and the piece stretches almost 50 percent in length. The width narrows at the same time. Shove your thumb into the middle. The cloth will assume an even curve without distorting.

This is the magic of fiberglass. The loose weave allows the threads to shift slightly to accommodate compound shapes. As long as no major holes get opened, it'll come out just as strong.

Back to our layup. Hold the cut cloth near the layup area, and approximately form it to shape. You're merely making sure that the piece is large enough, not getting it perfect before applying the resin.

The instructions should specify the weave orientation; therefore, make sure that the cloth you've cut is correct. Uni is usually lengthwise, but multiple layers sometimes are laid at slightly different orientations. Follow the plans' directions.

If there is definite excess, trim it back a bit. Be especially concerned where the edge of the cloth will just dangle over an edge. If there is too much cloth past the edge, the dangling weight will tend to lift the cloth still on the part.

When glassing two parts together, the cloth should overlap each part by at least 1 inch in order to develop full strength.

BASIC LAYUP

If you are applying the cloth to a fiberglass kit part, apply a thin layer of resin to the layup area. It fills the rough surface and helps saturate the down side of the cloth. Hidden air bubbles thus are reduced. Don't paint the area with resin if the cloth is being applied to aluminum or another smooth surface. Oversaturation results.

Next, lay the cloth in place (Fig. 8-22). Use your (gloved) fingers, the mixing stick, or anything else cleanable or disposable to push the cloth into all the nooks and crannies. It must lie flat against the surface, or the gap is a starting point for delamination.

Stretch and push the cloth as required to eliminate folds and wrinkles. Make sure that the weave doesn't, well, *weave.* It should run in a straight line across the part. This is vitally important. A wiggle in the weave is like slack in a rope. Get it as straight as humanly possible.

Brush resin atop the cloth, starting in the middle and working out toward the edges, to chase the bubbles away. Take care not to break them into smaller bubbles because that actually makes the situation more complicated.

If possible, use a squeegee to further spread the resin and eliminate bubbles. Don't use too much pressure—that might shift the cloth. If the layup is in a position that can't be reached with a brush, be especially careful not to apply too much resin.

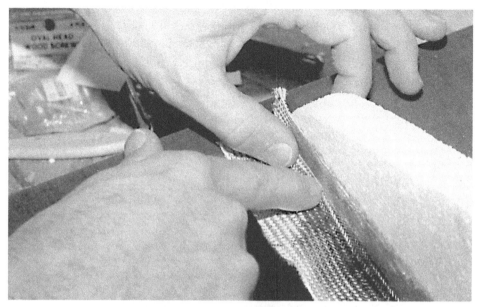

Fig. 8-22. *After applying a layer of resin mixture, lay the cloth down. Make sure that it makes solid contact on the fillet, leaving no air gaps under the cloth. The weave should be straight because waves reduce the strength of the layup. Gloves or barrier cream should be worn.*

One way to eliminate air bubbles is by *stippling* with the brush. The heel of the brush is pressed down on the cloth to force resin through to replace the air (Fig. 8-23). Note that this counters how you use a brush while painting, where you don't shove the brush down hard.

Take special care when applying a strip of cloth over a fillet. The cloth must set down in direct contact with the fillet material and not "bridge" over it.

Vinylesters have a unique problem. Where epoxies harden gradually, vinylester resin gels suddenly and with little warning. If a part gels before it gets saturated, it must be reworked. Once it has cured, sand away the unsaturated area with 80-grit or coarser sandpaper. Be careful in this process—you don't want to damage the previous layer. Once the unsaturated area is gone, cut a piece of cloth wide enough to cover the area with some overlap, and resin it into place.

A well-done layup should look wet, with no grayish white air bubbles or puddles of resin. The cloth pattern still should be visible. The color is a nice, rich gold. Take care to wipe up any resin dripped on the surface, especially if the drops land in a place where glass will be applied later. Once cured, they're hard to remove without damaging the surface.

If another layer of fiberglass is to be applied, lay it down while the previous layer is still wet. If it has cured, you'll have to sand it lightly and then apply resin. Take care with the weave orientation. Again, all layers usually are oriented the same way.

If the area is hard to work in, let the area get a little tacky before adding the next layer of fiberglass. Otherwise, your efforts might shift the previous layer, throwing off the layup and adding air bubbles.

280 *Composite construction*

Fig. 8-23. *Stippling with the brush ensures that the cloth is thoroughly soaked with resin. The force always should be perpendicular to the surface; any slant might shift the cloth.*

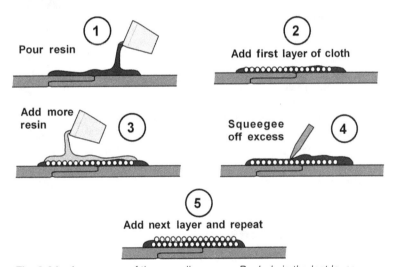

Fig. 8-24. *A summary of the overall process. Peel ply is the last layer.*

The general process is illustrated in Fig. 8-24. Figure 8-25 shows a completed Glasair rib. Note the basic rib shape is foam, with layers of fiberglass applied to the outside. Note also the fillet and fiberglass where the rib joins the lower skin.

Some multiple layups are awkward to apply to the airplane. Figure 8-26 shows a sneaky little trick you can use. Cut the layers of glass to size, and tack a large piece

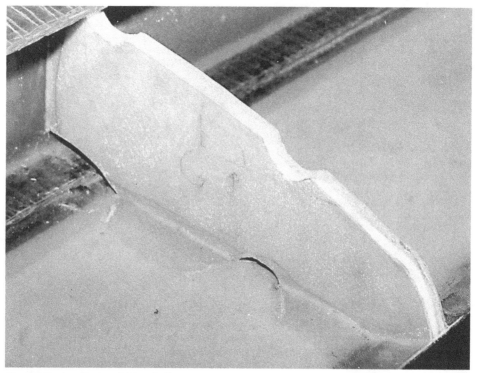

Fig. 8-25. *A completed Glasair rib.*

of plastic wrap or Mylar to your benchtop. Make the layups on the plastic instead of on the airplane. Make sure that the edges of each layer of cloth are lined up. When the last layer has been added, cover the layup with another layer of Mylar. Then run a roller back and forth over the outside of the plastic.

This ensures that the resin is entirely forced through all the plies. Excess resin and air bubbles ooze out the sides. Use your roller blade to cut the exact shape you need, and then prepare the target area. Lift the laminate off the table (the bottom sheet of plastic should stay tacked down. Look and make sure!). Apply the laminate to the airplane, press it down, and then take off the last piece of plastic. *Make sure that no plastic remains on the aircraft!*

Trimming and shaping

At some point, the resin-soaked cloth becomes tacky enough to stay together without yet hardening, called the "green cure stage." At this time, the cloth can be trimmed using a very sharp tool such as a single-edged razor blade. When trimmed, press the loose ends of the tacky cloth into the surface.

You can shape the fiberglass once the resin cures, but you won't like it. It's very hard and stiff. Use a grinder or files and coarse sandpaper. Careful, though—don't grind too much away.

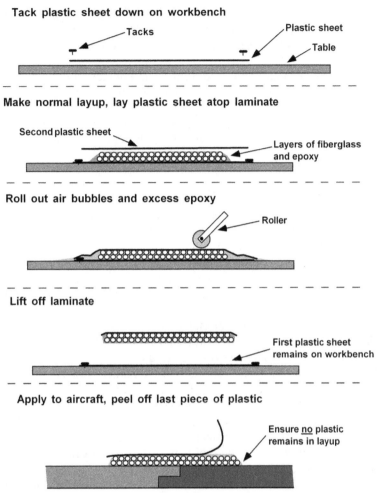

Tack plastic sheet down on workbench

Tacks

Plastic sheet

Table

Make normal layup, lay plastic sheet atop laminate

Second plastic sheet

Layers of fiberglass and epoxy

Roll out air bubbles and excess epoxy

Roller

Lift off laminate

First plastic sheet remains on workbench

Apply to aircraft, peel off last piece of plastic

Ensure <u>no</u> plastic remains in layup

Fig. 8-26. *Rather than attempt to make a layup in place on an awkward surface, make the layup on the workbench and transfer it to the appropriate location. Make sure that no plastic remains anywhere in the layup.*

Peel ply

Strips of Dacron cloth, called "peel ply," have several uses in fiberglass work. Peel ply doesn't bond structurally with the fiberglass. It can be applied, wetted out with resin, and left to cure. Even when the cure is complete, the peel ply (and the resin atop and within it) can be removed with a simple tug. The top layer of fiberglass is left with a uniformly rough surface.

Its major use is to ensure a good bond with a layup to be applied later. One problem with layups is that the brush strokes sometimes remain in the cured part. The location has to be thoroughly sanded first to eliminate the imperfections. And as you'll find, sanding cured resin is a tedious task.

An associated problem is comfort. The brush swirls are sharp, called "spikies" in the trade. They might not draw blood, but handling the cured part gets uncomfortable. Apply peel ply to all external layups and any internal ones that you'll be working around.

Peel ply is also good for holding resin in contact with cloth in awkward positions. For instance, a sharp corner might allow resin to flow downhill, starving the bond area. A piece of peel ply laid on the starved area will keep the resin in place.

Be sure to remove the peel ply after the resin cures. Especially don't try to glass another part down over peel ply. It will not bond but might stick just well enough to fool you. Peel ply can be identified by its close, almost invisible weave pattern.

No-stick application

Peel ply is great for limited areas, but occasionally you don't want the glass to stick to a surface. Examples include removable fairings or covers. A mold is made from foam or other materials, and fiberglass is applied. When the resin cures, you want to be able to pull the completed part away from the mold with little bother.

The secret is to give the surface of the mold as slick a surface as possible. If the mold could be made from aluminum, you'd be all set. The resin won't stick to it. However, if the mold could be made from aluminum, so could the part.

Because the usual mold material is foam, the surface must be made impervious to the resin. The first step is a heavy coat of slurry. You aren't interested in saving weight because the mold stays on the ground.

The slurry's surface isn't especially slick. Sand it down as well as possible, and then add a couple of coats of varnish. Finally, give two coats of ordinary paste wax and shine it up. The paste wax *cannot* contain silicon.

For you belt-and-suspenders types, a quart of mold release sells for under $10. This is the same stuff the manufacturer uses on the kit parts. Just brush or spray it onto the mold after the last coat of wax. After the part has been peeled away, use acetone to remove any release residue.

Other options abound; remember, all you need is a smooth surface. For instance, don't bother with the sealing; simply wrap the mold with wax paper or plastic. Glue it in place, or apply cellophane tape to the exterior. Wax paper is dandy for removable parts such as fairings. Cover the area with it, and form the part. When the resin has cured, remove the part and peel off the wax paper.

Hard points

Composites are great, but they have one weakness—poor crush resistance. You can run a bolt through a metal or wood part and tighten it down practically to your heart's desire.

Try that with your typical foam-and-fiberglass sandwich, though, and the bolt head will crush the foam between the fiberglass sheets.

So your plans will identify places to install *hard points*. This consists of imbedding a harder material within the composite sandwich. It could be aluminum, steel, or some high-tech chemists' nightmare. Often, though, it's a piece of one of the oldest composites out there: wood.

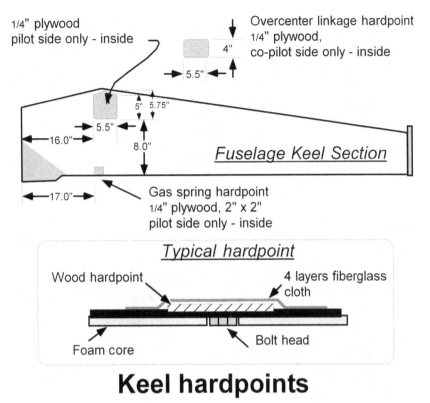

Keel hardpoints

Fig. 8-27. *Because foam-and-fiberglass structures can't withstand localized crushing loads, hard points are bonded in at bolt locations. The plans should show where the hardpoints should be installed.* Velocity, Inc.

The typical process, as shown in Fig. 8-27, is to first identify the size of the necessary hardpoint and to mark its location on the skin. Next, cut away the fiberglass on one side of the composite sandwich, typically on the inside. Then dig out the foam in the area, either by hand or using a power tool such as a router. Then epoxy the piece of wood in place, and make several layups over the modified area to finish the job.

Wing skin joining

Probably the most critical part of building a composite aircraft is the bonding of the upper wing skins to the spar and ribs. The instruction manual will cover this subject in great detail. It's important to follow the steps to the letter—let's see why.

Typical construction of a molded-composite wing is shown in Fig. 8-28. The wing spars are bonded to the lower skin, ribs are bonded to the lower skin and the spars, then the upper wing skin is bonded to the ribs and spars.

As we have discussed, it's important to get enough of the bonding agent into any joint. Typically, a flox mixture is applied to the top of the ribs and spar, which then spreads out when the top skin comes down. Figure 8-29 is a cross section of a

Fig. 8-28. *A typical molded-composite wing internal layout.*

wing coming together. Note that one rib is slightly shorter—its flox barely comes in contact with the upper skin, which leaves the joint weak.

You'll be doing *a lot* of test-fits of the upper wing skins. Some manufacturers have you temporarily piano-hinge the upper and lower skins together at the leading edge to facilitate proper development of the rib-spar-skin interfaces. Follow the instructions, and talk to the factory if you have any questions.

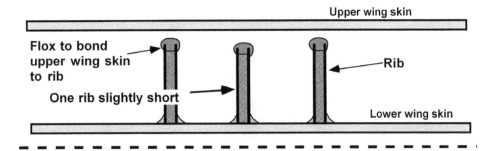

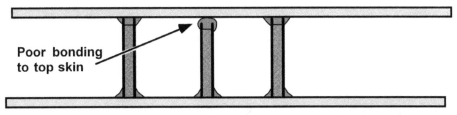

Fig. 8-29. *Build up all ribs to the same height before permanently bonding the wing skin in place.*

While it's a critical operation, it's not an impossible one. Thousands of other builders have managed it. So will you.

FINISHING FIBERGLASS

Just about every kit comes with several fiberglass parts. Usually, they aren't ready to paint, and some surface finishing is necessary. An exception is gel-coated parts. Gel-coating is white and shiny and needs only a light sanding before painting (Fig. 8-30). Some parts might require trimming before installation, as shown in Fig. 8-31.

Most kit parts are made in female molds that produce a reasonably smooth surface. The glass and resin produce the mold's smooth texture. Not so with the shop-made layups—the resin hardens in whatever shapes, patterns, and swirls the builder's brush applied. Peel ply is intended to produce the optimally rough surface for applying additional layers, not a smooth exterior finish. And even with a perfect application of resin mixture, the weave of the cloth will still be showing.

The following section addresses how to get the hand layup to a reasonable state of smoothness and then how to get a glass-smooth finish on either the layups or provided parts.

Smoothing the shop layup

If a fiberglass layup is located inside the structure, smoothness doesn't make any difference. Externally is another matter. The excess resin can be sanded down to the surface of the cloth without loss of strength. But the resin generally doesn't fill in level with the weave; the weave pattern remains, and painting makes it even more noticeable.

Fig. 8-30. *The fiberglass fuselage shells of these Europas come gel-coated, which requires only minor sanding should the builder decide to paint it.*

The solution is to sand and then fill with a dry micro mixture. This is a pretty horrible job over large areas, like those homebuilts that the builder totally covers with fiberglass. A KR-2 has about 200 square feet that must be sanded, filled, sanded again, and so on.

Fortunately, modern kitplanes reduce the area that must be filled. An example would be the horizontal stabilizer joint to the fuselage (Fig. 8-32). Fiberglass strips are used on the exterior to strengthen the bond. The weave in the strips must be filled, and the edges must be faired to make the strip invisible once the area is painted.

Begin by sanding the area with coarse sandpaper, 30 to 80 grit. This knocks down the roughness of the cured resin. *Do not* sand into the fiberglass. If you do, the typical repair method is to make another layup, overlapping the damaged area by 2 inches or so. Check with the kit manufacturer for exact repair methods.

The weave will be filled by covering the surface with dry micro and sanding it flush to the weave. Low areas or spots that must be built up should be prepped first

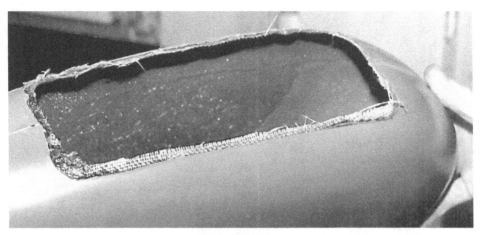

Fig. 8-31. *Note the rough, raw edges of fiberglass left on this wheelpant. These can be filed or sanded away or even cut with a saber saw if the part can be supported properly.*

Fig. 8-32. *Here's where the builder spends his time when building a molded-composite aircraft. The weave of the fiberglass cloth applied to this horizontal stabilizer must be filled prior to painting.*

by adding a thin coat of resin mixture. Then make a batch of dry micro—five parts or more of glass balloons to one part epoxy or catalyzed resin (mix by volume, not weight). Essentially, just enough bonding agent is added to hold the mass of balloons together. By limiting the amount of resin, dry micro doesn't add much weight.

Use a putty knife to apply. Fill the low spots, and then smear on a coat of dry micro. It's not especially easy to work with because it tends to crumble and ball up. Don't be tempted to add more resin—that will just make it harder to sand. When the micro has cured, don your respirator, and sand the surface smooth with 150-grit paper. The process is shown in Fig. 8-33.

You won't get it smooth in one application. It probably will take at least two coats.

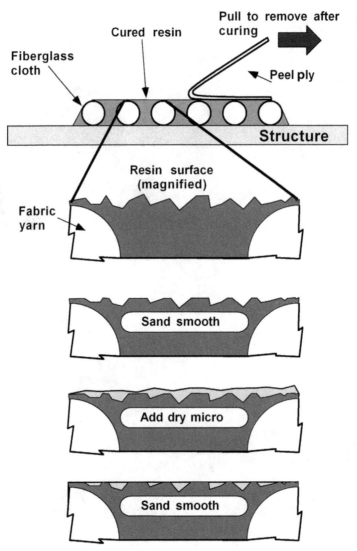

Fig. 8-33. *Filling the weave using dry micro. Following this process, the part must be sprayed with a filling primer to fill pinholes too small for the micro.*

Most of the micro will end up on the floor and on you. Wear old clothes and a shop coat. Have a hand-held vacuum convenient to slurp up the dust and to clean yourself off before going into the house. A floor mat—damp if you can arrange it—reduces tracking of the dust into the house. Cover your supply of fiberglass cloth before starting.

As you might gather, it makes sense to wait on your dry-micro activities until they all can be done at once.

Fine filling

The surfaces of smoothed shop layups and supplied fiberglass parts still are not perfect. Pinholes can be left by both the mold and the dry micro/sand process.

The traditional way of eliminating pinholes is with the application of Feather Fill or a similar product. It must be mixed carefully according to the directions and applied with a brush or spray gun. Let it cure for two hours or so, and then sand with 150-grit paper. Repeat as necessary, and again, don't forget your respirator.

Feather Fill is one of several products that can be used to fill the pinholes. Ask for a *filling primer*. These aren't aircraft-specific products; in fact, Feather Fill is sold at auto-parts stores. Check with the kitplane manufacturer or fellow builders for alternatives.

However, whichever product you use must be compatible with the primer and topcoat with which the aircraft will be painted.

On first glance, composite construction is the simplest: "All ya gotta' do is glue the parts together." As you've seen in this chapter, it just isn't true. It's hard, dirty, smelly work. But done right, it does produce the slickest airplanes this side of mach 1. And nothing is *difficult;* it is just a series of logical, easy steps.

Compare it with the alternative. Tube and fabric airplanes fly slowly. Wood airplane kits have far less work already done for you. So it boils down to either slinging a lot of resin or banging 15,000 rivets.

Composite kitplanes have nothing to apologize for. They're fast to build and fast in the air. With all the glass airplane kits out there, they must be doing something right.

CASE STUDY: TACKLING A BIG ONE

It certainly takes guts to tackle a homebuilt airplane project. It takes more guts to choose a huge six-seater instead of the usual two-seat planes for your first project. It takes even more guts to start such a project in a single-car garage. It takes even more to decide to be the first one to install a nonstandard, Russian-designed, Czech-Built engine in your plane.

Imagine the guts it takes to do all four.

Oh, one more thing: The builder has just started pilot training.

Selection and delivery

Neil Bryant had been messing around with boats for years, including time in the Coast Guard and ownership of a series of increasingly larger boats (Fig. 8-34).

"We were between boats when Neil was glancing through some borrowed *KIT-PLANES* magazines," remembers his wife, Marty. "He mentioned to me, 'Well, for the same amount of money, we could buy a kitplane.'"

After a test ride at a local kitplane manufacturer later, they were hooked. Neil had a love of the old flying boats, but none of the available kits met both their mission needs and fit their pocketbook. They wanted a plane with utility, good bush-flying capability, and the ability to share the fun with friends. Living near Seattle and having owned land on an island in saltwater Puget Sound, they knew that an aluminum

Fig. 8-34. *Neal and Marty Bryant took on a massive radial-powered six seater for their first home-built project. They're holding one of the propeller blades for their license-built M14P engine.*

plane wasn't the right approach. Their boat experience had taught them that more room eventually would be necessary, so they knew that they'd probably need a pretty big plane. This plane was to last them through retirement, not just until the start of the next project.

Scanning the for-sale ads revealed an Aerocomp Comp Air 6 project for sale not too far away. The owner had taken delivery of a kit (including floats quick-build wings), but a car accident made him unable to start work. The price was right, but as a nonpilot, Bryant couldn't evaluate the flying characteristics. The feedback he got from other builders and owners was overwhelmingly positive, so the Bryants bought the plane and trucked it home.

Unfortunately, their home had just a single-car garage—and the Comp Air more than filled it. The Comp Air is a six-seat fiberglass airplane, and they had to lengthen the garage temporarily to even *fit* the fuselage in the garage. The rest of the parts were stored in nooks and crannies around the house, but it still didn't leave much room for working on the plane. "I had to go outside to change my mind," says Neil. The garage didn't even have enough room to enable them to install both horizontal stabilizers at the same time.

Still, Neil plunged ahead. He and his wife knew that they'd need a larger facility eventually, and they started looking for a hangar to rent. In the process, they'd stopped by a home for sale on a local airpark. The owner had a buyer, but the Bryants casually dropped another offer on the table if the buyer backed out.

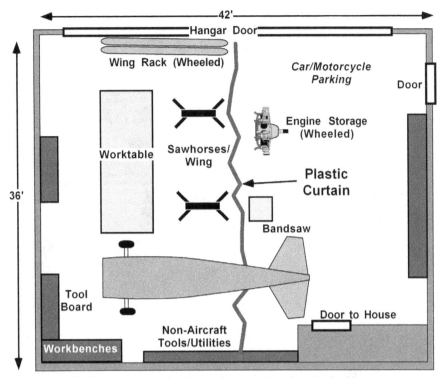

Fig. 8-35. *The Bryant's hangar/workshop is larger than the attached house.*

And sure enough, it happened. A year after starting the project, the Bryants went from a home with a single-car garage to one with an attached 44 × 36-foot hangar. The hangar, in fact, had more floor space than their new house (Fig. 8-35).

Engine selection

The Comp Air is a big airplane, and to fly off floats and do the bush flying the Bryants had in mind, they needed a 300-hp engine. Engines in this power range are very expensive—$60,000 or more. Many people building the bigger Comp Airs opt for a turbine engine, but even those are expensive.

Using the Internet, Neil eventually honed in on the Czech version of the Russian M14P radial engine. It is being installed in dozens of homebuilt aircraft types and has received good reviews. Murphy Aircraft has its successful Moose, a Super Rebel, with the engine, and it has been favorably compared with a DeHavilland Beaver. Since the Aero Comp is about the same size and configuration as a Super Rebel, Neil figured the engine would work.

Besides, the price was right: less than $20,000, including all accessories, the necessary tools, and a constant-speed propeller.

He confesses to a little trepidation when he actually bought the engine—he had to wire the full amount to a dealer in another country—but everything was as advertised.

Fig. 8-36. *Since the smaller conventional-engined Aero Comp is called the Comp Monster, the Bryants felt that theirs should be named "Compzilla."*

Aero Comp originally had named the smaller Comp Air 4 the "Comp Monster." The huge radial engine on the nose of the larger six-seat airplane led the Bryants to dub their project, "Compzilla" (Fig. 8-36).

Building experience

While the Seattle area has mild winters, the temperatures still hover around the mid-40s for several months—too cold for vinylester to cure properly. The single-car garage should have been easy enough to warm up, but the need for ventilation led to a thermal tug-of-war.

"Compzilla's" move into the new shop had its own problems, such as trying to heat that huge hangar without going broke. Bryant bought a variety of heaters, from home-sized electric furnaces to kerosene and electric space heaters, but it was still tough getting the temperatures stabilized. He hung a large sheet of polyethylene plastic across the center of the hangar like a shower curtain. This cut down the heated area into something more reasonable (Fig. 8-37).

Neil Bryant had no aircraft building background, but his previous experience as a boat owner had given him some fiberglass experience. However, his purchase of an uncompleted project led to problems with the vinylester resin. This resin has a shelf life, and by the time he tried to do his first layups, the vinylester had gone bad. He had to grind off some of the work he'd done and start over using new resin.

Since he intends to fly the plane with floats, Bryant has increased the size of the elevators and the rudder. Fortunately, Aero Comp publishes the instructions for this task.

Like all fiberglass-airplane builders, Neil has become an aficionado of sanding tools. His favorite is a RotoZip, originally designed for cutting drywall. However, with the available blades and bits, it can handle a wide variety of tasks. With a

Fig. 8-37. *It's expensive to heat their entire shop, so the Bryants use a large plastic curtain to isolate the current work area for localized heating.*

grinding pad, he finds that it is perfectly suited to taking off the outer gel coat without damaging the fiberglass below.

With all the sanding/grinding going on, Neil generates a lot of dust. He built a portable air cleaner based on a design in a magazine article. It's a wooden box with three small box-type fans arranged to suck the air out. A standard furnace-type air filter is mounted just on the inside of the only opening to the box, with a bit of grating on the box surface itself (Fig. 8-38). Wheels on the bottom allow him to move it close to wherever he happens to be working. When working on smaller parts, he sets them directly atop the intake grating.

One tool he found useful is a laser level. For just $40, it's perfect for leveling the work table prior to starting critical work.

Recently, he's had problems keeping going on the project. Health problems and family holidays tend to break his rhythm, as has a recent job with significant evening and weekend overtime. "Once I get going on regular weekends, I develop the building habit," he says, "But like inertia, it's sometimes tough to start. Maintaining momentum is easy to say but hard to do when working full time."

He and Marty have started flying lessons, but with several years' building time left, there's no rush. He realizes that he probably won't be ready to make the first flight himself and is looking at his options.

Fig. 8-38. *Neal Bryant's homemade shop air filter uses a furnace filter within the box. The grating allows parts to be worked on directly atop the intake.*

Advice

"'Modification' is now in the four-letter word category," declares Neil Bryant. The changes he's instituted, especially the engine change, have greatly increased his work load. "The Aero Comp team has been very helpful," he says, and he adds that he takes full advantage of the e-mail network of Aero Comp builders.

Living on an airpark now also gives him a nearby assistance network. Several Experimental Aircraft Association (EAA) members live on the field, and the local chapter meets right across the street. "Whether or not you're new at this, join EAA right away."

"Be honest and realistic about your mission goals. We'll finish the Aero Comp and eventually put floats on it. But the grass seems greener elsewhere sometimes, at least until we get this in the air."

"Stick to your dream."

QUALITY CONTROL

As you work with fiberglass, keep the following quality control points in mind:

- Keep all tools and surfaces at stable, warm temperatures.

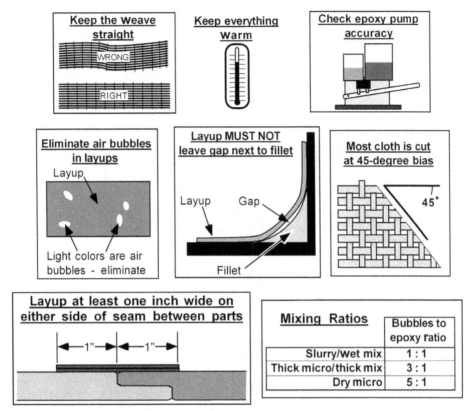

Fig. 8-39. *Crucial considerations when working with fiberglass.*

- Keep dust and dirt from fiberglass cloth and away from the layups.
- When bonding premade composite parts, roughen the bonding surfaces first.
- Spread out the force of clamps; localized pressure can damage parts.
- Regularly check the accuracy of your epoxy pump.
- Keep your fiberglass-cutting tools sharp.
- Most cloth will be cut at a 45-degree bias cut.
- Keep the fiberglass weave straight.
- Layups joining parts must go at least 1 inch past the seam between the parts.
- Fiberglass properly saturated with bonding agent will be a uniform color. Air bubbles produce lighter-colored areas and must be removed.
- Make sure that no gaps appear beneath layups applied over fillets or other concave shapes. Fillets with generous radii are less likely to produce gaps.
- Don't forget to remove the peel ply. This goes double for any layers of plastic.
- Trim the fiberglass during the green cure stage.

Figure 8-39 summarizes the key points.

9

Metal-monocoque construction

METAL-MONOCOQUE comes closest to being the "modern" traditional method of building an airplane. Most factory aircraft are built this way: thin metal skins, bent and curved for strength, held together by rivets. Construction usually isn't true monocoque, but a modified form that retains some features (longerons, for instance) from truss-type construction. A better term perhaps is "stressed-skin construction," meaning that the exterior skin itself adds to the strength of the aircraft.

This chapter covers those aircraft constructed primarily of aluminum with the metal skin absorbing flight loads. It's amazing the amount of strength a puny piece of sheet aluminum gains from clever design and a rivet or two. Skins and bulkheads typically are 0.040 inch thick or less. Certain smaller metal-monocoque homebuilts are built from a couple of 4 × 12-foot sheets of aluminum. Cut out and form bulkheads, ribs, and spars; cut out, bend, and rivet the skins; and the result is an airframe good for more than 6 Gs.

The metal-monocoque RV series (Fig. 9-1) is the most numerous type of kitplane irrespective of construction method. The line became popular for a number of reasons, including good engineering, excellent performance in the low- and high-speed areas, a reasonable kit cost, and an enthusiastic customer base. The company claims that over 4,000 kits have been sold worldwide. In the year 2004, almost 40 percent of the homebuilts added to the FAA register were RVs.

Even with the RV juggernaut, other all-metal designs have carved out a niche. Most emphasize the use of simple pulled rivets ("pop rivets") in their construction, short-field/bushplane performance levels, and/or lower construction costs than the Vans' designs.

Zenith Aircraft Company offers a combination of cruiser and STOL-type designs (Fig. 9-2). Murphy also shoots for the short-field/bush airplane crowd with the Rebel, Super Rebel, and Maverick. Sonex has a range of low-cost kits offering good

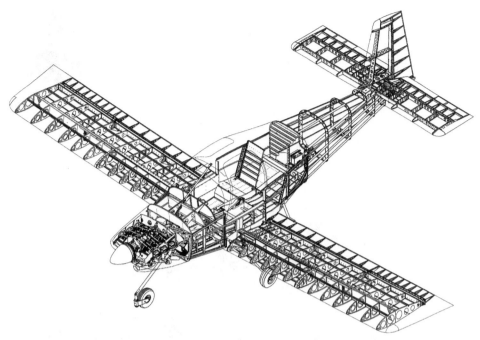

Fig. 9-1. *The Van's Aircraft RV-10 four-seater was one of the most anticipated new homebuilts.*
Van's Aircraft.

performance on small engines. The GlaStar, with its composite/steel-tube fuselage, has a completely conventional all-aluminum wing (Fig. 9-3).

This chapter is mostly concerned with sheet-metal techniques:

- Making long, straight cuts
- Making smooth bends in long sheets
- Riveting
- Corrosion-proofing aluminum

Installation of pulled or pop rivets is covered in Chap. 10.

ADDITIONAL TOOLS

In addition to the basic tools specified in Chap. 6, additional tools are necessary:

- If your plane uses conventional (i.e., driven) rivets, you'll need an air compressor of about 2 hp with a 20-gallon capacity. You are going to be operating tools with the compressor, so the little $100 tankless models won't do. A good compressor will cost between $250 and $500, less if bought used.
- Two useful compressor accessories are a moisture trap and an oiler. The moisture trap separates water from the air, which is necessary for paint spraying. Pneumatic tools, such as rivet guns, need a bit of oil with their air, hence the addition of an oiler.

Fig. 9-2. *The Zodiac XL appeals to those looking for good performance on a budget in an all-metal kitplane.*

- Needless to say, the two items are incompatible. If you install an oiler, remove it before painting, and use a different air hose because the old one is probably soaked with oil. Or just install a moisture trap and manually add a drop of oil to the tools before connecting them to the line.

- Size your pneumatic rivet gun based on the largest-diameter rivet used in the aircraft. A 2X gun is limited to ⅛-inch rivets; a 3X, to ³⁄₁₆-inch; and a 4X, to ¼-inch. A 3X is probably what you'll need; good kit instructions should tell you which to get. Cost is $150 to $250.

While the rivet gun supplies a hammering force to drive the rivet, it isn't the same as a pneumatic or air-impact hammer. They look identical, and the

Fig. 9-3. *While the GlaStar has a composite/steel-tube fuselage, the wing is riveted aluminum.*

hammer costs only $25 or so. But a rivet gun's trigger operates like a variable-speed drill. The air hammer's trigger starts the tool at full force, while the rivet gun can be started gradually. You can rivet with an air hammer, but it's a tricky operation. A typical kit has 5,000 rivets, and a gun is worth a penny per hole.

- You also will need bucking bars of various sizes and shapes. Buy two or three different ones to start. They run around $15 to $40.
- You also will need a rivet cutter ($20 to $50), plus rivet sets, dimpling dies and blocks as applicable, countersinks, fluting pliers, hand seamers, deburring tools, and a number of other small hand tools costing less than $15 each—and lots and lots of clecos. Find out the sizes of rivets you'll be driving, and buy at least 100 clecos per size. Buy an extra cleco pliers as well.

  One larger-cost option is a hand rivet-squeezer (about $100). They're easier to operate than a pneumatic riveter but can't rivet more than a few inches from a sheet edge.

- You'll need drift punches in the same sizes as the rivets. These are really nothing more than solid steel rods with one end slightly pointed.

One could go broke buying every tool useful for sheet-metal work. The preceding should get you started. Others can be picked up as needed.

Several companies offer tool kits for RV-brand aircraft builders. These kits include the rivet gun, rivet sets, bucking bars, and the like. Whether to buy one depends on what tools you already own and what kitplane you're putting together. Zenith and

Murphy builders, for example, use pulled (pop) rivets (explained in Chap. 10) instead of driven rivets and don't need the rivet guns and bucking bars. For as many rivets as will be installed, however, you'd end up buying a pneumatic rivet puller anyway.

Optional tools are suggested in the context of the various operations described.

MATERIALS, FASTENERS, AND SAFETY

Recall that aluminum-monocoque construction uses traditional materials and methods.

Materials

The primary materials you'll be working with are sheets of aluminum, either 2024-T3 or, less often, 6061-T6. The basic rules, such as edge margin and bend radius, were given in Chap. 7.

Few kits supply parts that are ready to be installed. Most have rough edges that first must be deburred. Details are also given in Chap. 7.

Because you are going to use the material for primary aircraft structure, there are a couple of additional precautions. The first is cosmetic. Many pilots like the appearance of a polished metal aircraft. If you are planning on leaving the exterior unpainted, take special care not to scratch or ding the skin during construction.

This seems like a trivial precaution, but it isn't. By definition, building a metal airplane means extensive use of metal cutting tools. A moment's inattention while carrying a drill might ruin a skin. The difficulty of the task is indicated by the relatively low number of metal homebuilts left unpainted. If you aren't going to paint, get used to covering portions of the aircraft that aren't being worked on with old sheets or blankets. Often, kit companies apply a sheet of protective plastic to the aluminum. Leave this in place until removal is necessary.

There are other aspects to handling sheet aluminum. Most references will tell you that a corrosion-resistant primer must be applied to every square inch of aluminum within a closed structure. But look inside a lot of old production planes; I had a 1965 Cessna 150 without internal protection, and it didn't have a speck of corrosion.

The exterior can be left unpainted because it can be inspected and polished easily. But the inside of the wings, tail, and fuselage can corrode unobserved. If you live in a high-corrosion environment, such as near salt water, you'll probably want to apply primer to the inside.

The techniques for applying corrosion protection are given later in this chapter, but there's one point to remember now: The presence of finger oils on the sheet reduces the primer's adherence. You'll wash and etch the metal prior to application, but the more oils on the sheet, the more chance there is that some might get missed.

Consider wearing thin cotton gloves or rubber surgical gloves when you are working on unprepared aluminum. Not only does this keep your hands cleaner, but it cuts down on the irritating little scratches and nicks you can get from rough metal edges. Once the primer coating is prepared, you can operate much more freely—all the more reason to apply the primer early.

Fasteners

The fastener of choice for metal aircraft is the rivet. There have been a few experiments with other means, but the lowly little 0.01-ounce pieces of aluminum are used in everything from Teenie-Twos to 747s.

What else would you use? Small nuts and bolts would be heavy and expensive. Welding aluminum is difficult and is impossible to undo. Similarly, bonding is permanent and uses dangerous chemicals besides. Rivets are cheap, light, nontoxic, and easy to install and remove.

A rivet is like a bolt without threads that is stuck through both pieces of metal to be joined. The end of the shank is then distorted so that the rivet remains in place while pinning the components tightly together.

This distortion can be applied in two ways. Sometimes the rivet is hollow with a shaft inside leading to a bell-shaped end. A special tool pulls the end into the hollow shank, widening the hollow shank. When the maximum pressure has been applied, the shaft breaks at a predetermined location. This is a *pulled,* or *pop, rivet,* and its use is discussed in Chap.10.

The other type of rivet is the *driven rivet,* and it's the one used most commonly on aircraft. Unlike the pulled rivet, driven rivets are solid metal. The end of the shank is distorted by repeatedly ramming it into a hard chunk of steel, called a "bucking bar." The rivet is softer and yields under impact with the bar, flattening its end. Steel rivets used on buildings and ships are installed red hot to ensure that they are softer than their bucking bar.

The distortion of the shank end forms what is called the "shop head." The "manufactured head" is the one the rivet came out of the box with. Two types of manufactured head generally are used: the MS20470 (AN470) with a "universal head" (sort of a flattened oval) and the MS20426 (AN426-100) with the "countersunk head."

For most applications, the standard MS20470 universal-head rivet is used. The MS20426 rivet is the *flush* rivet; when installed properly, no portion sticks into the slipstream to add drag. This requires dimpling or countersinking the skin before installation.

Aluminum rivets come in two alloys, 1100 and 2117-T4. The 1100 alloy is soft and can be used only in nonstructural applications. The 1100 alloy rivets have an *A* suffix on the specification (MS20470A, for example), and the 2117-T4 use an *AD* suffix. Unless you have a definite call for the softer rivets (they are easier to drive and work fine for interior trim), always order MS20470AD or MS20426AD rivets.

What if the labels fall off your rivet boxes? The structural rivets will have a small dimple in the middle of the manufactured head. This isn't just for identification; it's an aid to removing the rivet should installation go wrong.

While AN bolts with a variety of head markings can be interchanged, only rivets with *dimples* on the heads should be used for structural purposes. Other markings mean other alloys or other degrees of heat treatment.

Specific rivet sizes are specified by a dash number indicating diameter and length. For an MS20470AD3-2 rivet, for example, the 3 indicates the diameter in one-thirty-seconds of an inch, and the 2 indicates the length in one-sixteenths of an inch. So

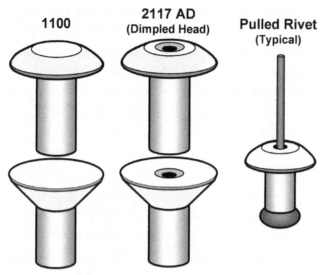

1100

2117 AD
(Dimpled Head)

Pulled Rivet
(Typical)

Fig. 9-4. *Typical homebuilding rivets. 1100 rivets are nonstructural and shouldn't be confused with the 2117 units.*

an MS20470AD4-10 rivet has a standard head, is made of 2117-T4 alloy, and is ⅛ inch in diameter (⁴⁄₃₂ inch) and ⅝ inch (¹⁰⁄₁₆ inch) long.

Common rivet types are illustrated in Fig. 9-4.

Safety

You can always recognize an old-time riveter's convention because they stand around yelling "What?" and "Speak up!" or "Say that again!" In other words, riveting is a noisy business. Wear ear plugs at a minimum; hearing protectors are preferred. Wear the plugs for normal work, and put on the earmuffs as well when you pick up your rivet gun.

Don safety glasses before cutting, drilling, or grinding. Metal work produces dust and a lot of sharp edges, so take whatever other precautions you prefer. I wear a shop coat but no gloves, for example. The aluminum dust goes everywhere, so it's nice to have the coat to shrug off before entering the house.

While aluminum work doesn't involve the witches' brew of chemicals that composite construction does, it has its own problems. Most primers are hazardous; the most common, zinc chromate, is suspected of causing lung cancer. You can't even get true zinc chromate in hardware stores anymore, so buy it though aviation suppliers if you decide to use it. Other less dangerous treatments are available—check with your kit manufacturer to see what alternatives it recommends.

METALWORKING

The basic rules for metalworking were given in Chap. 7. The following sections concern cutting and forming operations that go beyond making simple fittings as described in Chap. 7.

Fig. 9-5. *The Sonex can be built from a complete kit, from subkits, or from plans only. The builder must decide if the time saved is worth the additional money.*

How much you'll have to do depends on how much money you're willing to spend. Complete kits include all major cuttings, formings, and bendings (Fig. 9-5). But you can save a considerable amount of money by doing these operations yourself. It's the age-old trade: bucks versus building time.

Long, straight cuts

One advantage of building a metal-monocoque airplane is that the metal used is thin—0.040 inch (¹⁄₂₅ inch) or less. Aluminum this size is easily cut by a variety of means.

However, what is easy and simple to do on a 6-inch piece gets to be a hassle when the cut is 4 feet long and when the cut must be a good, sharp, straight line.

Of course, with enough money (or the right friends), the solution to long, straight cuts is simple: a *bench shear*. Similar to a bending brake, a bench shear makes large cuts with ease. All it takes is money; these units are expensive. Check with your EAA chapter to find out if someone has one. You probably can do most of your long cutting in a single evening.

With no bench shears, the problem boils down to three factors: finding a believable reference, transferring that reference to the metal, and cutting to that line.

What's straight? Or rather, what do you have around the shop that you can trust as being straight? You might think that the factory-supplied sheet of aluminum could be trusted, but you shouldn't. It can waver ⅛ inch or so over several feet. This might not sound like much, but an uneven edge can really cause a problem if you're trying to butt two edges against each other. But then again, if you can't cut any straighter, you might as well trust the factory edge.

You can buy long straightedges, which are objects with a guaranteed linear edge. Carpenter's squares can be trusted, although they might be too short. With care, you can draw part of the line and slide the square down and extend the line.

But there's one incontrovertible reference: the taut string. When referenced lying flat, it's a fast, easy, dead-accurate tool. Note that it cannot be measured from the side—gravity tugs the string into a bowed shape called a "catenary." Tension can reduce the amount of bow but never eliminate it completely.

So we have our references: taut string, carpenter's squares, straightedges, and the like. Now comes the problem of transferring the reference to the metal.

The archetypal taut string is the carpenter's chalk line, which not only provides a straight reference but also leaves a mark on the surface when "twanged." It might work in some cases, but usually the line of chalk is a good ⅛ inch wide.

Instead, get some "½ A control line" from a local hobby store. This is fine, extremely strong string for controlling small gas-powered U-control models. Stretch it tight across the metal, and then spray machinist's ink or paint over the string and metal. Not a heavy coat; just a dusting of color. The string acts like a stencil. When removed, it leaves a fine line of bare aluminum.

For various forms of straightedges, sharp pencils or a scribe work adequately. Apply machinist's ink first so that the line is easy to see. If you use a scribe, don't make any marks on "good" metal. The Sharpie brand of felt-tip pens is one of several that leave a good line on bare aluminum.

Once the line is marked, you're ready to cut. The use of aviation snips, nibblers, and various power saws is explained in Chap. 7. Larger pieces of sheet metal might be difficult to feed through a band saw, and properly supporting the metal for a saber saw might be tough, too.

Probably the best tool for long cuts is *power shears.* These are essentially electric- or air-powered aviation snips. They cost around $150, yet they might be worth it if you plan on cutting a lot of metal.

Eventually comes the moment of truth. How close do you cut to the line? The farther from the line you cut, the more additional work you will have to trim it closer. If you cut too close, the entire piece may be ruined. It depends on how much you trust yourself. For thinner metal, you might prefer to use a hand tool for better control. It's up to you.

Don't forget to file and deburr the edge. Special deburring tools for sheet edges make the job easier (Fig. 9-6).

Cutting large holes

You'll have occasional need to cut largish holes in aluminum sheet. Typical uses are lightening holes for weight reduction or access panels for inspections and maintenance.

It's best to cut round holes, if possible. Square corners, even if rounded, can concentrate stress and give cracks a place to start. In the second (and better) place, there are a number of tools that greatly simplify cutting round holes.

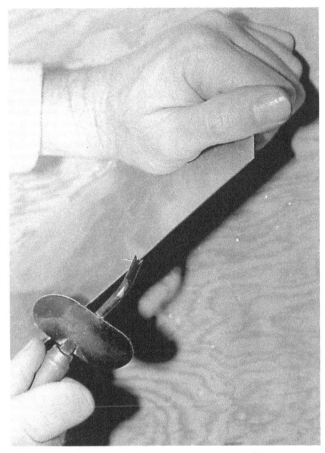

Fig. 9-6. *Edges of thin sheets can be smoothed with a deburring tool with a V-shaped notch.*

The ordinary hole saw is the first option. This is essentially an open cylinder with one edge covered with teeth; sizes are available to 2 inches or so in ⅛-inch increments. Larger sizes are sold for specific tasks, such as cutting instrument holes. Read the labels carefully when you buy hole saws at the hardware store—many are intended only for wood. They'll cut holes in metal but won't last very long.

One drawback is that a saw must be obtained for every size hole to be drilled. For holes up to ¾ inch or so, a *step drill* might be preferred. This is a cone-shaped bit with steps on the exterior for various sizes of holes. There are a number of other custom hole tools on the market that are able to cut or punch holes up to 4 inches or so. A fly cutter (Fig. 9-7) works well on thin aluminum.

When using a hole cutter, add *cutting fluid* to reduce friction and heat and extend the life of the tool. In a pinch, straight oil will work, but cutting (*tapping*) fluid is cheap and widely available. There are some formulations intended specifically for aluminum. Apply it to the tool-metal interface while cutting, and lift the tool away occasionally, and squirt a little on the cut itself.

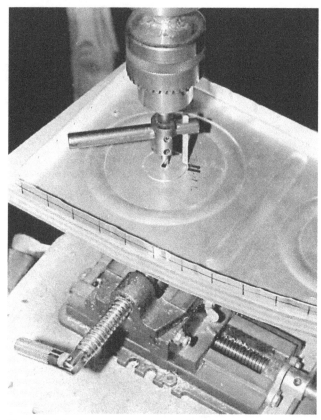

Fig. 9-7. *The fly cutter works best in a drill press. This rib is screwed onto a piece of wood to hold it steady while cutting the lightening hole.*

Access holes aren't always round. Sometimes a rectangular hole gives better access to a particular area. In this case, remember that you don't *want* nice square corners. The hole must have rounded corners, or it'll be a starting point for cracks. An inch or so radius should be about right.

To make a rectangular access hole, mark a square with dimensions equal to the final size minus the desired corner radius. In other words, for a 5 × 3-inch hole with 1-inch radius corners, mark a 4 × 2-inch square. Drill a pilot hole at each corner, and follow up with the corner-radius hole saw.

Lines drawn tangent to the outer edges of each hole will define the correct final size. A saber saw with a metal-cutting blade is a good tool for removing the metal within the rectangle. Take care when starting each cut—the saw can cross the line quite quickly.

Bending and forming

Aluminum sheet has many advantages as an aircraft construction material: It's strong, light, and fairly cheap, and it's also easily worked, up to a point.

Sheets of aluminum and paper share similar workability characteristics. It's easy to put a simple bend in the material, and the bend adds strength as well.

Take a sheet of paper, and draw a clock face on it. Now try to shape it into a bowl. It doesn't work very well, does it? As the "rim" of the bowl rises, the 12 and 6 o'clock positions come closer together. But the distance between the two points *around* the rim (from 12 to 1 to 2 through to 6) stays the same. Something has to give.

A simple bend wouldn't be any problem. The crease would go from 9 to 3. No matter how close the 12 and the 6 were brought to each other, the distance around the rim stays the same.

But we're making a bowl. As the 12 and the 6 are brought closer together, the 1 and the 7 are coming closer, too. So are the 2 and the 8, the 9 and the 3, and so forth.

The net effect is that the diameter of the rim decreases. But where does all the extra paper go? Into ugly folds and wrinkles. You can make a bowl out of paper, but only by cutting pie-shaped wedges from the rim to the center to get rid of the extra material. That's okay for paper but not for a person-carrying aircraft.

"Complex" or "compound" shape is the aircraft builder's term for curves beyond a simple linear bend. The stiffer the material, the harder it is to curve. Fiberglass cloth is limp until soaked in epoxy and allowed to harden. Composite airplane builders don't soak the cloth until it's ready to be placed into the final shape desired. Molten aluminum is limp as well, but it can be cast into all sorts of convoluted forms.

Aluminum sheet doesn't like to form complex shapes. (I didn't like school either, but Mom made me go.) With a little work, you can force aluminum into the shapes needed to build metal aircraft.

We form metal in two ways: stretching and shrinking. Take stretching first. If you take a piece of ¼-inch 6061-T6 plate and bend it 90 degrees around a ½-inch bend radius, the metal on the outside of the bend has to stretch more than ⅜ inch. Where did the extra metal come from?

If you could fit a micrometer in the middle of the bend, you'd find that the metal is now thinner than the original ¼ inch. Stretching is nothing more than converting thickness into length. Thin sheet doesn't have that much thickness to give up, but at the same time, the difference in radius between the inside and the outside of the bend isn't very much. Minimum bend radius tables are based on the amount of stretching the metal can withstand before tearing.

Stretching is simple enough, then. Now shrinking should be obvious as well: Excess metal is absorbed by a slight increase in the thickness of the material. In other words, if a small bulge in 0.032-inch aluminum is pounded flat, the metal in the region of the bulge might increase to 0.033 inch to absorb the extra metal.

Your mind might boggle slightly at the concept of actually forging the metal by hand. But it's nothing more than blacksmiths have been doing for thousands of years. Our high-tech materials are stronger, but don't try to make a horseshoe out of your kitplane's ribs. The hand workability of aluminum has its limits.

Actually, there are simple and effective ways around the shrinking problem. The excess metal can be used up in other ways, such as forming deep wrinkles. These wrinkles add strength as well. Look at a paper cupcake holder. These are circular pieces of paper that can take a bowl shape owing to numerous radial creases.

Fear not—you don't have to become a topologist to make metal shapes. You don't even have to be able to *spell* topologist.

But how does all this apply to the kitbuilder? The manufacturer says that all parts are prebent.

Even if you have a "complete" kit, however, the manufacturer may not completely form all the parts. Rib flanges, for instance, may not have been bent to 90 degrees. The initial crease is there, but the manufacturer expects you to shrink or stretch the material as required to produce the desired angle.

Or maybe you didn't buy the complete kit. It's a lot cheaper to buy the sheet metal and form the parts yourself. Or perhaps you've ruined a kit component and don't want to go through the delay of ordering a replacement.

Let's step through a metal-forming exercise.

Assume that you have to build a fuel tank antisurge baffle, as shown in Fig. 9-8. The overall shape is that of a half-oval with a flange on the bottom edge to allow it to be riveted to the inside skin of the fuel tank. A few holes are drilled through it to keep the fuel from surging during rapid maneuvers.

The flange isn't the only possible attachment method. Short pieces of angle could be riveted to the baffle and then to the tank (Fig. 9-9). But this would require several additional parts and more rivets. In other words, more weight, more complex, and more expensive—all four-letter words to homebuilders.

If the flange were to be added along a straight line, a bending brake would quickly do the job. But the flange is added along a curve. Bending it over means eliminating a bunch of excess metal.

The easiest solution is shown in Fig. 9-10. Just use a pair of hand shears to cut a series of notches along the flange, and then bend the tabs to the right angle. Or, depending on the degree of curvature, a series of slots might be sufficient. Either method is acceptable in some cases. But cutting the metal might weaken the baffle. Use this method if allowed in the plans; otherwise, do it another way.

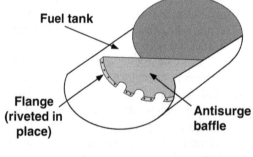

Fuel tank

Flange (riveted in place)

Antisurge baffle

Fig. 9-8.
Fuel tanks require baffles to reduce sloshing of the contents. These are often riveted in place.

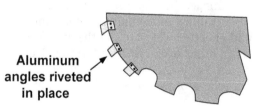

Aluminum angles riveted in place

Fig. 9-9.
Riveting angles along the edge of the baffles requires the least metal-working but adds excessive weight and complexity.

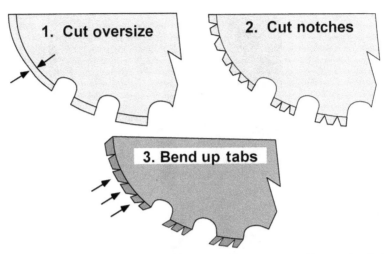

Fig. 9-10. *Cutting slots or V-notches in the flange area produces tabs that can be bent over and riveted.*

The next method is to use up the excess metal by inducing dents, or *flutes*, in the flange. Clamp the baffle in the vice between two pieces of wood, with one of the pieces having the proper curve for the part and the right bend radius for the metal.

Tap the metal lightly with a rubber or plastic hammer to bend the flange portion. Because of the curve of the baffle, you won't be able to work on more than a small section at a time. Do the actual bending in stages; bend one section over 20 to 30 degrees or so, and then you can work on the next section. Bend it the same amount and repeat. When the whole flange is bent to the same angle, bend one section over a bit farther and repeat the process. Eventually, the whole flange will be bent to 90 degrees.

But the flange and the baffle look awful. Both are wavy and distorted because of the excess metal on the flange.

This is where the fluting pliers come in. They look very similar to ordinary pliers except for the jaws. Looking end-on, one jaw has two bumps, and the other has one, which fits between the other two when the jaws are closed. The bumps are smooth and taper toward the tip of the tool (Fig. 9-11). Don't buy fluting pliers anywhere except at an aviation supplier. The nonaviation variety are often intended only for heating ducts, where damage to the metal is acceptable.

When the fluting pliers' handles are squeezed with the jaws pressed on the edge of a metal sheet, it makes a tapered U-shaped dent. This dent isn't made by stretching the metal. Rather, it's formed from available material.

So the fluting pliers are used to straighten the flange. Crimp carefully on the edge of the flange, and you'll see it start to work straight. A series of small flutes rather than a few large ones makes a nicer-shaped part. However, don't put a flute anywhere near where a rivet or a bolt will be installed. If you use a series of small flutes, make sure that they don't interfere with later rivet installation.

Eventually, the baffle will straighten out, and the flange will be completed (Fig. 9-12). The strength isn't affected. It looks a little ugly, but if it's on the inside of the wing, no one will see it (Fig. 9-13).

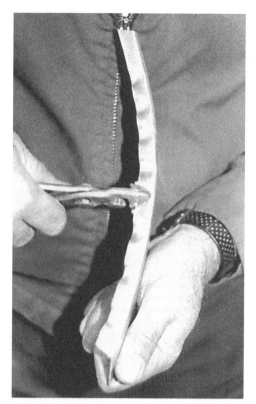

Fig. 9-11. *Note how the flutes in the bent portion curve the entire structure. This tendency also can be used to straighten a curved sheet.*

Another method uses a combination of shrinking and stretching to end up with a smooth flange without flutes. This uses a *backup block* and pointed hammer to shrink and stretch the metal as necessary as the flange is bent. It's a lot of work and not recommended for 2024-T3. It's a rather complex procedure, best learned in person from an experienced metalworker. Check with your local EAA chapter, and get some "dual instruction."

For extremely complex shapes, the "English wheel" has become the tool of choice. If you're building a modern kit, you shouldn't have to worry about one of these. But if you've got a scratchbuilt metal project, an English Wheel might be just the ticket for you. The EAA offers books and videos on the use of this tool.

RIVETING

At last we get to the nexus of metal aircraft building. You might be able to make pretzels out of ½-inch 2024-T3; punch a round 3.125-inch instrument hole with one blow of your mighty fist; or cut a 10-foot sheet of aluminum into $^{25}\!/_{64}$-inch strips with only your steady arm and a penknife.

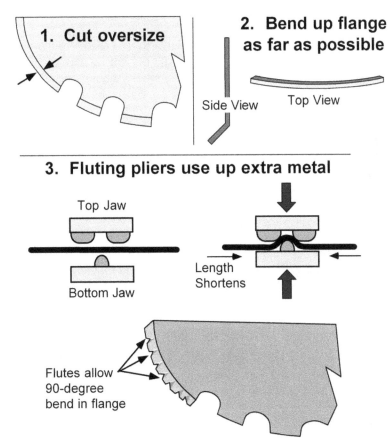

1. Cut oversize

2. Bend up flange as far as possible

Side View   Top View

3. Fluting pliers use up extra metal

Top Jaw

Bottom Jaw

Length
Shortens

Flutes allow
90-degree
bend in flange

Fig. 9-12. *The fluting process.*

Fig. 9-13. *Flutes have been added to this rib's flange to allow the top edge to curve. Rivet holes will be drilled in the straight sections of flange.*

But if you can't drive good rivets, your airplane ain't gonna' hold together. Period.

Riveting isn't that difficult. It requires good judgment and some attention, but with a little practice, it's easy.

One point to recall: Usually, the metal must be protected from corrosion before being riveted. Corrosion protection is covered later in this chapter.

Holding in position

Sometimes you have the luxury of being able to take the pieces to your workbench to rivet them. Most of the time you're trying to rivet an awkward-shaped piece of aluminum to another awkward-shaped piece that's already installed on the aircraft.

Obviously, when you start drilling holes, you want the new component to be sitting exactly in its final location. There are a number of items that help. C-clamps and spring clamps immediately come to mind, and there are several types of spring clamps worked by cleco pliers. Be careful, though—clamps can apply a surprising amount of pressure. Placed directly on metal, they'll leave marks. Never place metal directly against metal. Stick a largish scrap of wood between a clamp's jaws and the aircraft component. If only moderate pressure will be applied, wrap the clamp jaws with a couple layers of duct tape.

Other options include rope, string, tape, rubber bands, and the like. The important issue is to hold the pieces securely.

But only apply enough force to hold the piece in position. Too much pressure can cause subtle warping of the sheet, not enough to be permanent, but it slightly changes the component's relationships. If riveted in this position, the piece will retain an unsightly warp.

Figuring out where the hole goes

The first thing to do is determine the hole position. Many kits have pilot holes predrilled. In these cases, key rivet holes come already drilled, usually in a slightly smaller size. Cleco the part together, and drill out the additional holes. Or your plans may already indicate where the rivet holes should go (Fig. 9-14).

Otherwise, you'd hope that the rivet positions are supplied on the plans, in the form of a template to be taped to the piece. Otherwise, the plans might say, "Rivet the bottom edge to the bulkhead"—leaving the number of rivets and their exact locations up to you.

And in some cases you might not want to drill the hole where the template says. It's usually better to drill both pieces at once to ensure hole alignment.

The first thing to remember is the edge margin: at least three times the diameter of the hole between holes and twice the diameter to any edge. The equation for how many rivets can be installed is

$$N = 1 + \frac{W - 4 \times D}{3 \times D}$$

where W is the width of the riveted area, and D is the diameter of the rivet. Drop any fractional rivets from N; if the answer comes out to be 12.789, drop 0.789 for the resultant 12.

The actual spacing between rivets is given by the equation

$$S = \frac{W - 4 \times D}{N - 1}$$

Now, these equations are a *real* pain, aren't they? Relief comes in several forms. First, as long as you observe the edge-margin requirements, the exact spacing doesn't

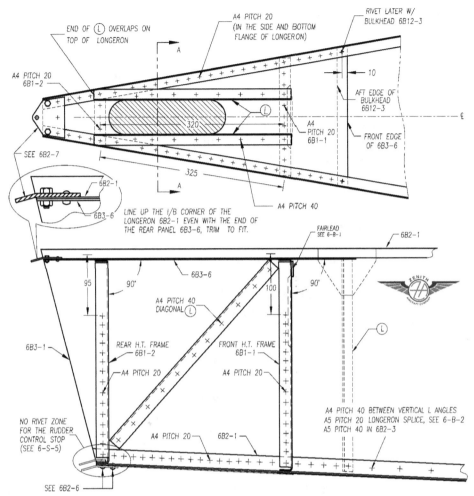

Fig. 9-14. *Rivet spacing is important to properly develop the strength of the part. In some cases, such as these Zenith 701 plans, the rivet locations are already specified.*

really matter. Even spacing looks better, but it's not necessary for strength. Put rivets in a nice row, and slight unevenness won't show.

Second, there's a handy little tool called a "rivet spacer," which is a bunch of metal strips pivoted together. It looks like a complex pantograph. Once you know how many rivets to install (and most plans tell you *that* much), lay the spacer against the metal, and it'll directly indicate the hole locations.

A cheaper way is to go to a sewing shop and buy a strip of white elastic for about $2. Lay out the elastic, and mark ½-inch intervals with a felt-tip pen. Then clamp one end to the start of the metal, and stretch the elastic until the marks are at proper intervals.

The usual first step is to draw a line on which the rivets will be installed. You won't use a scribe, of course, but pencil marks don't show up very well unless you've already applied a primer coat to the aluminum. Bare aluminum can be marked with a fine-line felt-tip pen and cleaned afterwards with MEK. Or you could apply a strip of wide masking tape and mark on that. Clean up the residue from the tape's glue afterwards, though.

Draw the line with a straightedge, and either calculate the spacing or use the rivet spacer to indicate correct locations. Put a small "x" at the location of the holes to be drilled. Position the automatic centerpunch at the center of the "x," and press down.

Sometimes you have to install a sheet atop components whose rivet holes are already drilled. In this case, use a hole duplicator. Select a duplicator based on the size of the hole already drilled. Slide the end with the stud under the sheet being added, drop the stud into the predrilled hole, and then drill the sheet through the bushing. A diagram of this process is contained in Chap. 7.

Drilling the hole

If the hole is too large, the rivet will fit sloppily. It might let the metal shift. Stresses could concentrate and cause premature failure.

The proper procedure is to drill an undersized pilot hole and then enlarge it to the final size. Reams can be used for this enlargement, but they work slower and are expensive. They leave a smooth hole, though.

There must be some small amount of clearance between the hole and the side of the rivet to allow it to pass. This clearance can't exceed 0.004 inch. So for the pilot drill, we'll use a bit of the same size as the rivet. The final bit then will be the next size larger. This works out as follows:

Rivet	Pilot	Final
$3/32$	$3/32$	#40
$1/8$	$1/8$	#30
$5/32$	$5/32$	#21
$3/16$	$3/16$	#11

When drilling the pilot hole, keep the drill bit perpendicular to the metal. Don't use too much pressure; there's no rush. If at all possible, support the metal on the back side with a block of wood. Otherwise, pushing on the drill might place an indentation around the site of the hole.

As each pilot hole is drilled, insert the proper-sized cleco. For the larger sizes, first drill and cleco all holes to $1/8$ inch. Then use the respective pilot bit, and replace the clecos with the larger size.

When the holes are drilled to the final size, remove the clecos and deburr the holes. If it isn't possible to disassemble the parts for deburring, you could use a *chip chaser*. It's a strip of metal with a hook at one end; slide it between the parts, and use the hook to pull away the raised metal. *But chip chasers can severely scratch the aluminum. Disassemble if at all possible.*

Countersinking and dimpling

If the plans call for flush rivets, the hole must be countersunk or dimpled. For counter-sinking, metal is cut away so that the sides of the hole match the cross section of the

Fig. 9-15. *A countersink in action. Note the tapering hole.*

rivet. Quite logically, this is done with a tool called a "countersink" (Fig. 9-15). It's like a wide-angled cone (actually, it's 100 degrees, to match the rivet) with flukes to do the cutting.

The biggest restriction on drill countersinking is the minimum allowable thickness of metal. It must be at least as thick as the rivet head is deep. This makes sense; otherwise, the bearing surface (the portion of the metal actually in contact with the rivet) is drastically reduced. The approximate minimum thickness values are

Rivet	Metal
$3/_{32}$	.036 inch
$1/_8$	.042 inch
$5/_{32}$	.070 inch

Note that only the metal to be countersunk must have this thickness. Thinner sheets must be on the same side as the shop head.

If the metal is too thin, you should dimple the metal instead of countersinking. Be advised, though, that one popular metal kitplane directs that the builder countersink a $3/_{32}$-inch rivet hole in 0.032-inch aluminum. It's not that big of a deal, but there is no margin for error. You'd really rather dimple instead. Talk to local builders.

After final drilling of the hole, use the countersink to remove enough metal to let a rivet sit flush in the hole. It's very important to apply the tool perpendicular to the metal—any angle will produce an oval shape that won't match the rivet head.

The major problem is cutting too deep. A little too far isn't critical; the rivet head is recessed instead of flush. But eventually the tip of the countersink will begin to enlarge the hole's diameter. The rivet will be loose. If this happens, ream out the hole for the next size larger rivet.

Countersink with a drill press whenever possible, and use an automatic-stop countersink. These units can be adjusted for maximum depth.

However, they are thrown off by curved surfaces. And most of the time a drill press isn't practical. Instead, use a regular countersink in a hand-held power drill at slow speed. Switch back and forth with the drill and a rivet until the proper depth is reached. With practice, you'll be able to eyeball the hole and won't have to check as often. When the hole approaches the proper size, finish it by hand using a countersink mounted in the end of a wooden dowel.

Dimpling is faster and easier than countersinking. Rather than carving metal away to match the rivet head, dimpling forms a calibrated dent in both sheets of metal. Equipment consists of a punch and a die (Fig. 9-16). The punch includes a projection of metal shaped like the flush rivet, and the die has a corresponding indentation. (The angles are actually a bit sharper, to allow for springback.)

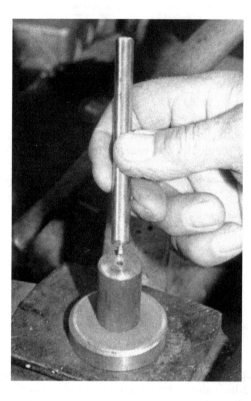

Fig. 9-16. *The dimpling punch and die. Note how the end of the die is shaped like a flush rivet. The die has an indentation to match.*

Don't final drill holes to be dimpled. Dimpling stretches the metal and actually increases the size of the hole. If you final drill before dimpling, the rivet will be loose.

Instead, pilot drill the hole to the size specified, and then place the die on the backside of the hole. The die must be solidly supported. Insert the punch and use the rivet gun to dimple the hole. Hand squeezers and pop riveters also can dimple with the proper dies. Don't use too much force because it tends to ripple the skins. Follow the instructions for the particular system you buy.

Then final drill the hole to the final diameter.

Dimpled rivet holes are stronger but don't look as good as countersunk holes. The dimpling effect isn't confined to the hole itself and tends to add a wide, shallow depression to the surrounding metal. Use whichever process the kitplane manufacturer recommends, or try each process on a piece of scrap and pick the one you prefer.

In some cases, a combination of methods is used. For instance, when riveting skins to heavier structure, the structure will have to be countersunk, while the skin must be dimpled. A summary of both operations is shown in Fig. 9-17.

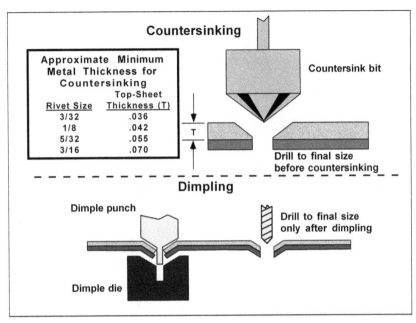

Approximate Minimum Metal Thickness for Countersinking	
Rivet Size	Top-Sheet Thickness (T)
3/32	.036
1/8	.042
5/32	.055
3/16	.070

Fig. 9-17. *Countersinking/dimpling summary.*

Making the shop head

Now comes the moment of truth. The shop head can be formed by either a pneumatic rivet gun or a rivet squeezer (Fig. 9-18). Squeezers make the job easy, but they can be used only near the edge of a sheet and are subject to other limitations. The "reach" or "throat" of these squeezers typically limits their utility. The bigger the reach, the more expensive, of course.

Actually, there's nothing magical about driving rivets. All you need is a continuous or momentary (shock) force and a hard surface to form the shop head. A carpenter's hammer can supply the force. But for building airplanes, pneumatic rivet guns are the tool of choice.

The rivet gun is like an electric drill; it supplies the motion, but interchangeable units supply the action required. Drills need bits; rivet guns need *rivet sets.*

The set has two functions: transmit the hammering action from the gun to the rivet and prevent damage to the surrounding skin. Standard and flush rivet sets operate in different ways. The sets for standard rivets are slim rods with an indented cup in the end that fits the manufactured head of the rivet. When used properly, this cup doesn't touch the metal skin. Because head dimensions change with rivet diameter, a separate set is required for each rivet diameter.

Flush rivet sets approach the problem differently. The end of the set is large, at least an inch in diameter. It's flat with rounded edges and has a polished, smooth surface. The flat set doesn't gouge the metal if slightly off-center. Some flush rivet sets cover the edge with rubber to reduce the possibility of skin damage. A variety of sets is shown in Fig. 9-19.

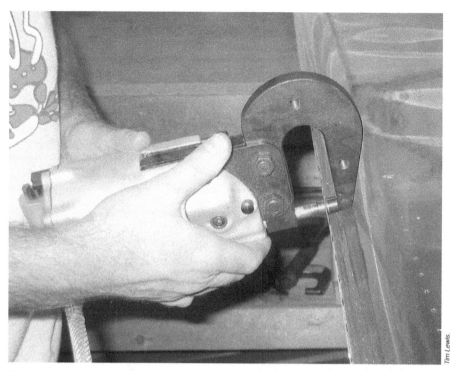

Fig. 9-18. *A hand-held pneumatic rivet squeezer in action. Manual fixed units are also available.*

The last necessary item is the bucking bar. All it really has to be is a piece of steel of the appropriate hardness and weight. If the bar will be hand-held, it should weigh a minimum of 25 to 30 times the rivet diameter. Hence ³/₃₂-inch rivets need a bucking bar of at least 2.5 pounds; ⅛-inch rivets call for 3.5 pounds; and so on.

If the bucking bar is too light, the edge of the sheet might distort. Bending its edge down before riveting will stop the warping, but it's not easy to do on long sheets. Select a heavy bucking bar instead.

All the bucking bar needs is a flat surface that can be held perpendicular to the rivet. One thing you'll notice about commercial bucking bars is their weird shapes (Fig. 9-20).

The weird shapes accommodate rivets installed in places with poor accessibility. The flange of a rib, for instance, might be only a half-inch wide. The rib probably curves, so only the corner of a wide bar could be used. The rest of the bar just gets in the way.

Instead, a manufacturer might design a tapering bar to make it more suited to those tight corners. The other end is usually designed for another kind of tight spot, one, for instance, where the bar can only approach from an oblique angle. Metal airplanes have a lot of nooks and crannies that the builder must snake a bucking bar into; hence metal-airplane builders tend to collect a variety of bucking bars.

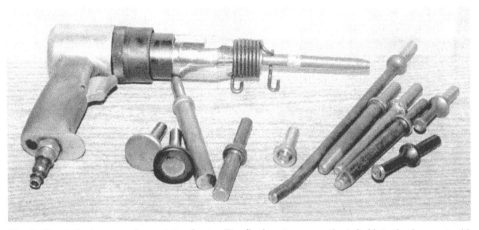

Fig. 9-19. *A rivet gun and a variety of sets. The flush sets are on the left. Note the long set with the angle on the end; it's designed to reach difficult places.*

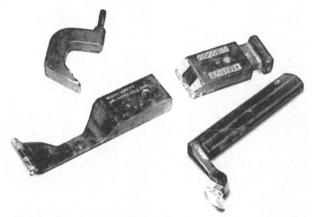

Fig. 9-20. *Bucking bars. Again, the odd shapes let the builder ease them into tight spots.*

Which leads to another point: Plans specify the order in which the rivets should be driven. Follow it exactly. Otherwise, you might wind up with a rivet in a corner that you can't reach. It's not the end of the world—the rivets could be removed, and you could start over, or the right kind of pop rivet could be substituted.

Similarly, lightening holes are often not *just* lightening holes. Sometimes they provide sole access to hard-to-reach spots. Don't arbitrarily decide not to make a lightening hole when the plans call for it. If it's listed as optional, fine. Otherwise, you might be hurting later.

Now you also might see why certain ribs' flanges face left and others face right. The flanges usually face away from the first rib to be riveted; therefore, make sure that you use the correct rib on the appropriate side. If all flanges face in one direction, the two wings are probably built with different procedures. The plans should make the flange orientation clear.

The plans also should specify the rivets or at least the general type (regular or countersunk) and diameter. If the plans just say, "Use ³⁄₃₂-inch AN470 rivets," it's up to you to determine length.

Rivet length should be at least one-and-a-half times the diameter of the rivet *plus* the thickness of the metal to be joined. So, if you're joining a 0.025-inch skin to a 0.040-inch rib with a ³⁄₃₂-inch rivet, the rivet would have to be about 0.025 + 0.040 + 1.5 × ³⁄₃₂ = 0.205 inch long, or a bit less than ¼ inch long. Because the dash numbers for rivet diameters are in ³⁄₃₂-inch increments and rivet lengths are in ¹⁄₁₆-inch increments, you'd need AN470AD3-4 (MS20470AD3-4) rivets.

If you don't have the correct size rivets, a rivet cutter can shorten longer ones. Some builders buy only a few different sizes and shorten them as required rather than maintain a stock of all lengths.

If the plans call for it, or if you desire, dip the rivet into primer before placing it in the hole. However, the paint acts as lubricant. The set and bucking bar tend to slide off the rivet.

With all this preparation, the actual driving of the rivet is anticlimatic. Remove the cleco, and insert the rivet. If you are using flush rivets, hold them in place with a strip of masking tape directly over the heads. The two pieces of metal to be joined must lie solidly against each other. If one sheet flexes away when the cleco is removed, use clamps or other means to rejoin them. Otherwise, the rivet material will flow between the pieces, resulting in a weak and unsightly join.

Insert the correct set into the rivet gun, and then check the gun's power-level setting. If it is too high, there's a danger of punching the rivet right through the skin if the bucking bar slips. If it is too low, the rivet will *work-harden* (become more brittle) and crack.

When adjusted properly, the gun should completely drive a rivet in two or three seconds. Practice with various rivet diameters, and mark the gun's dial at the correct settings.

Place the set on the manufactured head perpendicular to the metal's surface. For standard rivets, the set should make contact with the top of the head only. It shouldn't touch near the edges.

The riveter must apply the force directly along the length of the rivet. Any angle will tend to push the rivet sideways. An excessive angle might cause the set to contact the metal skin and place a little half-moon-shaped smile alongside the rivet.

Place the bucking bar perpendicular to the other end of the rivet. Hold it by hand if necessary, but it's easier if it can be clamped to a solid surface (Fig. 9-21). Squeeze the rivet gun's trigger.

The gun works by impact. You don't have to hold the gun and bucking bars with a death grip. Only push hard enough to keep the tools in place. If hand-held, the bucking bar vibrates in time with the gun; let it. Inertia does the work; by the time the bar is moving away, it's already done its job. Your hand pressure just moves it back into position in time for the next blow.

Some builders can buck their own flush-headed rivets, which is difficult because the set tends to slide unless held in place with the other hand. Rubber-edged sets

Fig. 9-21. *Basic riveting procedure. The bucking bar is clamped to the tabletop, although bars often must be hand held by the riveter or a helper. Note how the set and gun are directly in line with the rivet.*

can reduce the problem. Or if the riveting can be done at the workbench, clamp the bucking bar down and use both hands on skin side. Otherwise, count on needing some help.

Sometimes you can't get the gun to the manufactured head. In this case, you'll *back* rivet, which is, logically enough, riveting done backwards. You'll put a flush rivet set in your gun. If the manufactured head is the universal type (i.e., sticks up), you'll need a bucking bar with a corresponding dent; otherwise, an ordinary bucking bar will do. Then rivet normally.

Forward or backward, the rivet should be set in two to three seconds. The change in sound is a good clue. Release the trigger, and *then* remove the bucking bar. If you take the bar away before the gun stops, it'll bang the heck out of the skin and may even drive the whole rivet right through it.

It's in, but is it right?

Checking and removal

The shop head should be at least one-half the rivet diameter high and one and one-half times the diameter wide, as shown on Fig. 9-22. You can buy or make rivet gauges to quickly make this determination. If the shop head is still high but not wide enough, place the gun and bar back into position, and hit it a few more times.

The shop head should be uniformly flat. Neither the shop head nor the manufactured head should have any cracks or untoward deformation. Both heads should be aligned vertically with each other; if one is displaced sideways, either the set or the bucking bar was held at an angle.

If you have a lot of problems with cracking, check your technique. As mentioned earlier, if the rivet is hit too much, it work-hardens. However, if your technique

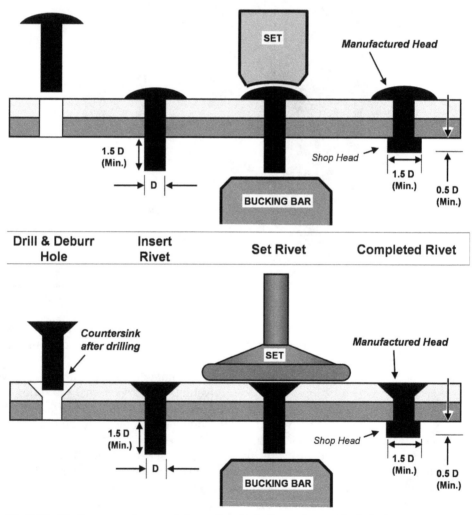

Fig. 9-22. *Riveting procedures and dimensions for both regular and flush rivets.*

seems right and you still have problems with rivets cracking, it may not be your fault. Rivets also age-harden. As they get older, they get harder and have a tendency to split. Kits that supply all the rivets assume that you'll be done with them in a few years. If you have a lot of problem with rivets cracking or distorting, buy a set of new rivets and see if they drive any easier.

Check flush rivets by placing a straightedge along the metal across the manufactured head. If it doesn't lie flat, the hole probably needs additional countersinking.

The joined metal shouldn't show any distortion. If the metal is bulged toward the shop head, the bar was too light. If it puckers around the rivet, you probably drove the rivet too far.

If a rivet is bad, it will have to be drilled out and replaced. The process is shown in Fig. 9-23. Structural rivets have a dimple on the head for guiding the drill. Use a bit slightly smaller than the rivet itself. Place it in the dimple, and drill until the head breaks free. Only friction is holding the rest.

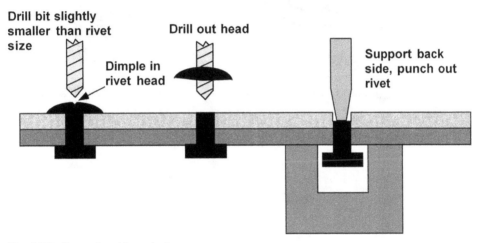

Fig. 9-23. *Removing driven rivets.*

With a good-fitting drift punch, the remainder will be knocked out. However, the metal around the hole must be supported, or residual friction might bend the skin before the rivet pops clear. Take a piece of wood with a hole drilled in it slightly larger than the shop head of the rivet. Place the wood flush to the back of the metal with the shop head inside the hole. Then knock the rivet through with a hammer and punch.

Note that this procedure takes at least three hands unless you can place the offending structure on a bench or other solid surface.

Touch up the hole a little bit with a file or deburring tool, and you're ready to drive another rivet.

Sometimes the rivet doesn't pop away cleanly, and the hole becomes distorted. It's allowable procedure to replace the rivet with the next higher size. Drill the pilot hole, ream, and drive the bigger rivet. Typical errors are shown in Fig. 9-24.

If you're building a metal airplane, you're going to drive thousands of rivets. The preceding information makes it sound like you're going to spend five minutes

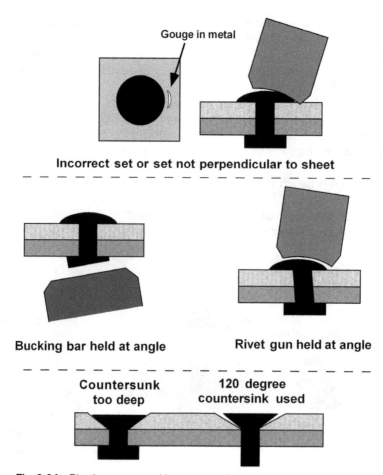

Gouge in metal

Incorrect set or set not perpendicular to sheet

Bucking bar held at angle   **Rivet gun held at angle**

Countersunk too deep   **120 degree countersink used**

Fig. 9-24. *Riveting errors and how to spot them.*

per rivet. You won't. Like many other tasks involved with building a kitplane, the preparations take more time than the actual operation. Ninety percent of your time will be spent preparing to rivet: priming, drilling, and countersinking.

In fact, most of that 90 percent will be spent making sure that the airplane won't corrode away. Let's look at the protection process.

PRIMING

Everybody loves to see pictures of unpainted kitplanes turned into grand champions by a few gallons of Imron. Painting is glamorous. It's the crowning moment of the kitbuilder's project—the airplane is painted just before it starts flying.

But if you're building an aluminum airplane, you'd be better off spending more attention on priming than painting. It's a dirty job that gets little publicity. While finish painting is often done in a paint booth (either permanent or temporary), priming takes place in the dirty, gritty environment of the workshop.

The need to prime surprises some people. Aluminum doesn't crumble away into reddish powder. Why prime?

Aluminum doesn't rust—or does it?

We tend to think that the major advantage of aluminum over steel is that aluminum doesn't rust. Well, that's true: When exposed to moisture and air, aluminum doesn't form ferric oxide. But aluminum does *corrode,* with the same visual and structural impacts. Fancy trim colors be damned—a properly prepared surface and a well-applied prime coat are going to make your airplane last.

Proof is available no farther than your local flight line. In the late 1970s, one major lightplane manufacturer changed its corrosion-proofing procedure, going to a simpler, cheaper, faster method. The procedure the company used was not approved by the paint manufacturer.

A few months after delivery, some of these airplanes' paint started cracking. Owners soon discovered that the problem wasn't in the paint—the aluminum under the paint was being eaten alive by *filliform corrosion.* "Filliform" means "wormlike"; the corrosion forms lines like worm tracks under the skin of paint.

The owners stripped the paint off the airplanes, primed them properly, and re-painted them. The problem returned. The corrosion was too deeply set; aftermarket fixes couldn't reach everywhere.

Within a few short years, a significant number of these aircraft were unairworthy—all from cutting corners, saving a few bucks on priming.

Your kitplane probably uses the same 2024 alloy. Without proper care, it can happen to you, too.

How much is enough?

The degree of corrosion protection varies from builder to builder. Most apply a primer to the outside of the aircraft prior to painting. They have to because paint won't adhere properly without it. But internal corrosion protection is another matter.

One local designer has built about six metal airplanes of his own design. He didn't apply internal priming on any of them. He had no problems while he owned the airplanes. However, one of his planes ended up in a local air museum. When they opened it up to restore it—you guessed it: corrosion.

Alclad is designed to resist corrosion. In this case, the restoration crew tackled the airplane nearly 20 years after the airplane first flew; the plane had been sitting outside and neglected for a number of years.

Most kit manufacturers demand a full corrosion-protection regimen. Kit builders follow these recommendations to various degrees. Some assemble the structure and then spray it with primer just prior to skinning. Others prime each and every piece before installation. Some even use separate kinds of primer on different parts (Fig. 9-25).

If the aircraft is made from 6061 alloy (which is highly corrosion-resistant), you might easily get by without priming. But otherwise it depends on a broad set of factors. Climate is the major player. An Arizona kitplane can get away with procedures

Fig. 9-25. *This builder used different primers on the bulkheads and skins.*

that would make a Florida kitplane crumble to dust within a few years. Builder care is another factor. The fewer scratches made during construction, the more Alclad the builder leaves in place, the fewer places major corrosion can start.

Most of us won't be that careful. And today's mobile society might end up moving you and your kitplane into warm, salty, corrosion-rife tropical breezes. Priming all the pieces before assembly has a few other advantages. The primer acts as a shield against scratches and minor damage. Most primers can be written on using a pencil, which makes marking rivet patterns far easier. Epoxy primers are impervious to most shop accidents.

There are a few caveats where the prime/no-prime decision is made for you. First, steel rusts quickly and always requires priming. Steel priming is covered in Chap. 10 under steel-tube fuselage preparation.

Second, if two different metals are to be joined, both must be primed. Dissimilar metals joined and exposed to salt air, exhaust, or even dew can build up an electrical potential. As mentioned in Chap. 7, the resulting current flow accelerates corrosion. Aluminum airplanes incorporate a lot of steel fittings, from landing gear parts to control bellcranks. Ensure that each piece is well primed.

Third, do not prime any surface to which something will be bonded or the inside of fuel tanks. Neither the bonding agents (epoxies, and so on) nor the substance used to seal the fuel tanks will bond to the primer. This is a definite danger in the gas tanks because sealant may come loose and clog the fuel lines.

Otherwise, you can decide for yourself, based on the kitplane manufacturer's recommendation and discussions with other builders. The following pages describe how to fully protect your airplane. It's your decision how much to implement.

The basics

Every item made from 2024 or 7075 alloy should be *cleaned, etched, alodined,* and *primed* before assembly. Steel parts need similar preparation, but they don't have to be etched. Aluminum alloy 6061 is more corrosion-resistant than other alloys; follow the kit manufacturer's recommendation. And don't take it into your head to substitute 6061 for 2024; 6061 is only two-thirds as strong.

Cleaning means the removal of surface grime and corrosion. The corrosion is the toughest. Don't attack it with steel wool, emery cloth, or a wire brush. It breaks the corrosion free but imbeds little pieces of steel into the metal. Try a mildly abrasive household cleanser and Scotchbrite pads. When finished, wipe it down with MEK or acetone to get rid of fingerprints. Don't handle the metal with bare hands—the whole purpose is to clean off fingerprints and similar impurities.

Paint doesn't like to stick to smooth, shiny surfaces. So we etch the surface using a weak acid solution. Apply the etchant, let it sit for a while, and then rinse it off. The metal takes on a rather dull, hazy appearance.

There are several quite satisfactory self-etching primers on the market that require only minor cleaning prior to application. Check with your fellow homebuilders and the kit manufacturer for recommendations.

The next step it to alodine the aluminum. Alodine is a chemical conversion coat; it chemically changes the outer layer of aluminum to make it more corrosion-resistant. You can actually see the result of this process because the end result is an attractive gold tint to the aluminum. Alodine is basically applied like the etchant—apply it, let it sit for a bit, and then clean it off.

Don't confuse alodining and anodizing. Anodizing requires immersing the parts in an electrolytic bath. It makes the aluminum practically impervious to corrosion and makes a wonderful base for painting, but isn't something you do in the average garage. Some of your parts might come anodized, though—the surface is a smooth, uniform black.

Once the metal is alodined, the primer can be applied. The primer is then sprayed on. Zinc chromate is the traditional primer, but several alternatives are available, such as zinc oxide and epoxy-based primers.

The preceding is just a general set of guidelines; whichever primer you chose, follow the manufacturer's instructions. Read the instructions and cautions carefully before you decide. For instance, one primer maker declares that its product doesn't require etching before application. But if you read carefully, you see the maker requires *another* process before priming, and *that* product requires etching.

The instructions are all easily followed, in theory. But the kitbuilder's workshop is a pretty nontheoretical place. Let's see how to apply primers outside the laboratory.

The real world

Cleaning and etching the metal are pretty easy to handle. Remember, one of the purposes is to eliminate finger oils from the metal. Gloves are a good start.

You'd rather not lay the parts flat on the floor during the painting process. Coathanger racks have been mentioned, but even small plywood easels would be adequate.

The nice thing about racks and easels is that they let you check if the paint is dry without touching the part itself. Touch the rack instead.

The etchant is generally a phosphoric acid solution; be sure to dilute it according to the instructions. Small parts can be dropped into a bucket of etchant. Brush it onto larger pieces, working the acid into the metal with a Scotchbrite pad. It stinks a bit. After five minutes, rinse the part under running water. My ecological conscience bothers me a bit about rinsing; it's impractical to catch the runoff. One could wipe off the etchant, but you're still left with the disposal of the paper towels or rags.

Once the metal has been alodined, you're ready for priming. It's great if your workshop is big enough to set up a permanent paint booth. It's not too likely, either.

There are two major problems to overcome. Anyone who's done any sort of spraying knows that the particles go everywhere. At the absolute minimum, a backdrop to catch the excess paint is necessary. For low-volume painting with spray cans, a large cardboard box makes a cheap and easy spray booth. But a compressor-type sprayer can kick out a lot of paint at high speed. You can "curtain off" your painting area using plastic sheets.

Painting in a basement means a houseful of paint odors. Even an attached garage seems to let the odors seep into the living areas. Some folks do their priming outside. It's not a bad idea, although spraying is restricted to windless days.

The second problem is the hazards associated with the paint itself. Wear a respirator specifically designed for painting, and ensure an adequate flow of fresh air. Note that I said a *flow* of fresh air, not just an open window. An open window allows the paint to linger; ventilation clears the air faster. Some primer manufacturers even specify a system to supply fresh air to the painter.

One common idea is to use a fan to move the air through the painting area, or build a paint booth using an old high-power fan. Perhaps, but the paint spray is explosive as well. At the right concentration, the fan's motor can ignite the suspended particles. Check supply houses for explosion-proof fans. If you rig up an exhaust system, run the exit pipe high enough so that the paint isn't deposited back on the house.

If any gas appliances that use an open flame (such as a furnace or water heater) are located in the workshop area, keep the painting area as far away as possible. Turn off the appliances until the air clears.

The actual primer application is no different from any other kind of painting. Primers are available in spray cans, but they aren't as cost-effective as using a spray gun and a compressor. The nice thing about the cans is the minimal cleanup afterwards. But a spray gun can be set to kick out a tight stream of paint to minimize overspray.

Avoid thick coats—they don't stick quite as well and take a lot longer to dry thoroughly than a couple of thin coats. Concentrate on the edges—they don't include any cladding. Also, corrosion tends to start between two sheets, so ensure good coverage on areas to be riveted together.

Let the pieces dry. If it's a calm sunny day, let the sun bake them. The coating always seems harder when I do this. While they're dry to the touch in just a few minutes, it takes 12 hours or so to reach full hardness.

It really doesn't take much primer to protect the metal. Look at the wing spar in Fig. 9-26. The builder applied just enough zinc oxide to ensure that every piece

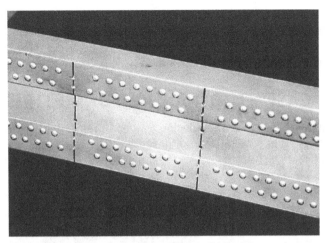

Fig. 9-26. *A primed spar for a Hummelbird. A heavy (nor an even) coat is not required.*

Fig. 9-27. *Properly designed all-metal homebuilts like this Mustang II are sturdy and long-lived.*

was covered. The coating isn't especially even—notice the mottled appearance—yet it covers the surface; it does its job.

Metal-monocoque airplanes have their drawbacks for the kitplane builder. The process was intended originally for factories, which can use expensive tooling to reduce personnel-hours and costs. A press with custom-made dies or a numerically controlled milling machine can crank out complex aluminum shapes a lot easier than a homebuilder's vise.

But the kitplane designers know this and make allowances. Traditional plans-built designs such as the Thorp T-18 Tiger and the Mustang II (Fig. 9-27) use simple curves throughout. Planes such as the RV series supply the curved components as part of the kit.

The skills aren't that hard to learn. Just pick up a few scrap pieces of aluminum and get some practice before starting on the kit.

CASE STUDY: AN RV FOR RETIREMENT

When many folks retire, they decide to enjoy themselves with a new recreational vehicle (RV). When Mike Dougherty retired, though, he and his wife, Arlene (Fig. 9-28), chose the *other* kind of RV—a Van's Aircraft RV-7A kit aircraft.

Fig. 9-28. *Arlene and Mike Dougherty picked a different kind of RV for their retirement—a Van's RV-7A.*

Selection and delivery

Mike had retired from Delco Electronics in Santa Barbara, California, and was working full time as a flight instructor. He and Arlene checked the used aircraft market, but, says Mike, "They were all corroded and overpriced."

Then they were at the airport one day, and a sporty two-seat airplane taxied in. Arlene, a pilot herself, said, "I want one of those!"

The plane? A Van's RV-6. They talked to the owner and were hooked. The Doughertys had been planning a vacation to Seattle and decided to drop in at the Van's Aircraft facility in Oregon on the way. After a test flight in the company demonstrator, they were decided. The RV-6 was being phased out, so they opted for the company's replacement bird, the RV-7. They chose the 7A tricycle-gear model.

A "standard" fuselage and tail kit was selected, but Mike and Arlene opted for a quick-build wing to jump-start the building process.

They still owned land and a house near Seattle from a previous period, and their vacation led to their decision to move back to the area. They bought an 8-foot-long trailer to move their household from California to Washington State, and the same trailer served to bring their kit home. The trailer needed some modifications to carry the approximately 12-foot-long quick-build wings safely.

The building experience

Mike and Arlene were experienced "homebuilders"—after all, they'd built their house in the mid-1970s. "With building the house and taking apart cars," says Mike, "I was handy with tools."

They hadn't worked with metal before, though. Mike and Arlene signed up for RV builder's classes offered by SynergyAir in Eugene, Oregon. The company had been founded by several RV builders and offers both a one-day "Fundamentals of Building" course and a six-day course designed to assist the builder in completing the empennage. The Doughertys took both courses.

They already had a shop for constructing the RV-7A. After building the main house, they'd built a smaller structure a few yards away. Over the years, the 22 × 22-foot building (Fig. 9-29) served host to spare bedrooms, a recreational room, and a shop for auto work. It included an eight-foot door leading to the attached carport (Fig. 9-30).

Since it has started out as living space, the shop walls and ceiling were fully finished, but the carpeting had been stripped out years earlier. Bare concrete is cold in the winter and uncomfortable to stand on for long periods anytime, so Mike started

Fig. 9-29. *Over the years, this outbuilding on the Dougherty property has served as a bedroom, a recreation room, and a garage. It now houses the Dougherty's RV project.*

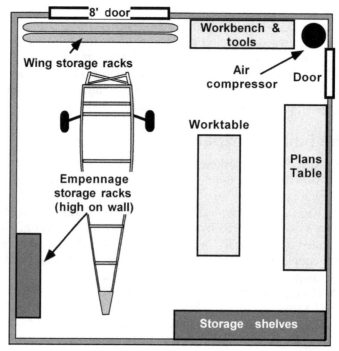

Fig. 9-30. *Layout of the shop.*

looking for floor covering. Buying new carpet made no sense; the building process would cover it with metal shavings and stain it with paint, grease, and oil.

Mike went to the local carpet supply store to try to buy some scrap carpeting. The store owner said Mike was welcome to dig through their trash and take what he needed for free. So Mike's building process began with "Dumpster diving!" He found enough large scraps to cover about half the floor.

Mike and Arlene limited their tool acquisitions. Arlene performed extensive research to find the best deals. They eventually bought an RV builder's kit from Aircraft Spruce and Specialty and an air compressor from Sears. The couple didn't splurge on the typical expensive power tools. Other than a bench grinder, they don't have any of the conventional shop power tools. Mike says, "I use a hacksaw for everything I should use a bandsaw for."

About a year into the construction process, Mike and Arlene have most of the fuselage structure complete and have started work on controls and cockpit systems (Fig. 9-31). They had anticipated initially needing a total of 18 months to build the plane, but they think that they've still got at least another year left.

While much of the riveting on an RV can be done solo, this is a joint project. "It's really handy having both of us working on this," says Mike. "Even one-person stuff goes quicker with two people working on it."

The amount of work involved in building their airplane has surprised them both. "It's amazing the amount of detail work necessary, even with a kit," says Mike. "It's a Godsend having two of us out here every day."

Fig. 9-31. *The Doughertys have plenty of working room around the fuselage of the RV.*

Mike and Arlene haven't completely decided on the engine yet. Initially, they were going to buy a kit for a Superior XP-360 engine, a noncertified version of the Lycoming O-360. Mike had planned to take Superior's engine-assembly course, where for $500 he would be able to assemble his kit with the assistance and monitoring of Superior's experienced staff. However, now they're considering a similar kit engine produced by Aero Sports Power of Kamloops, British Columbia, because of its better warranty. Ironically, the Aero Sports Power engine is built from Superior's components.

Advice

Mike and Arlene acknowledge that their minimal-tool philosophy may not have been the best approach. "If I were building again, I'd have a few more tools," says Mike. A bandsaw and a belt sander head the list of the tools he recommends. This extends to lights as well. While his shop seems well lit, he still has to dig out the halogen portable lights a little too often.

He feels that the money spent for better-quality tools is well worth it. "I wouldn't be as tight with tool setup. I would go for good quality, not the lowest dollar value."

The couple's final advice is simple: "Spousal buy-off is critical," says Mike. Arlene adds, "All wives should get out there and help; it makes things a lot easier."

QUALITY CONTROL

Most homebuilts require some aluminum work. Here are some quality control points to keep in mind from both this chapter and Chap. 7:

- Don't forget the edge-margin requirement: The centerpoint of all holes should be at least two times the rivet diameter from the closest edge.

- Similarly, don't space rivet (or bolt) holes any closer than three times the hole diameter.
- All bends should be across the metal grain (which is indicated by the printed labeling).
- Put a generous radius on all inside curves; cracks can start from too-tight turns.
- Smooth all metal edges to eliminate the "stress risers."
- Don't drill holes in metal directly to the desired size. Drill to the next size smaller, and then use a ream to finish.
- Deburr all holes.

Drilling Rivet Holes

Rivet Size	Min. Spacing	Edge Margin	Pilot Drill	Final Drill
3/32	9/32	3/16	3/32	#40
1/8	3/8	1/4	1/8	#30
5/32	15/32	5/16	5/32	#21
3/16	9/16	3/8	3/16	#11

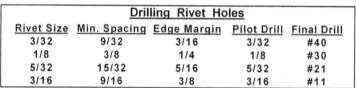

Countersink or Dimple for Flush Rivets

Use 100-degree Countersink or Dimple Die for Driven Rivets

Countersinking

Rivet Size	Min. Top-Sheet Thickness
3/32	.036
1/8	.042
5/32	.055
3/16	.070

Drill, then Countersink
Step 1 | Step 2

Dimpling

Dimple, then Final Drill
Step 1 | Step 2

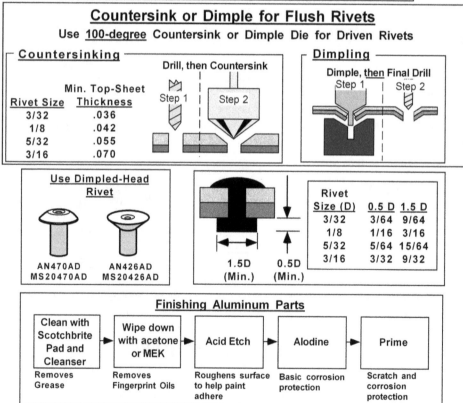

Use Dimpled-Head Rivet

AN470AD / MS20470AD

AN426AD / MS20426AD

1.5D (Min.) 0.5D (Min.)

Rivet Size (D)	0.5 D	1.5 D
3/32	3/64	9/64
1/8	1/16	3/16
5/32	5/64	15/64
3/16	3/32	9/32

Finishing Aluminum Parts

Clean with Scotchbrite Pad and Cleanser → Wipe down with acetone or MEK → Acid Etch → Alodine → Prime

Removes Grease | Removes Fingerprint Oils | Roughens surface to help paint adhere | Basic corrosion protection | Scratch and corrosion protection

Fig. 9-32. *Kitbuilders should find it easier to work with aluminum if they remember these key points.*

- When countersinking, drill and ream to the final size, and then countersink.
- For driven rivets, use a 100-degree countersink or dimple die. For pulled (pop) rivets, the proper angle is 120 degrees.
- When dimpling, drill the appropriate pilot hole for the dimple die, dimple, and then drill to final size.
- Use the proper rivets. Structural rivets have a dimple in the head to ease removal.
- The shop head for driven rivets should be at least one and one-half times the rivet size in diameter and at least one-half the rivet size high.
- Proper preparation of aluminum prevents corrosion and maintains an attractive appearance. Make sure that grease and finger oils have been removed prior to painting, and acid etch/alodine as the kitplane's instructions call for.

Figure 9-32 summarizes the key points.

10

Steel- and aluminum-tube construction

TUBE-AND-FABRIC CONSTRUCTION IS ONE TRADITIONAL AIRCRAFT DESIGN METHOD (Fig. 10-1). Steel tubing is cut into appropriate lengths, bent as necessary, and welded to form the fuselage structure. Any kit should be prewelded. But if you'd care to try it, there are still plenty of plans-built designs on the market.

Early homebuilts were mostly steel tube or wood. The drive to reduce building time begat more modern construction materials and methods. The 1960s brought the metal revolution with the Midget Mustang and the Thorp T-18, and the 1970s saw the composite stampede led by the VariEze, KR, and Glasair. Tube-and-fabric designs seemed destined to fade into history.

But the 1980s saw a resurgence. Two friends in Idaho developed the Avid Flyer, a light STOL design. The kit supplied a completely prewelded steel-tube fuselage and everything else necessary to complete the aircraft. Not only that, but it offered a simple wing-folding mechanism that didn't require disconnecting the aileron controls (Fig. 10-2).

This inexpensive kit led the comeback of tube-and-fabric designs. Production figures for this type of construction are larger than most realize, and the kits constituted 20 percent of the new homebuilts licensed in 2004. Tube-and-fabric kitplanes include the Skystar Kitfox, the RANS line (Fig. 10-3), the Murphy Renegade, the Sonerai, the CGS Hawk, and others. They've all grown, too—today's Barrows Bearhawk (Fig. 10-4) is a far cry from the early stages of tube-and-fabric homebuilts.

Popularity is fed by two factors. The first is cost. Steel-tube manufacturers don't need expensive composite molds or hydraulic presses. Because the initial investment is less, the kit sells for less.

The second factor is the ease of construction. A tube-type kitplane is the easiest to build. This is a bold statement. Right now, the composite, aluminum, and wooden airplane buffs are rising to refute it.

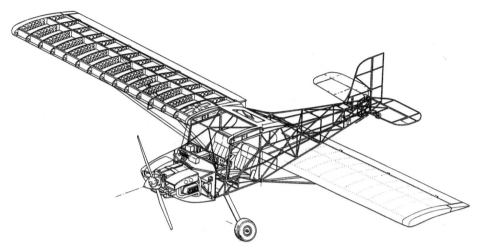

Fig. 10-1. *Today's tube-and-fabric kitplanes are a little different in design from the homebuilts of forty years ago.* ProTech Aircraft, Inc.

Fig. 10-2. *In the early 1980s, the Avid Flyer marked the beginning of the modern tube-and-fabric kitplane. Folding wings increased its appeal.*

Steel-tube airplane kits supply a structurally complete fuselage. No other type of construction does, unless one orders an expensive quick-build kit. Composite fuselages assemble fast, but you have to bond the fuselage halves together. Metal or wood structures are built from hundreds of separate parts. The kit might include prebent/precut bulkheads and precurved skins, but the builder still must rivet or bond them together.

Fig. 10-3. *The RANS S-7 is a great advance over the primitive tube-and-fabric kitplanes of the early 1980s.*

Fig. 10-4. *The Barrows Bearhawk is a four-seat steel-tube aircraft. It can be built from plans, or a AviPro kit can be ordered.*

But the builder pulls a complete steel-tube fuselage out of the delivery crate (Fig. 10-5). It's welded by the factory on jigs just like production airplanes such as the Citabria and Super Cub. A lot of work remains. The fuselage must be protected against corrosion. The wings must be built. But much of the work left for the builder is of the simple "bolt-in-place" variety.

Fig. 10-5. *Steel-tube kits come with the fuselage prewelded.*

Several tube-type kitplanes don't use welded-steel tubing. The Murphy Renegade uses aluminum tube joined by pop rivets and special fittings. The CIRCA Nieuport replicas (Bebe single-seater and N-12 two-seater) use aluminum-sheet gussets instead of the special fittings. Many ultralight-based homebuilts use the tube-and-gusset method as well.

In this chapter you'll learn some of the basic skills involved in building steel- and aluminum-tube airplanes:

- Checking and preparing welded-steel-tube fuselages
- Pop riveting
- Cutting, drilling, and bending aluminum tubes

Tube-type kitplanes require techniques from most construction disciplines. You'll probably be making some sheet metal structures; the details of this process are illustrated in Chap. 9. Composite work is shown in Chap. 8, and wood preparation and fabric coverings are discussed in Chap. 11.

Wing construction varies among types. Many use aluminum-tube spars; others build the wing from aluminum C-section and aluminum or wood ribs. A few have metal-covered wings, such as all-aluminum airplanes. Between this and the other construction chapters, you'll be prepared regardless of kit design.

ADDITIONAL TOOLS

The primary addition to the basic tool list in Chap. 6 is a *pop-rivet gun*. Prices vary from a few dollars at the local hardware store to $300 for a pneumatic gun.

Which model you buy depends on the amount of use. Builders of Murphy Renegades and Circa Nieuports construct most of the airplane with pop rivets, whereas builders of welded-steel-tube kitplanes probably can get by with a cheaper model.

Selecting a pop riveter is easier the second time around. I bought a $15 unit from the local hardware store, and it has worked well, but its ergonometrics are poor; the handles are too far apart to be comfortably worked by one hand. I'll know what to look for on the next one.

If there's a lot of riveting, consider springing for the pneumatic tool (Fig. 10-6). Zeniths, for instance, are entirely pop-riveted. There's only so much squeezing that your hands can take.

Fig. 10-6. *Pneumatic rivet pullers can save a lot of hand strain.*

Hand or pneumatic, make sure that the riveter can handle the variety of rivet sizes to be used. Most supply a selection of nosepieces that adapt the same gun. If the plans call for Cherry rivets, a special gun is needed.

Most other tools are trivial and cheap. You'll need a wire brush for your drill, a tubing cutter, and extra clecos. If there's a lot of sheet metal work, check the list of additional tools in Chap. 7.

MATERIALS, FASTENERS, AND SAFETY

Steel- or aluminum-tube airplanes are straightforward and traditional.

Materials

Tubing is the material you'll be working with, either steel or aluminum. Characteristics of each alloy were discussed in Chap. 7. Recall that 4130 steel is the most common. It's strong and weldable. It also rusts freely and must be protected.

Most aluminum-tube aircraft use 6061-T6 alloy. This isn't as strong as 2024-T3 but is significantly more corrosion-resistant. Wide-diameter 6061-T6 tubing is used for wing spars on several kitplanes.

Tubing is sold by diameter and wall thickness. The inside diameter can be determined by subtracting twice the wall thickness from the diameter. The alloy, diameter, and wall thickness are printed along the tubing and are repeated every foot or so.

Fasteners

The primary fastener in steel-tube aircraft is the *weld,* which you probably won't be doing. A steel-tube kitplane should come with a prewelded fuselage. I have no objection to those who want to learn to weld or build up their own fuselage. However, if the kit comes with the fuselage prewelded, everything *else* should come prewelded, too.

But you should know *something* about welding. You'd like to be able to check the welding on the fuselage. And if you ever have an accident and have to get the fuselage repaired, it's nice to be able to intelligently discuss the problem with the welder.

The welding operation consists of placing the two components to be joined in close proximity and then heating the junction into a molten state. The metal of the two pieces then flows together. When the joint cools, the two pieces of metal have been turned into one continuous unit. A filler rod supplies additional metal to the weld area to add thickness and ductility.

While you won't be doing any welding on your kitplane, you'll probably install a lot of pop rivets. These inexpensive fasteners, also called "pulled" or "blind" rivets, are ubiquitous on most tube-type airplanes.

The advantage of pop rivets over driven rivets (see Chap. 9) is that they can be installed one-handed in areas where a bucking bar can't reach. Conventional rivets form a shop head by ramming the rivet repeatedly into a steel block (the bucking bar). The rivets can't be installed into a tube—there's usually no way to get a bucking bar inside. Hence pop rivets are used instead.

The big disadvantage to pop rivets is their strength, or lack of it. They're typically about 60 percent as strong as driven rivets of the same size. They're a bit unsightly, too. The shop head is pretty ugly, and the stem leaves a hole in the manufactured head.

Stronger monel and stainless steel rivets are available. *Cherry* rivets are the most common. A $\frac{3}{16}$-inch stainless steel structural Cherry rivet is rated at 1,650 pounds in shear. However, the harder shank means more pull resistance. Thus they take more effort to install.

Other than pop rivets, the most common fastening hardware is standard AN nuts, washers, and bolts. Chapter 7 details the "nuts and bolts" of AN hardware.

Safety

Other than inflicting a few gouges and nicks from sharp edges, tube airplanes are fairly kind to builders. If any welding is done in the shop (e.g., you hire a welder to redo a bad joint), isolate flammables from the area, and keep a fire extinguisher handy. It's a hot flame, and a backfire can spit out bits of molten steel. The cheaper extinguishers don't turn off once started; spend a little extra for the multiple-use kind. Carbon dioxide works well.

If you're going to watch or help, wear old clothes, gloves, and safety goggles. Cotton shirts and denim jeans might be best. Polyester and other man-made fabrics can melt or ignite. No leisure suits in the shop, please.

The other safety hazard involves the health hazards of the anticorrosion primers applied to the tubing. Follow the instructions, wear a respirator, and ensure adequate ventilation.

WELDED STRUCTURES

As mentioned at the beginning of this chapter, most tube-type aircraft use welded-steel fuselages. The fuselage is structurally complete; all you have to do is attach fairings, controls, and other items that don't affect load-carrying ability.

But you can't just ignore the fuselage. You have to verify that it's been done properly and follow a few basic rules during construction.

Basic construction

Steel-tube fuselages begin as straight lengths of tubing. Pieces of the proper diameter and wall thickness are cut to length, bent as necessary, notched to match the mating tubes, and installed on a jig that holds the tubing in the desired shape. Figure 10-7 shows an example.

The main tubes that run fore and aft are called "longerons." Pieces that run vertically between them are called "uprights," and "laterals" run horizontally between longerons on opposite sides. "Diagonals" are mounted aslant between uprights, laterals, or longerons.

The tubes are welded together wherever they meet. Each tube must be shaped to make contact over the maximum amount of area. For example, on a simple T-joint, the end of the vertical bar must have a semicircular notch to fit snugly over the other tube. If there's a wide gap, the steel can't intermingle when the two pieces are heated. The filler rod adds some additional metal, but there's a limit.

The usual practice is to lay one side of the fuselage flat on the worktable, tack-weld all the joints, and then build the other side. A tack weld melts the material in just a small area to temporarily hold the components for shop handling. One could do the finish weld on the table, but all-around access is usually impossible without lifting the pieces.

Fig. 10-7. *Wood blocks nailed to the tabletop hold the steel tubing in position for tack welding.*

After the other side of the fuselage is tack-welded, the two sides are jigged upright, and laterals and diagonals are tack-welded between them. Once the entire fuselage is tack-welded, the builder does a finish weld on each joint.

In addition to welding up the structure, the builder adds tabs and plates in areas where items will be bolted later. The tubing used on the lighter homebuilts is pretty small, typically around ½ to ¾ inch in diameter. Drilling a ¼-inch hole through a tube for an AN4 bolt weakens the tube drastically. Instead, the kit manufacturer welds a small plate into place, bracing it with additional tubing or sheet as required.

Buying tack-welded fuselages

Sometimes the manufacturer offers a tack-welded fuselage as part of a lower-cost kit. This generally isn't an option for complete kits; rather, it's available as part of certain materials kits. Wag-Aero sells both fully welded and tack-welded fuselages for their ragwing Piper replicas, for example.

Cost is the main reason for ordering a tack-welded fuselage instead of a fully welded one. Typically, Wag-Aero charges several thousand dollars less for tack-welded Sport Trainer and Sportsman 2 + 2 fuselages. Depending on the complexity, a certified aircraft welder might want between $1,500 and $2,500 to finish weld the fuselage. A fellow EAA chapter member might do the job for half that.

The main point is to find someone experienced in welding aircraft tubing. Your best bet would be an FAA–certified repair station. Even if the station (as a matter of policy) doesn't do welding for experimental aircraft, the welders might take after-hours projects on their own. After all, you don't need the certification signoff, just someone who knows aircraft tubing.

Don't just page through the Yellow Pages under "Welding." Commercial welders normally work with far thicker materials than the 0.045-inch walls of 4130 tubing. You want someone with demonstrated aircraft expertise.

Or you could spend $100 for a community college welding course and learn how to do it yourself. The most time-consuming process is cutting and fitting all the tubes. If you buy a tack-welded fuselage, that's already done. You might make mistakes, but all you'd have to do is cut away the ruined tubing and weld in a replacement.

Welded-fuselage checkover

Whether you buy a tack-welded fuselage or the finished product that comes in a complete kit, check the fuselage carefully before proceeding. The first step is an overall examination for dented, bent, or crimped tubing. This type of damage shouldn't occur during normal shipment. If the exterior of the crate is broken, note the fact on the form when you sign for delivery. Immediately inspect for internal damage. The shipper is liable in this case and should pay the repair costs.

If the crate didn't show any sign of abuse, the damage probably was caused during assembly. Take pictures, and contact the kit maker.

Repair methods depend on the type and location of damage. Usually, a two-piece sleeve is welded over the damaged area. You'd prefer an experienced welder to do such repairs. Don't say, "Well, I'll fix it myself." Have it repaired professionally; an experienced aircraft welder not only will fix it properly but also might spot further damage. In severe cases, the fuselage might have to be shipped back to the factory and rejigged. Work out the details with the kit manufacturer.

After checking the tubes, examine the welds. I've seen some kits delivered on which the manufacturer missed finish welding some joints. Here's what good finish welds should look like:

- The weld should go all the way around the joint in a continuous bead. If there are just a couple of melted spots holding the tube in place, it's a tack weld that didn't get completed.

- The melted area should be smooth. Gentle, continuous ripples are normal and the sign of a good weld. Watch for sharp edges and generally rough and dull appearances. No matter how ugly a weld might look, don't file it. The metal is necessary for strength.

- The weld should taper smoothly into the surrounding metal.

- There should be no cracks in either the tubes or the weld itself.

- There should be no holes in the welds or tubes.

Figure 10-8 presents some of the characteristics of good and bad welds.

If you suspect a problem, call your EAA technical counselor or someone else knowledgeable. If that person confirms your suspicions, contact the manufacturer.

After the inspection, ream out all the drilled holes to the proper size and deburr (Fig. 10-9). Most will be for AN3 (³⁄₁₆-inch) or AN4 (¼- inch) bolts, but check the plans.

Internal corrosion-proofing

Steel-tube fuselages are strong and light. Properly protected, they'll last a long time. But 4130 steel tubing is prone to rust and must be protected. The outside is easy enough. But what about the inside?

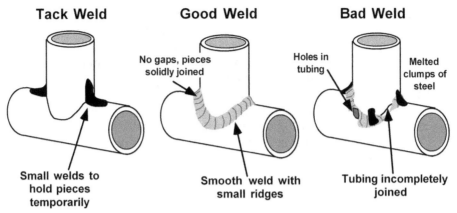

Tack Weld

Small welds to hold pieces temporarily

Good Weld

No gaps, pieces solidly joined

Smooth weld with small ridges

Bad Weld

Holes in tubing

Melted clumps of steel

Tubing incompletely joined

Fig. 10-8. *General characteristics of welds.*

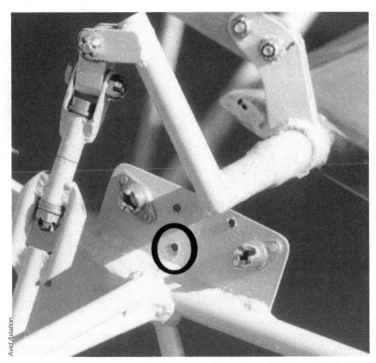

Fig. 10-9. *Kit manufacturers leave the deburring of holes to the builder. Note the burrs around the holes in the plate.*

Properly done finish welds make a complete airtight seal and stop exterior moisture from entering. The air inside retains whatever moisture it contained when the structure was welded. There might be enough to support corrosion. In any case, bolt holes and cracks can allow moist outer air to enter and accelerate the process.

Steel-tube structures are internally corrosion-proofed by a coating of *line oil*. The old-timers used hot linseed oil, but line oil combines corrosion-proofing with an ability to seal tiny holes. The oil is introduced into the lower longerons and then sloshed about by rolling the fuselage from side to side. Pour out the oil, and close the holes with self-tapping machine screws and a sealant, such as Permatex.

Should you line oil? It's a weird situation. Most of the old-time homebuilders I talked to were quite adamant about the necessity, yet the kit manufacturers generally don't make it a required operation. Most list it as optional.

I feel that the lower longerons need the protection, especially on amphibians and floatplanes (Fig. 10-10). These tubes are most likely to be rusted out on old production aircraft, so the same is probably true for kitplanes. It's a longevity issue that probably won't make much difference over the five years you'll probably own the airplane, especially if it's hangared.

Fig. 10-10. *The fiberglass outer shell of this GlaStar will mostly protect the steel tube inner structure from spray, but proper corrosion protection is still important.*

In any case, don't leave open holes into the structure. Atmospheric moisture is bad enough without allowing rain to get inside. Seal any bolts or screws that penetrate the tube wall.

Finally, dispose of the used line oil and any soaked material in an airtight container. The oil *can* ignite via spontaneous combustion. Here's a quote from an FAA Safety Alert:

> After using linseed oil as a corrosion inhibitor in a pair of aircraft wing struts, the soaked paper towels used to remove the residue were deposited in a trash barrel.
>
> Approximately one hour later, the odor of overheated material was noticed. While investigating the source of this odor, the linseed oil–soaked

paper towels were found to be near the point of spontaneous combustion. The entire barrel was warm to the touch, and the compacted paper towels were charred in the center.

The FAA's recommendation: "There are several precautions which should be observed when disposing of excess materials, and the best source of information on this subject is your local fire department. Their advice should be sought and adhered to strictly."

External corrosion-proofing

Applying a corrosion-resistant coating to the outside of the fuselage is one of the first things you should do. Alloy 4130 starts rusting almost immediately, and every day you delay priming just adds to the rust and scale that must be removed. The kit might have been shipped with an oil coating on the tubing to retard the rusting process, but many kits aren't shipped with a coating. In any case, the oil is easily disturbed during handling.

The degree of cleaning and the tools required depend on how rusty the fuselage has become. Sandpaper or steel wool might suffice if it's fairly clean. Or it might take a steel brush on a power drill.

The best way to clean the fuselage is to sandblast it, which presents its own problems—primarily sand everywhere. Usual practice is to pick a nice day and do it in the driveway or back yard.

Sandblasting setups for air compressors can be bought for $50 or so. If you don't have access to a compressor or yours isn't big enough, rent an all-in-one unit from the local U-Rent-It store. Or check with auto body shops, and then take the fuselage in ready to blast; they'll charge $200 or so. Paint it as soon as you get it home, or have the shop do that as well.

If you sandblast it yourself, set the unit for the narrowest stream possible. Because the fuselage is made of a bunch of thin tubes, most of the sand will be wasted anyway. Buy extra; you can always return unopened bags, and it eliminates the hassle of having to stop and go out for more.

Whatever the method, the intent is to remove all contamination. The nice things about steel compared with aluminum are that it doesn't have to be etched prior to priming and doesn't have an Alclad surface to baby. You can bear down on the sandpaper and the sand blaster. The rougher the surface after cleaning, the better the paint will stick.

Once the rust is knocked off, put on gloves, and clean the tubing with MEK or acetone. This gets rid of the residual grease, dust, and preservative oil. Until the primer has been applied, don't touch the fuselage with your bare hands.

Prime as soon as possible after sanding; otherwise, the tubing will just rust up again. Application is similar to priming aluminum, as given in Chap. 9. Note also the safety precautions given in Chap. 7; aircraft primers are rather nasty stuff, and you'll want to protect yourself properly. Make sure that the primer is compatible with the planned fabric-covering process.

Because the tubing isn't an external item, appearance of the primer coat isn't a factor. It can be applied with a brush. If you decide to spray, set the gun to the narrowest setting. If possible, rig up a rack that allows you to rotate the fuselage to be able to hit all the nooks and crannies.

Apply a light coat, let it dry, and then apply another coat. Then grab a flashlight or trouble light and go hunting for thin spots. Make sure that the bottom of the fuselage is well covered, especially the area around the tailwheel. These are the areas where water will collect.

An alternative is a professionally applied powder coating. This process applies an epoxy primer that is then baked until it is rock hard. A powder-coated part is almost impossible to scratch. This is available as an option from many manufacturers, and while it is a bit pricey ($500 on up), I recommend it. A typical primed steel-tube fuselage is shown in Fig. 10-11.

Fig. 10-11. *A primed RANS S-10 fuselage with wheels and engine attached.*

A lot of aluminum parts will be attached to the fuselage; these also must be protected against corrosion. Details on preparing and priming aluminum are given in Chap. 9.

Whenever different materials must be joined together, there is a chance of accelerated corrosion. Where steel and aluminum are to be attached, each must be primed thoroughly to prevent dissimilar-metals corrosion. This is one of the reasons for the cadmium plating applied to aircraft hardware.

And remember that wood retains a significant amount of moisture. When a wood component is joined to a metal one, varnish the wood in addition to priming the metal.

Steel-tube summary

The nice thing about steel-tube kitplanes is that once the fuselage is primed, you can forget about it. There are a few things to remember, such as only drilling holes specifically called for by the plans. But the fuselage just becomes a structure to hang other parts on. And many of those parts bolt to prewelded attachment points.

Building the wings is another problem. There aren't too many steel-tube-winged airplanes out there (the Dyke Delta is one). Most tube-and-fabric kitplanes' wings use either aluminum or wood construction. Working with wood is covered in Chap. 11. The following sections discuss aluminum structures.

ALUMINUM-TUBE STRUCTURES

Probably the biggest thing the ultralight fad did for homebuilt airplanes is popularize and prove aluminum-tube structures. Instead of a wooden spar, the ultralights used a tube of light aluminum alloy. Instead of a complicated longeron-and-bulkhead fuselage structure, they just ran a thicker aluminum tube from nose to tail (Fig. 10-12).

Fig. 10-12. *This Kolb ultralight is a modern example of aluminum-tube fuselage construction.*

Until the ultralight movement came along, the empty weight for the smallest conventionally built single-seat homebuilts was about 400 pounds. Ultralights cut this figure in half and more. When the first version of the Federal Aviation Regulations (FAR) Part 103 was due to come out, the rumored maximum weight was about 150 pounds. Many vehicles in production already met that limit. The actual limit turned out to be 254 pounds. A blizzard of comfort-type items such as fairings and closed cockpits then appeared in the sales catalogs.

The ultralight boom has subsided, but its effect on homebuilding lives on. The RANS series, the Kitfox, and many others use an aluminum tube as a combination

Fig. 10-13. *Many of modern tube-and-fabric airplanes use aluminum tubes as both spars and wing leading edges.*

main spar and leading edge (Fig. 10-13). Many bend ribs from ½-inch aluminum tubes. Those with conventional wood or metal ribs include them in the kit nearly ready to install.

For the homebuilder, aluminum tubing has several advantages over more traditional methods. It's light. It's easily cut with simple hand tools. It can be bent into graceful curves with little effort. It's more resistant to corrosion.

But its drawbacks also hit home. Aluminum isn't as strong or hard as steel, so some of the weight advantage is lost by the need for a beefier structure. It's tricky to weld, too. Steel's color depends on its temperature, which makes it easy to heat it to the proper condition. Set a torch to aluminum, and it looks the same right to the point at which it collapses into a molten puddle.

Because welding is difficult, aluminum tube structures are held together by other means, such as bolts and pop rivets. These labor-intensive processes are usually left to the builder. The following sections describe cutting, shaping, drilling, and riveting aluminum tubing.

Cutting

A horizontal bandsaw is the best way to cut aluminum tube. The clamp ensures straightness, and the cut requires only a moderate amount of cleaning up. Hand-held tubing cutters (Fig. 10-14) cost around $10 to $20 and consist of a movable cutting wheel positioned opposite a set of rollers. Get the kind for tubing ⅛ inch to 1⅛ inches in diameter.

Fig. 10-14.
The tubing cutter in action.

Place a pencil mark on the tube at the desired cut line. Turn the cutter's knob to retract the wheel until the tube slides between it and the rollers. Position the wheel just outside the cut line, and tighten the knob. Hold the tube stationary, and rotate the tool around it. Watch the tracking when you start—sometimes the tube doesn't seat properly, and the tool starts cutting a spiral.

As the wheel cuts a groove, the tool's resistance to rotation lessens. Tighten the knob every couple of turns to maintain the same level. Eventually, the wheel breaks through, and the tubing separates. This can happen suddenly; be careful not to drop the tool.

The operation leaves a pinched-in area at the end of the tube; the pinched-in area must be removed. Clamp the tube down and run a coarse file across the end a few times to get rid of some of the excess metal. A hook-type deburring tool (Fig. 10-15) takes care of the remaining thin lip with little fuss. Polish the end with emery cloth to eliminate any scratches or file marks. As discussed in Chap. 7, these can be starting points for cracks. A benchtop belt sander works well for smoothing aluminum.

A tubing cutter won't work in all cases. There must be enough tube past the cut line to allow the rollers to make full contact. The rollers on either side are of unequal length, and you sometimes can get closer to the end by flipping the tool the other way to put the shorter roller on the tube's end. If the amount to be removed is too small for the cutter, try a grinder.

Most tubing cutters can't open wide enough for tubes much greater than 1 inch. Without a bandsaw, you'll occasionally have to resort to a hacksaw. The problem is making a straight cut—hacksaws sometimes seem to have a mind of their own.

Fig. 10-15.
The tubing cutter leaves a pinched-in burr around the circumference of the tube. File it down a bit first, and then scrape it away with the deburring tool as shown.

The first step is to apply a cut line all the way around the circumference. A single pencil mark isn't enough because you need a definite reference to keep the cut perpendicular to the end of the tube.

Start with a piece of paper shaped like a grocery receipt. The length should be about four times the diameter of the tube, and the edges should be perfectly straight. Wrap it around the tube. When it starts to overlap, place the additional paper directly atop that previously applied. Tape the end down.

You've just made a paper tube with an inside diameter equal to the outside diameter of the aluminum one. By carefully overlapping the paper, the end of the paper tube is square and true.

Slide the tube to the cut line. The paper tube should cover the end of the good metal, leaving the cut-off end bare. Slather machinists' ink, spray paint, or a magic marker over the area where the paper tube meets the aluminum one.

When dry, remove the paper tube. The paper acted as a stencil, keeping the "good" tubing clean while marking the excess. If you cut away all the marked tubing, the tube end will be true.

Clamp the tube, leaving the end accessible to the hacksaw. Protect the tube from the vise jaws. Actually, vises aren't the best for holding round objects. A Black & Decker benchtop Workmate has triangular plastic jaws that are great for holding tubes.

Start the hacksaw on the painted area, about ⅛ inch from the marked line. You aren't trying to cut off all the end, you're just trimming it down to lessen the amount of filing.

Cut slowly and carefully. If the saw comes too near the line, apply a bit of twist to take it away. Keep close watch on the back of the tube, too, because it's very easy for the saw to take a bit of slant and cut the other side of the tube differently. When the end is removed, use files to trim away the remaining painted metal.

Where to drill holes

You can't drill a hole anywhere you want. A hole will weaken the tube, though, and how much depends on where you put it and how well it's aligned.

To help describe the right way to drill holes, I'll reference points around the circumference of the tube similarly to a compass. Looking at the tube end-on, it's a circle. North is straight up (or 0 degrees), east is directly to the right (or 90 degrees), south is straight down (180 degrees), and west is directly to the left (270 degrees).

Imagine a long tube supported by two sawhorses. Hang a bucket from the midpoint of the tube, and it flexes a bit. This flexing has similar effects as when aluminum sheet is bent, as described in Chap. 9. The south (bottom) edge of the tube has to stretch slightly, whereas the north edge compresses to match the curve. Add weight to the bucket, and the amount of stretching and compressing increases. If you add too much weight, either the south edge stretches too far and breaks open, or the north edge crumbles.

Looking at a cross section, there must be two points where the force changes from stretching to compression. The points are approximately halfway between north and south—in other words, east and west. At these positions, the force is neutral, and the metal isn't under stress.

If a hole is drilled in the top or bottom, the surrounding area is under stress, and the hole will cause the metal to fail early. But if the hole is drilled horizontally from east to west, no problems result. Because the area isn't under stress, drilling doesn't affect the tube's load-carrying ability.

As mentioned, many smaller kitplanes use aluminum tubes for spars. While the preceding description is drastically simplified, the lesson is unchanged: Do not drill holes through the top or bottom of the spar. They should go fore and aft, through the neutral portion. This is illustrated in Fig. 10-16.

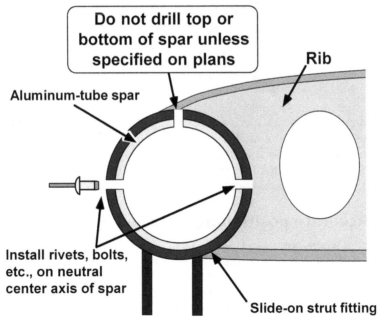

Do not drill top or bottom of spar unless specified on plans

Rib

Aluminum-tube spar

Install rivets, bolts, etc., on neutral center axis of spar

Slide-on strut fitting

Fig. 10-16. *In most cases, the builder should never drill holes on the top or bottom of the wing spar. The same holds true for other tubes as well.*

Some instructions call for drilling holes at various positions around the circumference. Anything closer than, say, 45 degrees to north or south might cause problems. But you'll find that the designers compensate for this in various ways. They'll nest another tube inside the drilled area, for example. Or the portion might not be under much load, like the tip of the spar.

Some aircraft install a couple of pop rivets on the bottom of the spar underneath the strut fitting (Fig. 10-17). The strut fittings themselves provide considerable reinforcement along the spar to compensate.

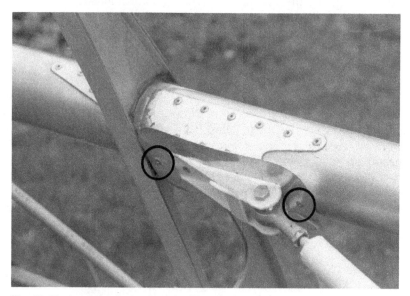

Fig. 10-17. *An exception to the rule. The thin sheet reinforces the fabric around the strut attachment and is pop-riveted to the bottom of the spar. But the large wing strut attachment plates riveted and epoxied to the tube provide more than enough compensation.*

Manufacturers can specify hole drilling in these seemingly forbidden locations because they have the facilities to analyze and test to ensure that strength isn't affected. You don't. So keep the holes out of the tops and bottoms unless the plans specifically call for them. If the situation feels funny, call the manufacturer. You might have misread something.

Accurate positioning of drilled holes

Knowing where holes can be drilled safely is one thing; placing them accurately is another. Drilling a hole in one side of a steel or aluminum tube isn't much of a problem. Dimple the spot with the punch, drill a pilot hole, and ream to the final diameter. This'll work for pop rivets, self-tapping screws, or any other fastener that doesn't pass all the way through the tube.

The problem arises when holes have to be oriented correctly with each other. Take the spars for the RANS S-10 Sakota, as shown in Fig. 10-18. Aluminum rib fittings and

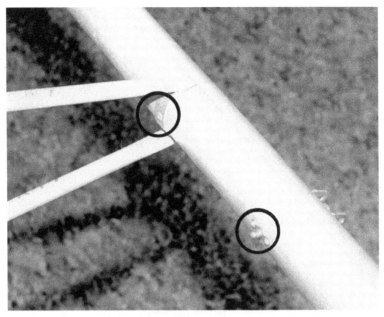

Fig. 10-18. *The rivet holding the rib and the nut plate for the aileron must be aligned exactly or weird things will happen to the wing aerodynamics.*

aileron nut plates are pop riveted at the nine o'clock position at given intervals from the root to the tip. Install a few at 10 o'clock instead and the wing aerodynamics suffer.

Even aircraft with different rib systems come against a similar problem. A fitting might be riveted to the root, and a fiberglass wingtip must be riveted to the other end. Get the holes off line, and the tip will have an odd twist.

Drilling the first hole is easy enough. But how do you position the rest of the holes at the same clock position? The labeling printed on the tube can't necessarily be trusted. The tube might have rotated slightly while the label was being applied, or the tube might have been painted already.

And what happens if the plans call for bolting through the spar? The bolt must pass directly through the center, perpendicular to the tube's long axis. The second hole must be drilled directly across from the first at the same distance from the tube end.

The farther the second hole is off, the weaker is the attachment. If one hole is at three o'clock and the other is at eight, the tube will warp as the bolt is tightened. If the bolt slants one way or the other along the length of the tube, other localized stresses can be induced. In either case, the bolt head, nut, and washers won't rest flat on the surface.

We need to establish a pair of reference lines along the length of the tube exactly opposite each other. It's easy enough for a pair of short lines because all you have to do is wrap the tube in carbon paper and close a vise gently on it. If the vise's jaws are parallel, their first points of contact are (by definition) exactly opposite to each other. But vises aren't really precision pieces of equipment, and this procedure wouldn't work for tubing longer than a few inches.

Instead, place the tube on a flat, smooth surface and hold it rigidly in place. One good way is to pin the tube between pieces of wood nailed to the surface. Holding it by hand won't do because with any little wiggle, the reference will be skewed.

Place a combination square on the table with the ruler vertical, as if you were measuring the height of something. Slide the square until the ruler touches the side of the tube. Take a look at the ruler line equal to half the tube's diameter. You'll find that line is the point in contact with the tube.

The actual dimension doesn't make any difference. If the base of the square is flat on the table, the ruler *must* make contact at the point 90 degrees around the circumference from the lowest and highest points.

Because we are only seeking two reference points exactly opposite to each other, the relationship to the six- and twelve-o'clock positions is moot. As long as the tabletop is flat, a square held to either side will make contact at two points exactly opposite to each other, as Fig. 10-19 demonstrates. The tube doesn't even have to be level. The only reason we attach it tightly to the tabletop is to make sure that it doesn't roll between markings.

Tangent points are exactly opposite each other

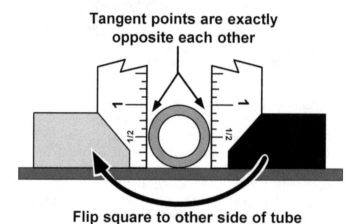

Flip square to other side of tube

Fig. 10-19. *No matter the diameter of the tube, a square held flat on the worktable on either side will touch the tube at directly opposite points. The measurement itself is immaterial because it is the points of contact we're interested in.*

But just marking single points of contact on either side doesn't do us much good. We need to define reference lines on both sides along the entire length.

Figure 10-20 shows one solution. Bring the square into position, then slide it along the length of the tube while remaining in contact. As long as the base of the square stays flat on the table, it will apply a long, straight scratch. This scratch and the one we immediately apply on the other side become our reference lines.

However, because you don't want to actually scratch the tube, paint the sides with machinist's ink first. Let it dry, and then slide the square with enough pressure to scrape the paint but not enough to excessively gouge the underlying aluminum. Then slide it along the other side as well.

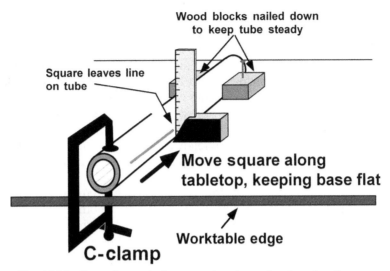

Fig. 10-20. *Opposite centerlines can be drawn by dragging the square along the tube on either side, keeping the base flat. Keeping the tube from moving or rotating during this process is vitally important.*

It isn't as damaging as it sounds. If the tube is 6061 alloy, it doesn't have an Alclad surface to disrupt. The scrape will be along the neutral, unloaded axis of the tube. And if you don't use too much pressure, the scratches should buff right out with emery cloth.

There's nothing magical about machinist's ink in this case. Just about anything will work: old spray paint, shoe polish, whatever. I use a wide-tipped marking pen. Use what is available, just as long as it can be cleaned off before priming.

The tube and whatever's holding it in place shouldn't be moved until both sides of the tube are marked. Holder placement must be planned so that the areas to be marked aren't obstructed.

With the two reference lines in place, any number of holes can be drilled exactly along the same axis. However, we still have a problem with bolts. The holes through both sides must be directly across from each other along the length of the tube as well.

I know what you're thinking. Why not mark on one side and then clamp the thing in the drill press and drill both sides at once? Go ahead. Everybody's got to try this once. Use a cheap piece of scrap tubing for the first attempt, though.

The problems are many. First, the tube has to be set up so that the drill point hits the exact center. Any inaccuracy, and the bit tries to crawl away, chewing the surface as it goes. You'll have to position the tube so that the point where you want the hole is the highest point, which is the point closest to the drill press.

Second, the tube has to be dead level when it's being drilled. Any sort of slant puts the opposite hole somewhere else.

Third, drill bits aren't stiff. They flex. The slightest off-center pressure between the first hole and the opposite inside wall will put the hole in some other location. If the drill's a little off center, the curve of the tube will change the aim. When it hits the opposite wall, the bit will walk around until it digs in at a random location.

A few of you might be able to do this perfectly every time. I hit about 50 percent correctly. Even if you can do it 90 percent of the time, every tenth hole will ruin the part. Do yourself a favor and drill the holes separately.

There are a couple of ways to find the opposite point on the other side of the tube. The best way is by measurement. If the bolt is supposed to be installed 30 inches from the end, measure the distance on both reference lines, and mark the points.

An alternate method uses a drilled pattern. Take a 2-inch piece of tubing with an inside diameter equal to the outside diameter of the tube to be drilled. Working very slowly and carefully, mark reference lines on the tube, and drill two small reference holes directly across from each other. This short bit of tubing then becomes a pattern for marking bolt holes.

Slide it over the tube to be drilled. Position one reference hole at the point to be drilled, and then insert the punch in the hole on the other side of the pattern and mark the location. You could even drill pilot holes through the pattern, but this tends to enlarge the reference holes.

With the positions marked, it's time to drill.

Drilling

Once the hole is marked accurately, drilling is a snap. Dimple the metal with the automatic centerpunch, and then drill a small pilot hole. Use the drill press, or hold a hand drill perpendicular to the surface while drilling. Then drill and ream to the desired size.

Before you reach for the larger drill bits, however, take a look at the pilot hole. Is it where it's supposed to be? If not, some degree of correction still can be made. When using the larger bit, drill at a slant. Point the drill in the direction the hole needs to be moved, and apply a moderate amount of pressure in that direction. Think of it as pushing the hole in the right direction; if the hole has to go to the right, the drill should be slanted to the left. Experience will teach you how much slant and force.

Large holes present a problem. The big bits have a tendency to grab and make wavy edges. A countersink makes smoother holes. For a ½-inch hole, for instance, drill to 3/16 inch, and then follow with a ½-inch countersink.

Dress the edges of drilled holes with a larger-diameter countersink to deburr and chamfer the edge. A hook-type deburring tool removes the burr on the inside of the tube. Or if the hole is near the end of the tube, slide a round or half-round file inside to knock down the inner burr. Polish with emery cloth.

It's especially important to eliminate the interior burr when another tube is going to slide inside the drilled one. A burr on the inside might stop the smaller tube. Even if you can force it past the hole, the sharp metal will scratch its exterior.

Other shaping

Sometimes the end of a tube must be shaped to fit other parts of the structure rather than being cut off square. Where one tube joins another, its end must match the other's curvature. For example, the end of a ¾-inch tube forming a T with a larger-diameter tube should be saddle-shaped to make contact with the maximum surface on the larger tube so as to spread out the compression loads (Fig. 10-21). If not, the

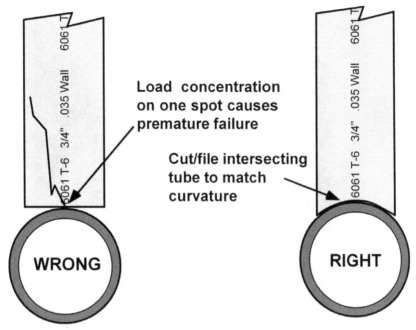

Fig. 10-21. *Intersecting tubes must be fishmouthed or loads will be concentrated.*

tubes would only make contact at one small point on either side. Either tube might then crumble under heavy loads.

This process is called "fishmouthing." Smaller or thin-walled (0.035-inch) tubing can be fishmouthed by a round or half-round file. Start a notch with a rat-tail file, and then enlarge it until it starts matching the tube to join. Then notch the other side and file it down until both sides are the same. Deepen each side sequentially, never letting either get too far ahead of the other.

Bench tools can help, especially when the intersection is at other than 90 degrees (Fig. 10-22). Fuselage diagonals are good examples. These can be done by hand, but it's a slow process. Instead, chuck up a reamer or a grinding stone of the same diameter as the longeron in a drill press. Mount the tube in a drill table clamp at the same angle (relative to the ream or stone) at which it is supposed to join the longeron. Start the motor, and slowly lower the ream to the tube.

If you have a bench press clamp with a table that cranks in and out, mount the longeron-sized ream in the chuck, and use the table to bring the diagonal into contact. This method also can be used with a metal lathe.

Whether fishmouthing by hand or with power, deburr and smooth the piece with emery cloth when the metal has been trimmed properly.

More complex-shaped notches can be made by a variety of methods. Both the Avid Flyer and the Kitfox need oddly shaped slots in the root ends of the spars. The companies supply paper templates to mark the area to be removed. For larger areas such as these, a Dremel Moto-Tool is probably best. Cut close to the line, and take out the rest with files.

Fig. 10-22. *The intersection of the two verticals with the longeron are easy to do, but the diagonal presents another problem.*

Hold the Moto-Tool firmly—if the cutter or wheel catches an edge, it'll jerk the tool, chewing up the surface. If possible, position the tool so that the tendency is to pull away from the spar instead of into it. As always, deburr and buff smooth when finished.

Fasteners

Two fastener types predominate aluminum-tube construction: pop rivets and bolts. Driven rivets form the shop head by repeatedly ramming the end into a bucking bar. Pop rivets do it differently.

As shown in Fig. 10-23, a pop rivet has a hole running its length. A stem passes from the shop end through the center of the rivet and sticks out above the manufactured head. At the shop end, the stem is wider to form a *mandrel* that is wider than the center hole and sits flush against the shop end.

The rivet tool grabs the free end of the stem and pulls while holding the rest of the rivet stationary. The force tries to drag the mandrel toward the manufactured head. Because it's wider than the center hole, the only way the mandrel can move is to crush the softer metal of the rivet shank. When the shank is completely compressed and can't collapse any more, the stem of the rivet breaks away.

Where a flat plate must be pop-riveted to a tube, the line of rivets should run straight down the centerline (Fig. 10-24). Any offset weakens the connection.

Countersunk pop rivets are used on some airplanes. However, be advised that the head angle is different from that of standard rivets. Countersunk pop rivets require a 120-degree tool. Countersinks are available in this angle, as are sheet-metal dimplers designed to work with the pop-rivet tool. The countersinking/dimpling procedure is identical to that of driven rivets, which was discussed in Chap. 9.

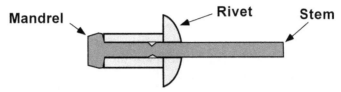

Fig. 10-23. *The pop, or pulled, rivet.*

Fig. 10-24. *Where a plate must be riveted to a tube, the holes should be drilled on the tube's centerline. (This photo shows at least three violations of proper workmanship as described in Chap. 7. Can you spot them?)*

There are MS/NAS standard pop rivets, but they don't seem to be marketed widely. Thus it becomes even more important not to diverge from the plan's specifications. The standard pop rivet is the Cherry rivet, sold by most homebuilder's supply outlets. These are available in different styles, sizes, and materials. Use only approved substitutes—rivets of the wrong material might be only half as strong as required.

One advantage of some types of pop rivets is the ability to expand and match misdrilled holes. Don't depend on this too much because the stronger rivets specified for aircraft generally don't expand as much.

Preparation is similar to that for driven rivets, as discussed in Chap. 9. The hole is pilot drilled and then reamed to the final size by using the correct bit. As with driven rivets, pop rivet sizes are exact, so this final drill must be slightly larger than the rivet diameter. The pilot and final drill sizes (incremental inches) for popular rivets are

Rivet	Pilot Hole	Final Hole
$\frac{3}{32}$ inch	$\frac{3}{32}$ inch	#40
$\frac{1}{8}$ inch	$\frac{1}{8}$ inch	#30
$\frac{5}{32}$ inch	$\frac{5}{32}$ inch	#21
$\frac{3}{16}$ inch	$\frac{3}{16}$ inch	#11

Deburr the holes, and cleco the components together in the usual fashion. When all the holes are ready, remove a cleco and insert a pop rivet. If the plans specify, dip the rivet in corrosion preventative first. Some types of rivets don't have smooth shanks and resist sliding all the way into the hole. A piece of wooden dowel with a stem-sized hole drilled in one end can be used to apply a little force.

The rivet's manufactured head should rest solidly on the surface. The components being riveted must be held tightly together during the process—the rivet might try to expand into any gap. This results in wrinkled aluminum and a weak joint.

Make sure that the correct head or nosepiece is installed on the rivet gun, and slide it down the stem until tip is solidly against the rivet head. Some types of Cherry rivets require a special tool.

Squeeze the handles of the tool together (Fig. 10-25). The jaws inside the riveter will clench the stem and then start drawing it back. The mandrel crushes the shell of the rivet. When it has moved as far as it can, the stem suddenly breaks away. Figure 10-26 summarizes the process.

Unless the material is especially thick or the rivet is short, the tool's handles will meet before the rivet pops. Release the handles, slide the nosepiece down the stem until it rests on the head again, and repeat.

When the stem breaks away, lift the tool away from the metal, release the handles, and twitch the tool in a safe direction. The stem should release from the riveter, and you're ready to go again.

Pop rivets are designed to work only in shear, so never install one in a position where the load tries to pull it directly out of its hole. Also, some pop rivets have rather soft heads; don't lift the top sheet away from the material once the first rivet is in place. Lifting the top sheet away bends the head.

If a pop rivet must be removed, drill out the mandrel first with a small drill, and then punch out the remainder of the mandrel. Then drill the center with the pilot

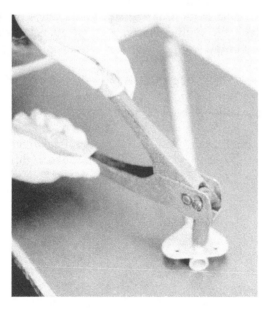

Fig. 10-25.
The rivet puller in action.

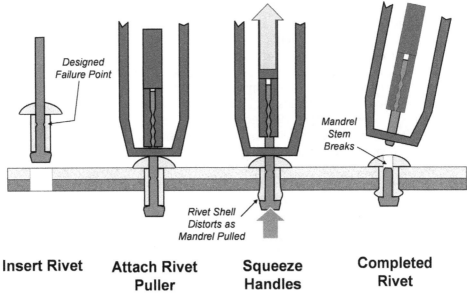

Designed Failure Point

Mandrel Stem Breaks

Rivet Shell Distorts as Mandrel Pulled

Insert Rivet   **Attach Rivet Puller**   **Squeeze Handles**   **Completed Rivet**

Fig. 10-26. *The mechanics of installing a pop rivet.*

drill. The driven-rivet method of using a smaller drill to remove the head and punch out the remainder doesn't work because of the shank's tendency to exapnd.

Installation of bolts is perfectly straightforward and follows the same basic rules explained in Chap. 7. However, there's another factor to consider when bolting aluminum tubes.

Screw threads are force multipliers; they convert a small rotational motion into high compression. The old hand-operated printing presses are a prime example of this effect. A few footpounds generated via the handle resulted in hundreds of pounds of force at the platen.

Bolts work the same way. However, most commonly used sizes of aluminum tubing aren't strong enough to withstand the force. As the bolt tightens, the tubing pinches in. You could completely squash a 0.049- or 0.058-inch aluminum tube using normal hand tools. Of course, an aircraft bolt would bottom out on its threads first. But the tube would still distort, and recommended torque levels couldn't be reached.

The solution is to increase the tube's wall thickness in the area of the bolt. This can be done in two ways. Some kits apply reinforcement plates to the exterior. This is especially common on kitplanes with aluminum-tube spars. Installation of these plates varies with the type of aircraft.

The other method is to install reinforcement inside the drilled area. For this to be effective, the outside diameter of the insert must be nearly equal to the inside diameter of the tubing.

The easiest way is to use a short piece of the next-size-smaller tubing. This works out fairly well; a 1-inch-diameter tube with a 0.058-inch wall has an inside diameter (I.D.) of 0.884 inch, which is just a tad larger than the outside diameter

(O.D.) of ⅞-inch (0.875-inch) tubing. If the ⅞-inch tubing also has a 0.058-inch wall, the total wall thickness has been doubled. Twice 0.058 inch is 0.116 inch, just a tad less than ⅛ inch. Therefore, for tubes with 0.058-inch wall thickness, select an insert with ⅛ inch shorter diameter.

Other reinforcement methods are possible. A solid aluminum plug will be strong, albeit heavy. Probably the best insert is solid plastic. Wood is cheaper and easily worked but should be varnished before installation. Steel has dissimilar-corrosion problems when in contact with aluminum.

The absolute minimum length of the insert is four times the diameter of the hole to be drilled. This allows the appropriate end margin in both directions. I like to make them a bit longer—their resistance to compression also depends on length. Mine are usually about twice the minimum, about 1 ½ inches long for AN3-sized bolt holes.

On the tubing to be reinforced, mark the position of the bolt holes. Drill a pilot hole on both sides. Deburr as well as possible. Drill a single, same-sized hole midway on the insert. Slide the insert into the tube until its hole matches one of the pilot holes. Cleco the pieces together, and then pilot drill the other side of the insert through the other pilot hole in the tubing. Drill and ream the holes to final size, and then disassemble (if possible) and deburr.

This method works when the insert can be rotated to match its pilot hole with the tube's. This isn't possible on square tubing or the ovals used on some ARVs. To install, drill the pilot holes in the outer tubing. Mark the center of the insert using a felt-tip pen. Draw a line all the way across.

Watch through a pilot hole as the insert is slid into place (Fig 10-27). When the centerline appears, drill the insert's pilot holes. Then drill and ream the holes to final size.

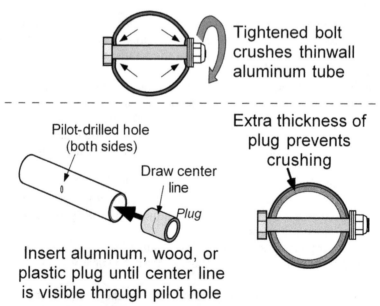

Fig. 10-27. *When bolts run through aluminum tubes, a reinforcement plug is usually necessary to support the pressure of the bolt.*

Bending tubing

Bending aluminum tubing is very much like bending a thick aluminum plate. The inside of the bend radius stays pretty much the same, whereas the metal on the outside stretches to remain continuous.

There are a couple of differences. First, you're manipulating less metal. A 1-inch-diameter tube is far easier to bend than a 1-inch-thick piece of aluminum plate. Believe me.

Second, the tube has a lot of ways to react if it doesn't like the bend. As mentioned earlier, a tube under a bending load is compressed on one side and stretched on the other. The center remains neutral. When a piece of tubing is overbent or bent incorrectly, the outside of the bend can tear open and/or the inside can crimp.

How the tube reacts depends on the bending procedure. Proper fixtures and practice minimize problems.

To begin, let's understand how a crimp is formed. Take an ordinary drinking straw and slowly fold it in half while watching the inside of the bend. The material under compression has to go someplace, so it folds inward. But because the surface is curved, the material expands sideways as well. When the straw is fully folded, the crimp is one-half the circumference of the straw across, and there *is* no inside diameter at the fold point. And because tubes depend on their circular cross section for strength, the area of the crimp is very weak.

The goal when bending aluminum tubing is to avoid crimps. One way is to maintain the cross section of the tube throughout the bending process. There are coiled-spring tube shells that slide over the outside, but they usually aren't available for the larger sizes. Or you can fill the tube with sand prior to bending. This works, but long or wide tubes need a lot of sand.

There are custom tools on the market for bending tubing. Before you buy one, make sure that it'll handle both the diameter and material of the tubes you wish to bend. Some are intended for soft annealed aluminum brake lines. An adequate tool to bend a single size of large-diameter aluminum tubing costs about $100; those that can handle a wide range of tubing sizes cost four or five times that.

The cheapest solution is procedural: Bend the tube gradually around a radius in such a manner as to stretch the outside surface rather than crimp the inside edge. This can be done in the shop with a few tools.

Two types of bends are used on kitplanes: *radius bends,* where bends of particular characteristics are required, and *shape bends,* where the final shape is given rather than a set of radii. We'll look at radius bends first.

A benchtop jig is necessary. The kitplane's instructions will specify the radius, and the total angle of bend can be eyeballed. Take a bit of scrap plywood, draw a circle of the proper radius on it, and then cut it out with a saber or band saw.

A full circle, not just 90 degrees or whatever. The tube will have a bit of spring-back, meaning that the disk must continue the radius a few degrees more than the angle the plans call for. You'll typically have to draw nearly a full circle on the wood anyway, so you might as well cut the whole thing out; you might need a greater angle at the same radius later. Smooth the edge a bit with sandpaper, and nail the disk to the worktable.

Of course, it would be even better if the edge were grooved like a pulley to match the tube. It's not easy to do, but if you've got a router table, it might be worth a try. Another option is to keep your eyes open for cheap pulleys of the proper sizes.

When ready to bend, place the tube against the disk, and nail a piece of 2 × 4 against the free end on the opposite side. Position the tube so that the point at which the bend is to begin is the point touching the disk.

Draw an imaginary line from the point at which the tube contacts the disk through the disk center; call this the "axis of contact." To make the bend, grab the free end of the tube, and pull *parallel* or slightly *outward* to the axis of contact, as shown in Fig. 10-28. *Never* apply any sort of pressure *toward* the disk. If you do, the metal on the inside tries to compress and starts a crimp.

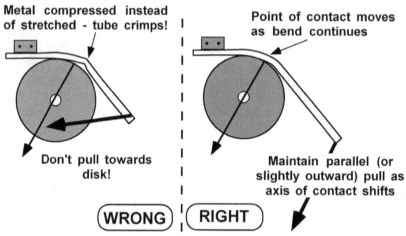

Fig. 10-28. *Smooth tubing bends are made if the pressure is applied parallel to the axis of contact.*

Bend the tube to the desired angle while watching carefully for crimping. If it looks like the tube is beginning to flatten, relax pressure and slide the tube slightly to move the point of contact. If table space becomes a problem, move the scrap holding the free end and resume until the desired angle is reached.

Heavier tubing requires a lot of pressure. The longer the work end, the more force you apply with less effort. If possible, don't cut the tube to length before bending. If the tube is already to length, slip a smaller-diameter tube or a wooden dowel into the work end to give you more leverage.

Shape bends are usually easier. Begin by stapling a piece of brown wrapping paper to the tabletop. Transfer the shape shown in the plans to full size on the paper.

The shaping process consists of bending the tube and then comparing it with the desired shape. Bends are applied where necessary and as necessary. If a piece gets bent too far, you can bend it back as long as it hasn't started to crimp.

The same rules still apply. Bend all tubes around a radius. Too sharp an edge will start a crimp. But these shapes usually involve gradual curves that don't start problems. Slight bends at many points are less angular than a few big ones.

A very common shape is a rib formed of bent ½-inch aluminum tubes. A common bending tool in these cases comes in pairs: your knees. These smaller tubes can be shaped easily by hand.

Where heavier tubes must be shaped, a combination approach works best. Bending jigs can make gradual curves by making closely spaced small bends. For very wide-radius shapes in heavy tubes, a car's full-size spare tire can be used. Prop up the tire, and place the tube on top. You and a friend then push downward, see-sawing back and forth over the tire. It sounds ugly, but it works.

Building structures

Drilling, pop-riveting, and bending are all steps in the construction of aluminum structures. Aluminum structures usually are held together with gussets—pieces of sheet or plate aluminum that rivet to both pieces of the structure.

Figure 10-29 illustrates the three primary methods for building aluminum-tube structures. Thin wrap-type gussets (typically 0.025-inch 2024-T3) are used in many light aircraft, such as the CIRCA Nieuport replicas. Other aircraft, such as the RANS series, use heavier plate-type gussets. Finally, a few, such as the Murphy Renegade (Fig. 10-30), use custom extrusions that are lighter than the plate-type and easier to use than the wrap-type.

The first step in building a full structure is to draw a single long reference line atop the worktable. This is your basic reference line for measuring positions and angles of all elements. You could use a chalk line, but go over it with something more permanent. Or pull an ordinary string tight, mark positions every foot or so, and then connect the dots with a straightedge and pencil.

Define one end of the line as being the firewall or wing root (depending on what's being built), and use a carpenter's square to draw another line perpendicular to it.

The plans will give the distances at which verticals, compression struts, or ribs must be installed. Transfer these dimensions to the table, and draw perpendiculars at the points with the carpenter's square.

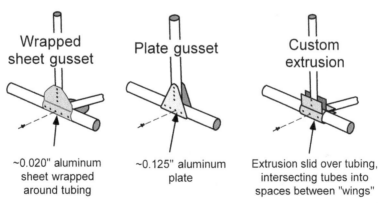

Wrapped
sheet gusset
~0.020" aluminum sheet wrapped around tubing

Plate gusset
~0.125" aluminum plate

Custom extrusion
Extrusion slid over tubing, intersecting tubes into spaces between "wings"

Fig. 10-29. *Aluminum tube structures generally are assembled in one of three ways.*

Fig. 10-30. *The Murphy Renegade's fuselage is aluminum tubing. The builder assembles the entire structure using special fittings and pop rivets.*

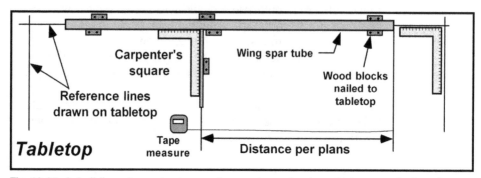

Fig. 10-31. *In building a large structure such as a wing or fuselage, two perpendicular reference lines are first drawn on the surface of the worktable. Then other components are put in place in reference to the dimensions shown in the plans. Parts are held in place with short pieces of wood nailed in place.*

Cut the components to size, and install them in relationship to the patterns drawn on the table. Hold them in place with pieces of 1 × 2s or 2 × 4s (called "jigging blocks") nailed to the tabletop. Don't pound the nails all the way in; leave a half-inch or so sticking up so that they'll be easy to remove. This process is shown in Fig. 10-31.

Corrosion protection

If the fuselage is made from 6061 alloy, you might be able to forego the corrosion protection regime required for 2024 aluminum or even 4130 steel. It depends on where the plane will be stored, your proximity to salt water, and how strongly the kitplane manufacturer recommends it.

Priming aluminum is detailed in Chap. 9. To recap, the metal first must be cleaned to eliminate grease and finger oils. Then it's etched with a mild phosphoric acid wash to roughen the surface to enhance paint adhesion ("engineerese" for making the paint stick well), and the primer is sprayed on.

Tube-fuselaged kitplanes are about the fastest to build. They're intended as fun airplanes, so you don't get involved with large amounts of subsystems work. As such, they're generally pretty light as well, which makes them easier to manhandle around the shop.

Almost all of these aircraft are fabric-covered. This process is addressed in the second half of Chap. 11.

CASE STUDY: AVID MAGNUM

It was probably the coziest homebuilding scene you could imagine: a snug work-shop, a crackling wood fire in the corner, and an Avid Magnum under construction on center stage. Kirk McCarty (Fig. 10-32) retired a few years back after a career as an engineer and manager with a major aerospace firm. His hasn't been an idle re-tirement—bicycle touring through Ireland and Spain with his wife; getting his in-strument, commercial, and instructor flight ratings; flying his first homebuilt; and building a second one, the Magnum (Fig. 10-33), in a snug single-car garage.

Let's take a look at how he did it.

Selection and delivery

The Pacific Northwest is an ideal environment for water birds of both the feathered and manufactured varieties. Kirk's first homebuilt was a plans-built Osprey amphib-ian that took him about eight years of work. It first flew in 1989.

Years later, he still enjoyed flying the Osprey. As time went on, though, he found the Osprey lacking in two characteristics. First, the all-wood amphibian has a rather tight cabin with insufficient cargo space. Second, McCarty wanted a plane with bet-ter short-field capabilities.

"I had a heck of a time choosing between an RV-6 and the Magnum," Kirk remem-bers. "But I decided I wanted a decent floatplane." The Murphy Rebel also was a contender, but Kirk liked the looks of the Avid better. A test flight in a friend's new Magnum helped settle the decision.

He ordered the Magnum kit in August 1994, and it was delivered directly to his garage workshop the next month. The kit was delivered in a single huge box, 16 feet by 4 by 4 feet. He'd known that he wouldn't be able to unload the crate. Instead, the shipper agreed to leave the trailer overnight, and Kirk called his two adult sons to come over and help. That night, they broke down the crate inside the trailer and carried the parts into the garage. They inventoried the contents as they went. "Everything was there, and nothing was damaged," Kirk remembers.

He was quite happy with the shipping firm, too: "A really nice bunch."

Kirk's Lycoming O-360 has an unusual history. Originally mounted on a brand-new Piper in 1978, the airplane went into the water a year later. The engine was

Fig. 10-32.
Kirk McCarty built two small homebuilts in the same single-car garage.

Fig. 10-33. *McCarty's Avid Magnum took about four years to build.*

tied up in an insurance dispute for a number of years, finally being rebuilt and sold to an Oregon homebuilder. The builder lost his medical certificate, and Kirk answered the man's advertisement in the *General Aviation News*.

Building experience

Kirk built the Magnum in the same single-car garage where he constructed his Osprey. By most standards, things were tight, but he was used to it. Since it's a stand-alone building, it's probably a bit bigger than the typical attached garage. His shop layout is shown in Fig. 10-34.

With such tight quarters, storage space was at a premium. Wings hung on the walls when not being worked on. The flaperons were hung from the ceiling, next to the cowling parts on their own shelf. Shelves and cabinets lined the far wall with hardware and various parts such as fairings and wheel pants.

The Magnum uses a traditional welded-steel-tube fuselage truss with the aluminum-tube wing spar concept introduced by the ultralighters of the early 1980s. The main spar doubles as the leading edge. The two spars are joined by compression struts and bracing, with wooden ribs epoxied to the spars.

Kirk ordered his Magnum fuselage powder-coated, so he didn't have to worry about protecting bare steel. Holes drilled through the steel usually have to be reamed because they are partially closed with paint. Some holes in the mounting tabs also have to be drilled.

Again, having built a previous aircraft, Kirk didn't have a major outlay for additional tools. As far as building the Magnum, though, he says, "A hand-held power grinder is an absolute must." The edges of the Magnum's fiberglass parts were a bit rough, and the hand-held grinder makes quick work of shaping or cutting.

Kirk completed his Magnum after about four years of building and had many happy flights. Like many builders, though, the bug bit him again. His newest project is a Murphy Moose. Fortunately, he's moved to a new home with a more spacious shop!

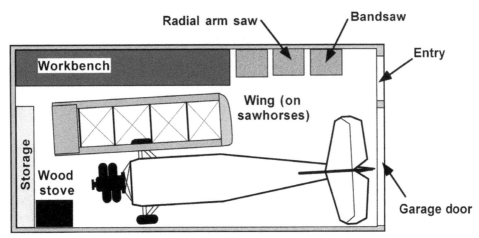

Fig. 10-34. *Careful planning makes the most of available space for kitplane building.*

Builder advice

Kirk McCarty has definite advice for those considering a kitplane: "Think about it. Make sure that you want to go through the agony of building."

If you do decide to build, "There are a lot of people around who can help." Kirk has sought advice from numerous experts over the years, picking up tips on welding, fiberglassing, and engine installation.

"Stay the course," he advises. "It can be boring as hell sometimes, but it's very fulfilling."

QUALITY CONTROL

A lot of the quality control for steel-tube fuselages is out of your hands. Still, there are a number of points you should remember. In any case, most of the following summary applies to aluminum-tube structures as well:

- When attaching anything to a steel or aluminum tube, all the holes should be aligned with the tube centerline.
- The centerpoint of all holes should be at least two times the hole diameter from the closest edge.
- Bolt holes shouldn't be spaced any closer than three times the hole diameter.
- Don't drill holes in the tops or bottoms of your wing spars unless the plans specifically call for them. This doesn't apply out near the tip because there's little load on the spar at that point.
- Since the lever action of the bolt threads can overcome the stiffness of most aluminum tubes, the tube should be reinforced wherever a bolt is installed. Some instructions call for installation of external plates at the attachment point. Most often an insert is slipped into the tube with holes to match the bolt holes. This insert can be a smaller segment of tubing or a solid plug of plastic or wood. Varnish wood plugs before insertion to protect the metal from the wood's retained moisture.
- When a tube intersects another, its end should be shaped to closely match the surface of the other tube ("saddling).
- Smooth the ends of all cut tubes and metal plates to eliminate the "stress risers."
- Put a generous radius on all inside curves; cracks can start from too-tight turns.
- When bending tubing, minimize crimping on the inside of the turn by stretching the tube while bending.
- Deburr all holes, whether you drill them or they come predrilled in the kit.
- Don't drill holes directly to the target size. Drill one size smaller, and then use a ream to enlarge to the final size.
- Obtain your pulled rivets from an aviation supplier.

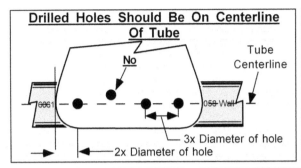

Drilled Holes Should Be On Centerline Of Tube

No

Tube Centerline

6061

.058 Wall

3x Diameter of hole

2x Diameter of hole

Edge Margin/Spacing		
Hole Size	Edge Margin	Minimum Spacing
3/32	3/16	9/32
1/8	1/4	3/8
5/32	5/16	15/32
3/16	3/8	9/16
7/32	7/16	21/32
1/4	1/2	3/4

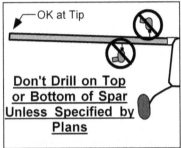

OK at Tip

Don't Drill on Top or Bottom of Spar Unless Specified by Plans

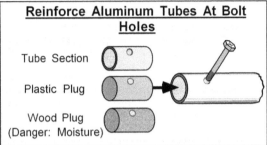

Reinforce Aluminum Tubes At Bolt Holes

Tube Section

Plastic Plug

Wood Plug
(Danger: Moisture)

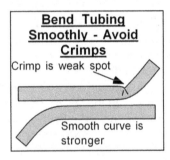

Bend Tubing Smoothly - Avoid Crimps

Crimp is weak spot

Smooth curve is stronger

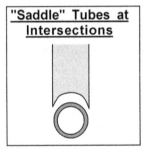

"Saddle" Tubes at Intersections

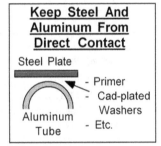

Keep Steel And Aluminum From Direct Contact

Steel Plate

- Primer
- Cad-plated Washers
- Etc.

Aluminum Tube

Fig. 10-35. *Key tube construction techniques make the job easier and maintain that essential margin of safety.*

- Don't let raw steel and aluminum make contact; accelerated corrosion results. Prime the pieces and/or use cadmium-plated or stainless-steel hardware to separate them.

Figure 10-35 summarizes the key points. Also, don't forget the basic workmanship rules given in Chap. 7.

11

Wood-and-fabric construction

WOODWORKING HASN'T BENEFITED MUCH FROM THE TECHNOLOGICAL REVOLUTION. Wood is passé. Industry spends millions of dollars on developing faster, more precise methods for cutting aluminum or on formulating new composite construction materials and methods. Mainstream aviation was the catalyst for some of these activities, and the kitplanes industry reaps the benefits.

Wood suffers from other faults as well. Manufactured materials can be checked easily for flaws—samples of 2024-T3 aluminum should be identical. Any difference is bad. It's easy to design equipment to detect imperfections and reject faulty products.

But wood is a natural material. Two samples from the same spruce tree will show differences, yet each might be perfectly adequate aircraft material. Computers can't tell minor variances from dangerous flaws. It takes a trained human eye. And if you can't detect the bad pieces with a computer, modern manufacturers aren't interested.

And as a natural material, wood is subject to nature's course. Everyone has seen rotted wood fences and other structures. Knute Rockne's death resulted in wooden box spars being banned for commercial aircraft. As with the *Hindenburg* disaster, a single, well-publicized failure negated many years of proven service.

When talk turns to wooden homebuilts, our mental image is that of an old geezer in bib overalls surrounded by ancient saws puttering happily along in clouds of sawdust. Somewhere in the corner of the shop, a Fly Baby or a World War I biplane replica is slowly taking shape.

You might think that wood has become a dusty footnote in aviation history. There's one problem: Wood is a perfect material for aircraft. It's strong, it's light, it forms complex shapes with ease, it doesn't get weaker from fatigue like metal, and it can be cut and shaped with inexpensive hand tools (Fig. 11-1). Even Burt Rutan started out with a wooden airplane (Fig. 11-2). A Rand KR-1 uses foam for shape and

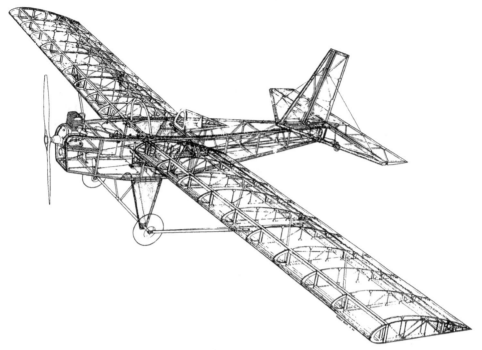

Fig. 11-1. *Although an ultralight, the MiniMAX is built of wood in the classic manner.* TEAM, Inc.

Fig. 11-2. *Burt Rutan gained fame with composite airplanes, but his first homebuilt, the VariViggen, featured a wooden structure.*

fiberglass for surface toughness, but underneath is a wooden structure Glenn Curtiss would recognize. The Lancair and the RV-4 might be battling for the small-engine speed crown, but the all-wood GP-4 and the Falco are breathing down their necks.

Wood rot hasn't been eliminated, but modern finishes provide far better protection. And who leaves a brand-new homebuilt out in the rain, anyway?

Wood is probably the safest and most satisfying medium from which to build a kitplane. Sawdust smells good and won't kill you. Varnish might make you high but won't give you hives. Buy your tools at the hardware store. Dip a wood strip in hot water, and it loosens up enough to tie a pretzel. It reacts predictably when carved or sanded, and mistakes are corrected easily.

Worried about strength? Pound for pound, wood has double the tensile strength of aluminum. Tools? You don't have to order exotic framistats from homebuilder's supply outfits—good woodworking tools are sold at most good hardware stores. At decent prices, to boot.

Wood is nice.

But there are a few disadvantages for the kitbuilder. Your pieces must be protected from temperature and humidity extremes. Some airplanes come with prerouted and preglued ribs, but often the kitbuilder is expected to do most of the cutting, shaping, and gluing. Thus a typical wooden kitplane will take longer to build.

No matter. It's fun and easy. And you'll have all those dusty Fly Baby–building geezers coming to your shop to give you free advice and to lend you all sorts of useful tools. If there's one thing the EAA builder network can offer, it's expertise in building wooden aircraft.

This chapter has two parts. The first half presents the techniques of building wooden aircraft:

Cutting wood parts

Shaping by hand and with power tools

Gluing components

Preserving wooden structures

The second half of the chapter discusses the basics of fabric coverings. Fabric isn't just a wooden aircraft operation, but the techniques are similar, no matter the structural material.

The following sections discuss the basics of working with wood.

ADDITIONAL TOOLS

Additions to the basic tool list given in Chap. 6 depend on the kit being built. The smaller kits such as the Fishers and the Loehle P-40, as well as classic scratch-built aircraft like the Pietenpol (Fig. 11-3), can get by with the basic list, with an emphasis on wood-cutting blades for the power tools. As always, a bandsaw is practically required. Typical wood-cutting speed is about 2,000 feet per minute; find one that also can run at half that speed for cutting aluminum.

Larger, more complex wooden aircraft might need a table saw. Or you'll need one if you decide to build a simpler aircraft with minimal reliance on precut parts. A table saw is irreplaceable for turning thick lumber into thin capstrips or stringers.

Fig. 11-3. *Wooden homebuilts such as this Pietenpol need only basic woodworking tools to build.*

Long, straight cuts are its forte. Benchtop table saws can be had for $200 or so, and floor models start at about $100 more. Good-quality tools reduce the frustration factor in kit building, and for a wooden aircraft, the table saw is even more important. Check your recommended tool list; if a table saw is necessary, break the bucks loose and buy a good one.

If you're tight for space in your shop, go to the tools department at Sears and buy a set of retractable wheels for your table saw. I've got them on mine; they let me shove it into a corner when it's not in use.

Power sanders are must-have tools. A benchtop model, definitely. A small hand-held belt sander, too. The pad-type sanders are gentle and allow good control for delicate shaping. I don't buy the special sanding strips for my pad sander; I buy ordinary sandpaper sheets and cut them to size.

There is a wide variety of common woodworking tools that might come in handy for use on your airplane. A router, for instance, can be useful for rounding the edges of wood. A radial arm saw greatly simplifies cutting longerons to a particular angle. Check your kitplane's instructions to see what power tools the kit manufacturer recommends.

One "trendy" item is laser guides for various saws—small devices that shine a red stripe on the work to be cut. These are neat, but they can take a bit of practice. The one on my radial arm saw actually displays about $1/16$ inch to the left of the actual cut line.

Lay in a stock of blades for all your saws. There is nothing more irritating than breaking a blade just after the hardware stores close.

Another thing you'll need is clamps—a *lot* of clamps. One or two types predominate depending on the kit. You might be able to get by with just a few, but you'll spend a lot of time waiting for glue to dry so that you can remove the clamps and use them on the next component. It's not farfetched to need 25 or more of each variety.

Spring clamps work like clothespins—squeeze the ends, and the clamp opens; relax, and it shuts and holds fairly tightly. If you end up making laminations, you'll need dozens. However, these aren't very strong. Some of the types of glues used on homebuilts need a lot of pressure; C-clamps are about the only way to go. A related device is a bar clamp. A bar clamp is like a C-clamp that works over a wide distance.

Add the usual assortment of inexpensive hand tools: chisels, saws, and files. The final item is a shop vacuum. Maybe the little hand-held unit works great for other types of construction, but woodworking generates tons of sawdust. You might as well bite the bullet and get an adequate vacuum from the start. Professional models of some hand tools (such as sanders) come with an inlet for hooking up a shop vac. This cuts down on the airborne dust but jacks up the price.

MATERIALS, FASTENERS, AND SAFETY

The materials and fasteners for wood construction are quite straightforward, and the additional safety precautions are minor.

Materials

The premier wood used in homebuilt construction is Sitka spruce. The trees come from Alaska and British Columbia, and the timber is carefully sawed, dried, and milled to produce aircraft-quality material.

Make no mistake about it, aircraft spruce doesn't come from orange crates. Just as with aluminum and steel, there are extensive specifications that certified lumber must meet. Type A wood must have a minimum density of 24 pounds per cubic foot and a moisture content of between 10 and 17 percent, and the grain slope must be less than 1 inch in 15 inches (Fig. 11-4). Type A wood is approved for all applications. Type B specifications allow slightly less density and a bit more grain slope (1 in 15 inches) and shouldn't be part of the primary load path.

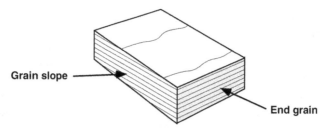

Grain slope

End grain

Fig. 11-4. *The grain slope must be less than 1 in 15 inches for wood used in spars and other structural elements.*

If you botch a part, you can't just run off to the hardware store for another piece of wood. Face it, aircraft-quality lumber does not end up at "Wood Is Us" at 49 cents a board.

This isn't to say that spruce is the only wood that can be used. Western hemlock, for example, is 14 percent stronger. FAA Aircraft Circular (AC) 41.13 allows direct substitution for spruce. But its quality varies more (good hemlock is more difficult to find), and it is about 7 percent heavier. Certified lumber in other than spruce is hard to find.

You can use the hardware store stuff for trim and interior purposes. For instance, I've used thin fir molding strips for turtledeck stringers and have installed small pieces of oak as backup blocks. But don't replace anything resembling a structural component with nonapproved wood.

The other wood commonly used in aircraft consists of multiple thin plies of bass, mahogany, birch, fir, or poplar bonded together with glue under heat and pressure—plywood, in other words. It comes in various combinations of thickness, plies, and grain orientation. Between three to seven plies are used. The "grain" refers to the relative orientation of the grain for each ply, usually either 90 or 45 degrees.

Plywood is used several ways. The most common use is to cover a structure while adding strength. Simple wood airplane fuselages are built as a wood truss. Gluing plywood to the exterior makes the truss far stronger. Plywood is often used to cover wings, even on aircraft that don't use wood for primary structure. The original Rotax-powered Pulsar, for instance, had plywood-covered wings.

Proper storage is vital. While metal corrosion is a chemical process, dry rot is caused by a fungus. It is an infection, pure and simple. A piece of contaminated wood will contaminate any nearby lumber. Infected wood shows whitish dots and streaks at first. If left unchecked, the wood turns black and crumbly.

Rot is stopped by keeping lumber moisture content below 20 percent. Keep the wood well ventilated, but you don't want it to become absolutely dry. A moisture content of about 12 percent is about right. Too little will result in embrittlement and overabsorption of glue.

Store wood in an area with a relative humidity of around 12 to 15 percent. Because the cut ends of lumber dry out faster, slap a quick coat of varnish or paint over the ends of stored wood. Keep it dry, and don't store it near a heat source.

When new wood is delivered to you, its moisture content is likely to be less than 10 percent. Store the new stuff with the old stuff for a couple of weeks until its moisture comes up.

If you're building in a heated shop during the winter, the heated air has a low moisture content, and your wood will dry out. Get a humidifier and a hydrometer.

To be sure of your wood's moisture content, you could buy the moisture meter shown in Fig. 11-5. In most cases, though, you can get by with monitoring your shop's humidity and giving newly arrived wood a chance to stabilize before using it.

Wood must be stored flat, not leaned against the wall. Any small but steady pressure will warp the pieces over time. Make sure that it's well supported; don't just lay an 8-foot piece of plywood across two 2 × 4s. This is an exasperating exercise—you want the wood supported over its entire length, but you still must allow air circulation or the unequal drying will cause it to warp.

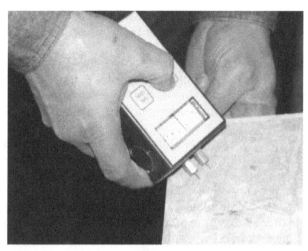

Fig. 11-5. *If your shop environment changes with the seasons, a moisture meter can be used to check the wood.*

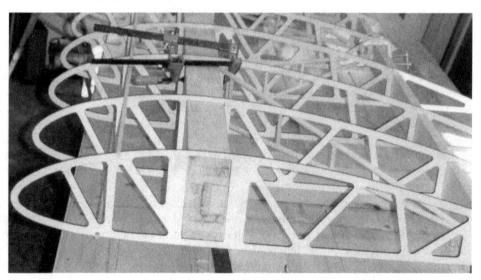

Fig. 11-6. *Wood kits don't have to be crude. These are laser-cut wooden ribs for a Warner Revolution.*

Plywood can be stored flat against the wall, but clamp it to prevent drooping and warping. A ceiling rack gets it out of the way but complicates access.

The preceding makes it sound like all you get with a wood airplane kit is a couple of baulks of lumber. This is not so—in many cases, at least. On most kits, you *can* build from scratch if you desire. But many kits have premade parts available, such as the laser-cut ribs shown in Fig. 11-6. Planes such as the Fisher R-80 Tiger Moth (Fig. 11-7) can be ordered with the fuselage sides and tail feathers already glued up. Again, the rules of storage are the same for these complete parts.

Fig. 11-7. *The sides of the fuselage and the tail surfaces of this Fisher R-80 Tiger Moth are available prebuilt as part of a quick-build kit. A welded steel-tube version is also available.*

Fasteners

The primary fastener is glue. There are several varieties. Casein glues are powders that are mixed with water. Resin and resorcinol glues are usually two-part synthetics that show superior resistance to moisture and thus are popular with amphibious aircraft. The most common resin component is urea-formaldehyde.

There are two main problems with both these glues: They don't fill gaps very well, and they require pressure to bond properly. The joints must be tight to begin with, and the pieces must be clamped while the glue sets. Usually, also, these glues require the shop temperature to be at least 70°F during the multiday curing period.

Again, modern epoxies come to the rescue, in the form of structural adhesives. They form a thick glop that fills gaps and bonds tightly under moderate pressure. They cure quickly and can be used at lower temperatures (although the cure time is affected).

This increased usability and strength have their costs, though. Just as with composite aircraft, these epoxy glues can lose strength as the temperature increases. Effects can be seen at as low as 125°F—a temperature that can be reached easily on wings exposed to direct sunlight.

Read the instructions carefully before using your glue. Some require different techniques for different types of wood. Also, don't mix types of glue—some resin glues use acid as a hardener, and the acid fumes can disrupt the epoxy hardening process.

No matter which kind of glue you use, one requirement is that it must form a bond stronger than the wood itself. If the joint fails, it must fail due to a splintering of the wood rather than a break in the glue bond. Make up test pieces out of scrap wood. Bond them together, let them cure, and then clamp one end in a vise and break them apart. Save the pieces—the FAA inspector may want some sort of proof of the glue's effectiveness. You might hang onto one or two unbroken test joints as well.

Other fasteners for wood aircraft are familiar to most home craftsmen: nails and wood screws. The nails look like what the home craftsman would call "brads," thin and less than an inch long. They are not used to hold the airplane together.

Rather, they hold plywood in place until the glue sets. The plywood can either be a large sheet, such as a wing covering, or a small gusset used to reinforce a truss joint.

Wood screws are either the AN545 round-head or the AN550 flat-head variety. They aren't used for structural purposes.

Safety

The main danger in building a wooden aircraft is from your tools—they all have nasty sharp edges. Even if you're using T-88, you aren't going to be awash in chemicals like composite aircraft builders are. The varnishes and other spirits used are pretty benign, unlike the primers necessary for metal airplanes.

However, the moisture content of wood can cause its own problems. Steel bolts through wood will tend to rust. In most cases, this won't cause a significant problem for quite awhile, but it's a good idea to coat the hole with varnish before inserting the bolt. Or use stainless steel hardware if your budget can stand it.

The nails mentioned earlier come with antirust coatings. That's a good reason to use genuine aircraft nails rather than hardware store brads.

Epoxy glues are occasionally sensitive about mixing ratios, so use a scale or a pair of syringes, as shown in Fig. 11-8. Read the label—some don't use a simple 1:1 ratio. I've seen a glue that requires a 3:2 mixture of resin and hardener. Read the label for safety information as well. Chapter 8 details the problems of chemical sensitivity.

Again, don't be tempted to substitute hardware store wood for aircraft materials. And use the size specified. Substituting $\frac{1}{16}$-inch plywood for $\frac{1}{32}$-inch plywood will double the weight. Builders have been known to use $\frac{1}{8}$-inch marine plywood instead of $\frac{1}{16}$-inch aircraft grade. While you might be able to find a break on price, it just isn't worth the extra weight and the reduction in flexibility.

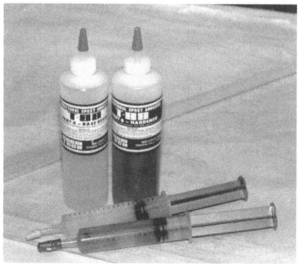

Fig. 11-8. *T-88 is a popular structural epoxy. The syringes aid in dispensing the proper ratio.*

As a last point, keep in mind that any airplane has a lot of metal parts. These have to be protected from corrosion, as described in Chap. 9.

CUTTING

Just about everyone has sawed a board or two. But building an airplane isn't like slapping a birdhouse together.

The grain

When cutting wood, the grain orientation is a key factor. The grain is a remnant of the annular rings caused by the variation in growth rates through the years. The long wood cells and the natural cellulose cement that binds them make the wood strongest in the direction of the grain.

The plans should specify the grain direction for each piece. Make sure to get it right—it makes a great difference in strength. If the plans don't give the direction, align the long axis of the part in the same direction as the grain.

The difficulty of the cut depends on its direction relative to the grain. The cement is actually harder than the wood fibers themselves. Cross-grain cuts are therefore hard work for the saw—it must cut through multiple cement lines. While the cut takes longer and all the cement tends to dull the blade, the saw is easy to guide.

Cutting with the grain is another story. The cement lines act like grooves on a roadway, they complicate lane changes. On the highway, the path of least resistance is along the same direction as the grooves. When met at a shallow angle, they try to force a car straight.

Thus it is with the grain. When a blade hits one of the cement lines, it tries to deflect the saw into a path of lower resistance along the grain. If the grain ran dead-straight in the direction you want to cut, this wouldn't be much of a problem. But even the most stringent certification standards allow a 1-inch-in-15-inch slope, approximately 4 degrees. If you're cutting a slot 1 inch deep, just following the grain will result in a $\frac{1}{16}$-inch gap, which is the maximum allowed when using high-tech epoxies. This gap would be intolerable with conventional aircraft glues.

Another by-product of this process is the tendency of the wood fibers to break away from the cement, forming splinters. Splinters can make a shambles of a cut line. Some woods are worse than others; pine, for example, splinters easily. Spruce's popularity is due in no small part to its splinter resistance—*resistance*, mind you, not immunity. It will still splinter.

While aluminum-airplane proponents have to watch the bend radius and composite-kit adherents watch the thermometer, the wood airplane builder must keep grain orientation foremost in his or her mind. There aren't any magic cures. You just have to take care to cut and trim to the line, fighting the grain all the way. A sharp tool is less likely to be deflected, so sharpen/replace as needed.

One way to reduce problems is to select the cutting direction carefully. If the grain is going to divert the cutting tool, cut so that the deflection will be *away* from the piece, not into it.

Laying out the shape

Before a part can be cut out, its outline must be marked. The usual practice is to use a pencil to draw the shape directly on the wood. Remaining pencil marks don't affect the varnish, but since the varnish is transparent, you'll want to erase the marks if the part will still be visible after completion of the aircraft. (I used a piece of hardware store hemlock for a seat spacer and neglected to sand off the rubber-stamped price before varnishing. The whole world now can see where $1.09 of my money went.)

Some plans include full-size templates. Photocopy the template, and glue the copy to the wood. If you botch the part, it's then easy enough to just grab another copy and start over. Or scan the image into your computer, and print the shape on sticky-backed paper. Check the scaling when you print, though. You might have to make slight adjustments to get the images to print full size.

Simple cuts

The vast majority of cutting on a wooden kitplane consists of cutting lumber to length or making shapes out of plywood. This process is little different from that for scale models. Thin plywood even can be cut with a modeler's knife. Run the blade along the same line a few times, and break the wood at the line.

For heavier pieces, a number of tools can be used: saber saws, drill presses, and table saws. Long, straight cuts are best handled by a table saw. Its unencumbered surface makes it one of the few tools that can handle large sheets of plywood. The blade housing on bandsaws often gets in the way.

Alternatives include the circular saw and the saber saw. Table saws have a *rip fence* to guide straight cuts, and it's a good idea to add one when using these hand tools. Commercial models are available, but a long piece of angle aluminum (or wood, if you can find a straight piece) and a pair of clamps work just as well. Measure the distance from the outer side of the saw's shoe to the opposite edge of the blade, and set the fence this distance away from the cut line. Clamp it in place. When cutting, maintain slight pressure against the fence.

Curving cuts generally are left to the band saw or the saber saw. Circular saws can't make tight turns. Really sharp turns might call for manual tools such as coping saws or keyhole saws.

While the cutting process is similar to that for aluminum, wood has a couple of differences that require changes in technique. The first is the rapid rate of cutting. Wood cuts easily; this is one reason why it's so popular. But a misdirected saw will do far more damage before your reflexes respond. One jiggle, and the part is ruined.

A factor in your favor is that wooden airplane building, even wooden kitplane building, isn't a manufacturing process. It's more a function of making pieces to fit other pieces. Even if you cut exactly to the line, you'll probably do some shaping to get parts to fit. Just don't cross the cut line.

The second problem is a saw's tendency to splinter the wood at the point that the blade exits the wood. As each tooth exits, friction pulls the surrounding wood with it. The wood lifts away and splinters (Fig. 11-9). Clamp the wood to a piece of scrap, and cut through both at the same time. The pressure from the scrap will hold

Fig. 11-9. *Wood tends to splinter on the backside of the cut; when using a saber saw as shown here, it'll splinter on both sides. To prevent splintering, clamp the wood to a piece of scrap, and cut through both pieces at once. Or cut far enough from the outline so that the splintering doesn't reach the part.*

the workpiece's wood fibers in place. This works well with a bandsaw, where the blade is moving in only one direction. A saber or scroll saw, though, still will see a little bit of splintering on top.

Again, don't cut exactly next to the line. Leave a little extra wood to splinter instead of the piece itself. Any splintering usually doesn't propagate far in spruce. There are also special plywood-cutting blades that have small teeth to reduce the problem.

Speaking of plywood, the glues used between the plies are tough on blades. Be prepared to replace them more often.

Fine cuts

While the information in Chap. 7 is suited for general cutting, frequently there are times when more exact trimming is necessary. The typical example is cutting a notch to join another piece of wood.

A simple L-shaped notch on a corner is easy enough. Use the same tools and procedures mentioned in the last section.

A square U-shaped notch is another matter. The sides can be cut with any convenient saw: saber, band, or razor. But how do you make the last cut across the inside?

It's pretty easy with wide notches. Cut the notch as a rounded U, running the blade along the marked lines on each side as far as practical. Then there will be enough room to cut out the corners.

If the notch is narrow, your options are many. Cut out the sides first in any case. This leaves an unsupported tab that is pretty flimsy. You can cut across the base with a razor blade. Or drill some holes near the corners, and slip a coping saw into place.

Or cut additional slots between the sides and cut up the pieces with a *wood chisel.* Wood chisels are wedge-shaped knives that come in a variety of widths. Set the chisel along the cut line, with the flat side toward the part and the sloped edge facing the tab to be removed. Hold the tool vertical, and then tap on the end of the handle with your other palm. A good, sharp chisel should cut right through, leaving few splinters (Fig. 11-10). Harder woods or duller tools may need a hammer and may leave the edge a bit ragged.

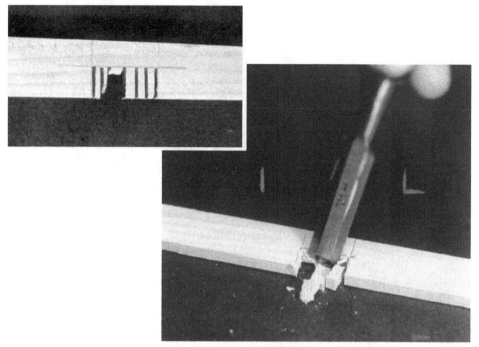

Fig. 11-10. *When cutting small notches, run the saw between the edges of the wood to make a sort of comb. Then cut away the teeth and use a chisel as shown to trim the notch.*

Chisels come in a variety of widths, starting at ¼ inch and working upwards. It's best to use a smaller width than the notch even if you have a chisel of exactly the right size. This gives you a little extra control.

Most cuts will have to be smoothed afterwards. Chisels work well with larger notches. Be careful with files—sometimes they're *too* effective. Emery boards are nice notch-cleaning accessories.

Holes

Back in Chap. 7 we discussed how to cut a hole in metal: Drill a hole to almost the right size, and then use a ream to enlarge it to the exact size.

If you're drilling in wood, forget all that. Drill the hole directly to the final size. Sequoia Aircraft recommends brad-point drill bits because they leave a very smooth hole.

Hole cutters for wood are widely available for those times when no drill bit is big enough. A drill press is almost a necessity, but use of the tool is straightforward. The main thing to watch out for is overheating caused by trying to cut too much too fast. It dulls the tools and chars the wood. If you smell something burning, back off on the press, and let the pieces cool.

Large-diameter holes, especially in plywood, are the saber saw's forte. Start with a drill bit that is slightly larger than the width of the saber saw's blade. Cut a hole near the cut line on the portion of the wood to be removed. Then insert the saber saw and cut toward the cut line at a shallow angle.

FINAL SHAPING

At some point you'll have to add the final shape to the wood pieces. This might amount to the last adjustment to joint surfaces, trimming away a blob of glue, or a variety of other actions. This requires hand tools of several varieties.

Smoothing the edge of a large piece of wood calls for a wood plane. Also, the tool can be used to make cut edges square with the top and bottom. The longer the plane, the smoother is the edge. The amount of wood removed with each stroke of the plane is adjusted by setting the height of the cutting blade.

The grain orientation has a great impact on the planing direction. Don't plane into the slope of the grain. For example, if the board's grain slopes downward to the right, plane from left to right. Otherwise, the blade will catch the cement lines and dig in.

A plane is nothing but a chisel on a carrier. Therefore, use a chisel in situations where lines of wood must be removed.

But chisels won't leave the smooth surface that a plane will. And some cases just require moderate shaping and smoothing. The power sanders (benchtop, belt, or pad-type) are the obvious choice. But they can't reach everywhere. In these cases, sandpaper and files are the obvious choices.

Files work best where inside corners must be sharp or where hard glue deposits must be eliminated. Their design lets them reach into tight crannies for those last-minute adjustments.

Sandpaper's various grit sizes allow better control of the amount of material removed. Unless you are sanding a curved surface, use a sanding block. Wrap a sheet of sandpaper around a piece of 2 × 4 or 1 × 2. This evens out the hand pressure; otherwise, the paper directly under each finger will dig in more.

A power option to consider is the small hand-held Dremel Moto-Tool. The company has a tiny sanding drum accessory that is dandy for tight, precision shaping.

Power sanders share a main problem with all power tools: If they are not watched carefully, they can strip too much material too fast. In addition, watch the edge effects of using power sanders. If you don't work on spreading out the action, the tool can leave indentations at the edge of the working area.

One type of final shaping is the rounding of all edges that won't be joined to other pieces. Nice sharp edges might look good, but they are a weak point for several reasons. They aren't supported very well, and a bit of impact will break off pieces.

This isn't dangerous from the point of view of structural strength, but it is unsightly. The part might stay just as strong, but broken-off edges make your workmanship look embarrassingly crude.

Also, the splinters that get started can propagate and disrupt gluing surfaces. And if the corner breaks away after varnishing, the exposed wood surfaces can be a starting point for deterioration.

The tools used for rounding vary with the size of the piece involved. Small pieces need only a bit of sanding. If you use too much tool, you stand the chance of breaking the piece.

Larger pieces leave a number of options. The power sanders work great. If you can, use a plane to cut the corner off, and then use the sander to round the remainder. This leaves a beautiful, durable edge. Depending on access to the part, the router might be an option as well.

BENDING

One of the reasons wood has always been popular as a homebuilt aircraft material is its ease in forming smooth curves. One has only to look at, say, a Falco and a T-18 to see the difference. Metal *can* form complex shapes, but such forming usually is beyond the capability of the average homebuilder.

It's easy to see that wood is a bit flexible. We'd expect a 10-foot piece of 2-inch-square spruce to be able to take the moderate bend required of a fuselage longeron. But some pieces must be bent more than the wood might take naturally.

The basic method is soaking and forming.

Soaking and forming

Did you ever see a piece of unprotected plywood left at the mercy of the elements? After awhile, it ends up severely warped. When wet, wood loses some of its stiffness. A little bit of pressure will bend it. When dried out, the wood regains full strength and tends to hold its new shape.

And this is exactly what you'll do with some of the wood pieces of your kitplane. Wood bends normally, of course, but for building your airplane, you're going to encourage it to bend even tighter.

The following factors make wood easier to bend:

- Decreasing the thickness of the wood
- Orienting the grain parallel to the bend
- Raising the moisture content
- Increasing the temperature

The bending process calls for soaking the wood, forming it into the desired shape, and then letting it dry. The first requirement is to get the wood wet. Hot water works best—the hotter the better. Boiling water works best, even steam if you can rig up a way to produce and deliver it. Small, long pieces can be placed in a metal tube fed with the exhaust from a teakettle.

This is pretty extreme. You might have some smaller pieces to bend, but the likely target will be plywood. The easiest soaking place is the bathtub. It makes sense—it's long, generally wide, and already set up to deliver hot water.

Household water heaters generally are set to around 130°F or so. More heat would be nice, but make sure that the system can take it. Make sure also that nobody is planning on washing dishes or taking a shower in another bathroom. Not only will they be robbing you of hot water, but the extra high temperature of the tap supply might be an unpleasant surprise. Family relations in homebuilder households generally are bad enough without broiling one's spouse or children.

Make sure that the sheet is in contact with the hot water on both sides. If not, it might try to take a contrary bend. It'll probably try to float; push it back down with sticks. If you weigh it down, have some scraps underneath it to keep from sitting flat on the bottom.

Soak time will depend on the temperature used and the thickness of the piece. Thirty minutes should be plenty. Then it's time to force the piece into the desired shape and hold it until dry.

Then you're ready to bend. The main thing is to do it *slowly*. As the wet wood comes under stress, it takes a moment or two to adjust and relax.

For simple bends, a former can be built from scrap 2×4s, as shown in Fig. 11-11. The plywood is bent by hand and slid between the two boards. The boards should be positioned so that the radius held by the plywood is slightly smaller than the final radius because the sheet will have a slight amount of springback when removed from the former. Leave it in place until it is dry, a process that might take a day or so.

On some airplanes, the wet pieces are glued in place immediately. This eliminates additional formers. The glues used by wooden kitplanes are still effective while the piece is wet. Details on gluing and clamping are provided later.

Another option is to form the plywood directly in place. Temporarily attach one edge, and lay thick, soaking-wet strips of towel along the area to be bent. Use an ordinary household iron to heat the area. As the plywood becomes flexible, wrap it around, and clamp it in place. This process is easier if there's a long piece of scrap wood attached to the free end of the plywood; this keeps the end from becoming wavy.

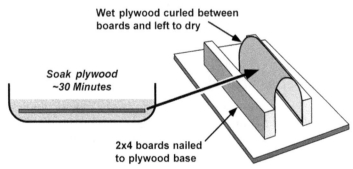

Fig. 11-11. *After soaking thin plywood in hot water for a half hour, curl it between two boards. When dry, it will retain the angle, less a little springback.*

Leave the part in place and clamped until it dries completely—this may take several days. When you remove it, apply straps to *keep* it in its flexed position. Otherwise, it will relax. Then prepare the glue line (or whatever additional steps are left to be done), and permanently attach the sheet.

Some kit manufacturers have builders do this in one step—*glue* the free end down, soak the plywood, apply glue, and wrap it around. It's your choice.

Laminating

The soak and flex method works for thin pieces and plywood. But what if you need a large piece of wood with a bend in it?

The solution is to take a lot of small pieces, bend them to the final shape, and glue them together. Typically, laminated parts come complete with the kit. Still, if you're trying to build for the utterly least cost, it's something you may end up doing. Let's take a brief look at the procedure.

It begins with a form in the shape of the desired final product. This might consist of a continuous surface or just a few blocks set up at key locations. The wooden strips that will be laminated together are coated on both sides with glue, placed together, and clamped into the form. When the glue has cured, the laminate is removed.

This doesn't sound too tough, does it? But getting the form right is the hard part. When removed from the form, the laminated assembly will spring back slightly. Thus the form must take this into account, relying on the builder's eye more than anything.

If your kit calls for laminating an assembly, the plans should give exact dimensions for the form. Clamping must be spread out evenly over the whole surface. Don't rely on just the shoe of the clamp; use blocks of wood between the clamps and the laminate.

The clamps must be applied to the centerline of the laminate. If they are located too near one edge, the laminations spread at the opposite end, an effect known as "fanning."

The more clamps you use, the better is the result. Figure 11-12 shows a lamination in progress, and Fig. 11-13 shows a close-up of the result.

GLUING

After all the work cutting and trimming the pieces, actually gluing them together is simple.

Joint preparation

The amount of allowable joint gap depends on the type of glue being used. *Casein* and *resorcinol glues* require tight joints, while *epoxies* such as T-88 can fill a gap of up to $\frac{1}{16}$ inch. When cut, the wood surfaces tend to dry out. If possible, glue the parts within a few hours of cutting.

If something gets between the two surfaces being glued, the glue will bond to the contaminant instead. Hence the surfaces of the joints must be clean. Sawdust is a major culprit. Sanding makes fine tight-fitting joints, but the dust tends to fill the

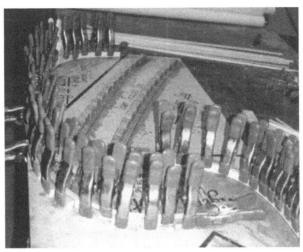

Fig. 11-12. *Spring clamps supply adequate pressure for epoxy adhesives. Here, the tip bows for a small wooden homebuilt are being laminated. The more clamps, the better.*

Fig. 11-13. *The result of the laminating process. Note the dark line of glue between each lamination.*

wood's pores and interfere with bonding. Don't sand the joint areas. Epoxies like rougher surfaces, so straight from the saw is fine. Otherwise, use a planer to shave just a tiny bit of wood from the joint area—just 0.002 or 0.003 inch is fine.

Get rid of oil and grease with acetone or lacquer thinner. Let the surfaces dry before gluing.

Glue application

If you've been through the chapter on composite construction (Chap. 8), you have read about the need to use glues sparingly to keep the weight down. This *doesn't* apply to wood construction. Wood soaks up glue. While this is bad from the weight standpoint, it's even worse from the bond-strength aspect. If the wood absorbs the glue, less is left to actually form the bond. Hence the builder must apply enough glue to ensure that enough remains on the surface.

The situation is relieved in one way: Wood is always glued under pressure, and the excess glue is squeezed out of the joint. It can be wiped away and removed. But if glue *doesn't* ooze out, open the joint up and add more glue.

To avoid glue "starvation," spread the glue thickly on both surfaces using a brush. If a piece of lumber has been cut directly across the end, the exposed "end grain" is extremely porous and thirsty for glue. Dab some glue onto the end, wait 20 minutes, and then add more glue and clamp the pieces together.

In some cases, the pieces aren't joined immediately. Sometimes the plans or the glue manufacturer requires that the glue air-dry for awhile before clamping. For instance, one popular brand of epoxy requires the user to wait until the epoxy mixture suddenly gets warm. It then hardens—at which point a second coat of epoxy is applied, and the pieces are joined and clamped.

There really is no one set of instructions for wood glues. With urea-formaldehyde glues, the user spreads glue on one side of the joint and hardener on the other. Many glues can be applied while the pieces are damp from the soak and bend cycle. Some can't.

Read the instructions.

Clamping

The parts must be clamped during the curing process. This is not quite so critical for epoxy glues, but casein and resorcinol glues require at least 125 pounds per square inch of pressure.

Clamping should make glue ooze from the joint. If it doesn't, separate the pieces and add more glue. Wipe up the excess with acetone or thinner. Smooth the joint with a fingertip or popsicle stick, leaving a small fillet.

Clamping methods vary. Ordinary C-clamps and spring clamps are the most common. C-clamps apply very localized pressure and can damage the wood, so add scrap pieces of wood under the shoes to spread the force over a larger area. Spring clamps can't apply enough pressure for traditional glues but are just dandy for epoxy-glued structures. (The lamination in Figs. 11-12 and 11-13 is being made with T-88.) There are other types of clamps as well; the only requirement is the ability to supply the required pressure without damaging the wood. Even clothespins can have their uses; they're cheap, too.

Any hardware store should carry a stock of various types of woodworking clamps. If the budget allows, pick up a sample or two in advance. Otherwise, keep their stock in mind as you face various jobs. It's always better to use the right tool than to struggle along with the wrong one. But sometimes exactly the right clamp just isn't made, and you have to come up with a solution (Fig. 11-14).

Large plywood pieces require pressure over a large area, which is tough to supply with conventional clamps. Enter the aircraft nail. The nail is used as a clamp in either a temporary or a permanent installation. Nails can be used instead of clamps in other applications as well. The nails should be four times as long as the sheet of plywood being glued.

Fig. 11-14. *Sometimes you have to clamp the clamps. Here, one clamp is used to help the other clamp maintain pressure. Note the thin strips of scrap wood under the pads of the clamps to protect the components.*

The nails don't add to the strength of the joint, so it's nice (but not necessary) to remove them once the glue has cured. Individually, they're devils to extract. But installing them through a thin strip of wood makes removal easier. Either lift up on the strip or break it away to expose the heads to a pair of pliers.

Staples are an alternative to nails. Don't use a standard commercial gun, though, because the staples are thicker than necessary. Buy one intended for aircraft use. The staples can be removed using the sharpened blade of a screwdriver (Fig. 11-15).

Scarf joints

One of the necessary evils of building a wooden airplane is the need for *scarf joints*. Often, two pieces of wood must be joined end to end. For instance, two 4-foot pieces of lumber are used to make a single 8-foot unit or two sheets of plywood are laid side by side.

An ordinary butt joint isn't strong enough. The contact area is too small, and just a teeny bit of glue can be used. Instead, aircraft use scarf joints. The edges of the two pieces are beveled and then glued. Because the bevel slope must be at least 10:1 (some sources recommend 12:1 or even 16:1), the contact area is increased dramatically. The overlap adds greatly to the shear strength.

Some kits supply components prebeveled; others require you make your own bevels. For boards, the slope can be marked along the edge and cut with a bandsaw. Plywood is a little more difficult. The basic requirement is to feed the sheet to a cutting tool that is set to the prescribed angle. Your plans should show how, or check with your EAA technical counselor. There are dozens of ways to scarf plywood.

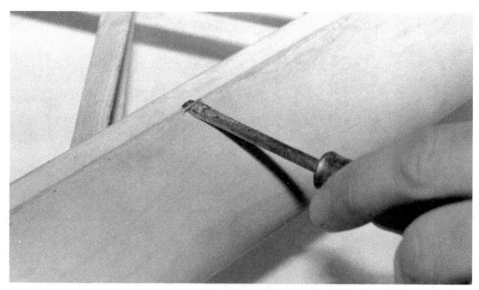

Fig. 11-15. *Loehle recommends holding plywood in place with small staples until the glue dries. Here, a sharpened screwdriver demonstrates how to remove the staples from the rudder of a 5151 Mustang.*

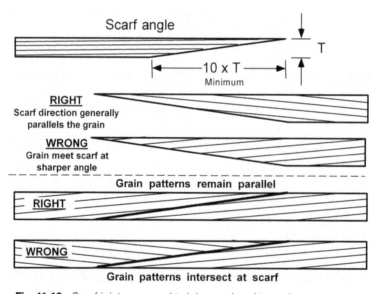

Fig. 11-16. *Scarf joints are used to join wood end to end.*

The biggest problem is to ensure that the scarf is oriented properly in relation to the grain. The scarf should approximately parallel the grain; when completed, only one or two grain lines should show on the cut area. The second piece should be cut in the same fashion. When joined, the two grains should appear continuous, except for a slight joggle at the scarf line (Fig. 11-16).

PROTECTION

As mentioned earlier in this chapter, wood's moisture content must remain at about 12 percent for maximum life. If it is too much wetter, rot can begin. If it is too dry, it becomes brittle and weak.

This is the reason you must varnish wood components. It's nothing but a water barrier to maintain the optimal moisture content. It sheds dampness, and it seals existing water into the wood.

If the varnish protects the wood, why do older wooden aircraft have rotting problems? The main problem is old-fashioned varnishes. Wood flexes under load; that's one reason it's strong. The older varnishes are not flexible, so they crack when the wood bends. Eventually, the cracks go though to the wood, and problems begin.

Modern urethane and epoxy varnishes are flexible, which results in less cracking. Most are two-part systems that must be mixed before use.

They're usually applied with a brush. The skeletal structure of most wooden airplanes makes spraying it on a waste of material. The goal is complete coverage because any gaps are a starting point for rot.

The best opportunity to apply varnish is before gluing. It's hard to get to some of the nooks and crannies before the structure goes together. However, keep varnish off any surfaces that will be glued. As mentioned previously, contaminants disrupt the bond and result in a weak joint.

The result is often a compromise. If access will be good after assembly, or if the area is one where appearance matters, glue first (Fig. 11-17). Otherwise, varnish the components, leaving a good margin around all areas to be glued. Test fit the components, and mark the limits with a pencil. You can keep a careful eye on the marked

Fig. 11-17. *The cockpit area of this Hevle Classic is easy to get to after assembly; hence varnishing can wait.*

areas while brushing on the varnish, or the glue areas can be protected with masking tape. Be sure to clean off any residue of the tape's adhesive.

In any case, don't shave the margins too closely. The varnish tends to ooze a little bit. Let the coating dry, and then glue. When the glue is cured, add a coat of varnish around all the joints.

Smaller components such as ribs can be dipped into a shallow pan of varnish and then hung to dry. This gives excellent coverage with very little additional weight. Again, place a strip of tape over the portions that will be glued later.

One of the major factors with varnish, even the modern epoxy types, is the number of coats. The level of protection rises tremendously with a second coat. One old airframe and powerplant (A&P) mechanic I talked to recommended initially applying a thinned coat; it would flow into the nicks and crannies better. Then back it up with two full-strength coats.

Sequoia Aircraft, the maker of the Falco, reports that the Forest Products Laboratory did a test on wood protection. They applied varying coats of a number of protectants and then exposed the wood to 90 percent humidity at 80°F for two weeks. The samples were then rated as to how efficiently they excluded moisture. The higher the number, the better they did:

Finish Type	One Coat	Two Coats	Three Coats
Spar varnish	0%	15%	30%
Polyurethane vanish	2%	23%	44%
Polyurethane paint	41%	61%	70%
Epoxy paint	40%	78%	83%
Enamel paint	50%	70%	80%
Aluminum-pigmented varnish	41%	77%	84%

Obviously, the more coats the better. Balancing the preceding results with the known durability of the finishes, Sequoia recommends the use of two coats of epoxy or polyurethane coatings.

Wood is the traditional aircraft construction material. While probably the slowest-building material, it's undoubtedly the most forgiving of mistakes. Personally, I find woodworking most satisfying. The components are smooth and warm to the touch, and rarely do they have dangerous sharp edges. Wood is the easiest to cut and shape, and the shop's smell rivals that of a kitchen on bread-baking day. There's nothing prettier than a wooden airplane just before the structure is closed up.

CASE STUDY: FALCO

If there's a pinnacle of wooden homebuilts, the Sequoia Falco F.8L is it. Designed by Stelio Frati, this aircraft is often called the "Ferrari of the air." The Falco proves that a homebuilt doesn't have to be made of fiberglass to look sporty and go like blazes. The Falco is unusual in that it was a certified production aircraft originally.

Dave Nason (Fig. 11-18) is a building contractor who decided years ago to get into another kind of "homebuilding." The complex all-wood Falco took him five and a half years to build, even with the pressures of business and family. He's taken several

Fig. 11-18. *Dave Nason during the construction of his Sequoia Falco.*

awards with his plane (Fig. 11-19), including Reserve Grand Champion at Oshkosh. The plane even appeared on the cover of *National Geographic Adventure* magazine.

Selection and delivery

Dave was no stranger to working on aircraft. Back in the 1960s, he restored a World War II Taylorcraft—and learned to fly in it. A few years later, he started work on a Midget Mustang, an all-metal plans-built homebuilt similar to the RV-3.

Dave's interest in finishing the Mustang eventually waned. He got bogged down with the need to build every single part of this aircraft—especially when a fire played havoc with the portions he'd already completed.

While continuing to own production aircraft, he got into large-scale RC planes. Eventually, he spent almost a year building a one-fifth scale Spitfire. He made a decision at that point: "If I'm going to spend this much time building an airplane, I'm going to ride in it!"

At about the same time, an article appeared in *Sport Aviation* magazine detailing the construction and flight to Oshkosh of a Norwegian Falco. The Falco used the

Fig. 11-19. *Nason's Falco. The Italian-designed machine has an unusual history; it's a homebuilt version of a production aircraft.*

same mode of construction as his Spitfire model, and he was sold. "It's easier to build the Falco," he says. "The parts aren't as small."

The Falco homebuilt has come a long way since its beginnings as a certified aircraft. In the 1970s, Alfred Scott of Sequoia Aircraft reengineered the design and produced plans. Over the years, Sequoia has provided more and more builder support. Currently, the company offers over twenty subkits.

Nason picked up the kits as he needed them. Most came by UPS. The biggest part was the wing spars; they came in a single crate *28 feet long*. Fortunately, the crate only weighed 250 pounds, and Dave was able to unload it without having to summon a batch of beefy friends.

The Falco plans give the full directions for building the spars from scratch. However, says Dave, "It would take at least six months to a year." Using the Sequoia subkits allows him to self-finance construction. Dave also saved time by buying a wrecked Falco for its wiring and electrical accessories.

Building experience

Dave lives on an airpark, and his property includes a large hangar as well as a large free-standing garage. This doesn't necessarily leave him with a lot of room, though. The garage is full of cars, boats, and a recreational vehicle, and the hangar is full of aircraft.

He built the Falco in a large room above the hangar (Fig. 11-20). A wood stove provided heat, and large windows and skylights augmented the fluorescent fixtures.

One problem, of course: He built the aircraft on the second story, and the Falco has a 26-foot one-piece wing—plus, the main fuselage is integral with the wing. He used a hoist to remove the Falco when it was almost completed.

Dave eschewed high-tech epoxies on his Falco, using Aerolite 306, a urea-formaldehyde glue. This type of glue was used on the famous Dehavilland Mosquito

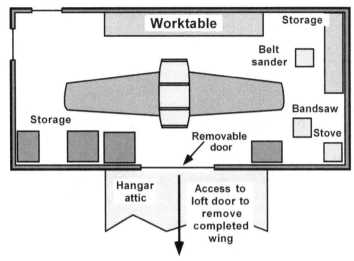

Fig. 11-20. *Nason's shop layout. Since it was on the second floor of his hangar, a crane was necessary to lower the plane to the ground.*

of World War II. The very complete Sequoia plans include details for using a number of other glues but recommend the Aerolite 306.

As with most glues, the working time depends on the temperature. It the temperature gets much above 75 to 80°F, the working time becomes too short. Dave arranged his building schedule by the thermometer. For instance, he skinned his wings on an early summer morning before the temperature started to rise.

The Falco's wing is covered with 2.5-mm birch plywood on top and 2.0-mm birch plywood on the bottom. "I didn't know you could bend wood like that," Dave marvels. He used heavy, wet towels and a steam iron to bend his plywood. He soaked the area to be bent, laid a wet towel along the bend line, steamed it with the hot iron, and applied bending force via clamps or rubber straps cut from an inner tube. He then removed the plywood, using clamps and straps to prevent the wood from straightening out while he prepared the surface and the glue.

Dave bought his plywood locally. He had to deviate slightly from the plans, in that the local supplier sells 61-inch-wide plywood instead of the 50-inch material the plans assume. It placed his scarf joints in different places.

Dave originally tried to set up a scarfing drum on his radial arm saw, but he found it easier to scarf by eye using a belt sander. "It took twenty minutes at first," he recalls. "But now I can do it in just a couple of minutes." He used the glue lines of the laminations to judge how the scarf joint was progressing. Starting out with a large belt sander, he would then transition to a smaller one and then finish up with a sanding block.

As far as small sanders go, he highly recommends the Skil Sandcat belt sander (Model 7102). It can be operated easily with one hand. Dave also considers a bandsaw and a benchtop belt sander as necessities.

The Falco's wooden skin is protected by a single layer of fiberglass. Dave used a slightly heavier cloth than the plans call for—he found it at a bargain price at a local aircraft manufacturer's surplus outlet. As with composite airplanes, the weave of the cloth must be filled in to get a smooth surface. In the beginning, he used microsphere filler, but he felt that he got better results from West Systems' Microlight.

Dave hadn't had to ask for any help locally. "I enjoy working on my own," he says. Probably his biggest source of aid and encouragement was the annual Falco-builder fly-in. There, he was able to take pictures to help answer his questions, and he had several opportunities to fly completed Falcos. He now regularly attends the fly-in with his own award-winning machine.

Advice

"If you want to get flying right away, the Falco is not the plane to build," says Dave Nason. "It's not something you can do overnight. If I get done in four to five years, I'll be happy."

"Don't look clear to the end of the road," he adds. "Look at it as building a bunch of parts, not a complete aircraft."

"You have to be patient. It's not a fast project, but it's very satisfying."

FABRIC COVERING

Fabric covering methods and materials have grown with aviation. Early pioneers used linen coated with ordinary varnish. Later on, various grades of cotton fabric became standard. But all these materials had one big problem—poor longevity. Even when kept indoors, a cotton-covered aircraft had to be re-covered approximately every 5 years and re-covered every year or so if the airplane is left outside.

It didn't bother people too much—it was just another cost of owning an airplane. Everybody had to do it. Fabric longevity didn't become an issue until the strides in aircraft design during the 1930s and 1940s, when aviation's explosive growth brought about lightweight aluminum alloys and the acceptance of monocoque construction.

After the war, the technology trickled down to the lightplane world. In 1946, the Cessna 120/140 aircraft featured a metal-monocoque fuselage and fabric-covered wings. Only the wings had to be recovered every few years. Later-model Cessna 140s metalized the wings as well.

Suddenly, one of the basic costs of ownership evaporated. In 10 years, a Piper PA-12 owner might have to recover and repaint three or four times. A Cessna might lose a bit of shine, that's all.

All-metal airplanes didn't replace the tube-and-fabric models overnight. Outfits such as Piper had a big investment in welding and fabric-working equipment. But tube-and-fabric construction is labor-intensive, and it's not very compatible with lower-cost assembly-line methods. In the long run, all-metal airplanes were cheaper to build.

Piper's steel-tube Colts and Tri-Pacers couldn't hold back the Cessna tide, so the Cherokee was introduced in the early 1960s. The Champion and Bellanca survived by filling specialized market niches.

While all this was happening, a fabric revolution occurred. Synthetic cloth had been around for a hundred years, but Dupont's polyester (trade name: Dacron) was the first to exhibit the peculiar qualities needed for aircraft use.

But the all-metal tide was unstoppable. While durable and easy to apply, Dacron couldn't change the fact that metal airplanes were perceived as the wave of the future. Under a variety of names, Dacron quickly took over the fabric market. Now cotton is used only by those antiquers interested in exact duplication of the original processes.

Why do some kitplanes use fabric, anyway? Why fiddle with obsolete technology? Why not make them composite or all-metal and be done with it?

In the first place, just because metal has advantages for assembly-line production doesn't necessarily make it best for a person building a single airplane. A factory worker gets lots of experience at riveting aluminum skins, but the homebuilder generally gets only one crack at it. A Kitfox builder applies glue to the structure, lays down the fabric, and then twitches it back and forth to eliminate any wrinkles.

Strength? The P-51 Mustang is a metal-monocoque design, but the rudder is fabric-covered. If it's strong enough for a 400 + -mph fighter, it's sufficient for a RANS. Even the early-model Beechcraft Bonanzas had fabric-covered control surfaces.

Weight? Fabric covering for a typical kitplane might weigh 10 pounds. The 0.040-inch aluminum skins on an all-metal airplane can weigh 20 times as much, although they also provide structural strength, for which the fabric airplane needs a beefier frame.

Drop a wrench on a metal airplane, and you'll get a dent. It'll probably just bounce right off a fabric surface. And if the surface gets damaged or must be removed to access internal components, fabric can be stripped away easily. Depending on the process used, patched areas are almost invisible.

Durability? During the 1960s, one local homebuilt had its tail surfaces recovered with nurses' uniform cloth from the local fabric store. The material was Dacron. The plane is still flying today, with the same fabric on the tail.

Some people might prefer the antiseptic sheen of a composite speedster or the mirror-like shine of an unpainted metal airplane. Personally, I love the satiny glow of doped fabric—and the delicate curves and dips as it wraps tightly around the underlying structure.

Pretty as it could become, it won't get that way without some work on your part. The following sections give a little background on the techniques and problems of applying a covering to your aircraft:

- Fabric selection
- Surface preparation
- Application
- Shrinking and doping

Fabric is applied to both wood and metal surfaces. For instance, the Fisher Classic and the Loehle 5151 Mustang (Fig. 11-21) have all-wood structures. The RANS Sakota is a typical all-metal tube structure. The Kitfox is somewhat of a crossbreed, in which the wing fabric must be attached to wood ribs and metal-tube spars. Some

Fig. 11-21. *Wooden airplanes such as this Loehle 5151 are usually (but not always) fabric covered.*

Loehle Aviation.

wooden airplanes use fabric as an exterior cover, with no aerodynamic purpose, because it reduces the need to fill nail holes and the like.

Whatever the underlying surface, fabric is installed nearly the same way. Let's take a look at some of the basics.

The Basics

Aircraft coverings, be they fabric, metal, or whatever, are used to define the exterior shape of the aircraft. They lessen drag by enclosing the fuselage structure and produce lift by assuming an airfoil shape. Therefore, aircraft coverings must be airtight (to be able to redirect the wind) and should tightly conform to whatever structure they are applied to.

This is easy enough with aluminum. But aircraft fabric is another matter. It's little different from the material in the shirt you're wearing—the breeze moves through it, and it wrinkles and sags at a whim. Obviously, builder effort is necessary to turn fabric into a proper covering.

The airtight aspect is the easiest. The weave is filled with a flexible coating called "dope." Flexibility is the key—the airstream will cause the fabric to vibrate. A stiff coating will crack.

Conforming to the underlying structure is a bit more difficult. Attachment is by a variety of methods from lacing to dope to glues. But to conform to the desired shape, the excess fabric must be removed between the attachment points. Once the fabric is attached to the skeletal structure of the aircraft, then the bagginess is removed by causing the fabric to shrink between the attachment points.

The shrinking method depends on the fabric material. Cotton is shrunk by the same dope used to fill the weave, while heat is used for polyester. It might seem that cotton is a more efficient system, but the polyester allows better control because it is attached, shrunk, and doped in separate steps.

Processes

The major polyester fabric systems (Poly-Fiber, Ceconite, and HIPEC) generally follow the steps described in the preceding section but get there in slightly different ways. Marketers offer low-cost videos on installing their product and often distribute installation instructions at no charge.

The major difference between the processes is the method of gluing down the fabric. Poly-Fiber (often referred to by its original name, Stits) and HIPEC use a single-part glue, while Ceconite has a cement activator that might allow better control. Any of them will produce an acceptable result. The processes allow some optimization based on aircraft type and mission.

Whichever way you decide to use, *don't mix processes.* They are not necessarily compatible. You might end up with a finish that develops cracks and flakes away. Materials might not be compatible between processes; you might apply dope and find that it dissolves the other company's fabric glue. If you decided on one of the major players, use only that company's components, except where substitutions are specifically permitted.

If your kit doesn't include the covering materials, count on spending at least $1,000 for the fabric, dopes, and accessories. You can save about 25 to 35 percent by buying generic Dacron and dope. But it's easier for a first-timer to work within the comforting framework of a well-designed commercial process.

Additional Tools

Primary tools will be brushes, mixing cups, *pinking shears,* and a 1,100-watt (or higher) iron. Pinking shears are scissors that cut a zigzag line. When cut straight, fabric edges start to unravel. When cut with pinking shears, the material can only unravel to the next zig. These tools are available through fabric stores or homebuilders' catalogs.

Irons are different. An ordinary clothes iron will do, but don't swipe the spouse's unless you replace it with a new one. The homebuilder's iron will get all grungy with glue and dope. If you're really being cheap, check the local Goodwill or Salvation Army stores for a used iron. That's what I did.

Units designed for aircraft use are sold for $20 or so. Hobby shops sell small irons (for applying plastic coverings to RC models) that work great for tight corners.

The primary requirement is the ability to hold a particular temperature. Clothing irons are labeled for material types, not actual heat setting. Buy a thermometer at a hardware store and a little tube of silicon heat-sink compound from Radio

Shack. Place the thermometer in a pan of boiling water and make sure that it reads 212°F (or whatever the boiling point is at your altitude). If the scale is off, adjust it to read correctly, and glue it in place.

Place a dab of the compound on the bottom of the iron, and insert the bulb of the thermometer. Turn on the iron, and adjust its dial until the thermometer reads 200°F. Make a mark on the iron's dial, and repeat for the other setpoints required by your covering process.

Some processes use a heat gun. These cost about $75.

Material, Fasteners, and Safety

There's very little high-tech involved in fabric operations. Most of it is traditional, but there are a few modern quirks.

MATERIAL

The primary material is the polyester fabric. You can still use cotton if you desire, but as mentioned earlier, it will need replacement regularly.

In addition to being strong and long-lasting, polyester has another trait that makes it well suited for covering airplanes: heat-induced shrinkage. The fabric is applied to the surface, and then the wrinkles are taken out by applying heat.

Most fabric-covered kitplanes include the covering materials. In a few cases, such as the RANS Coyote, the covering is a sailcloth slipcover similar to that used by ultralights. Otherwise, your kit includes at least a few bolts of cloth.

Fabric's strength and durability are determined by its weight in ounces per square yard. The heavier the cloth, the stronger and more durable it is. Typical weights run from about 1 ½ to 4 ounces.

The need for strength is obvious enough because you don't want the fabric ripping away at high speeds. Durability is another aspect. When first introduced, polyester was marketed as a lifetime fabric. Experience over the years has shown that "lifetime" in this case is about 20 years.

But there are other aspects to durability. Wear resistance is an important one. The typical fabric-covered kitplane is a taildragger, likely to be flown from rough grass strips. While a lighter fabric might be strong enough, the designer may pick a heavier grade to make the finished surface more resistant to puncture damage.

If you run out of fabric, two factors are important when resupplying:

1. *Order the proper weight.* One size heavier is OK, except from the weight standpoint, but don't install lighter fabric than the designer specifies.

2. *Order from the same fabric manufacturer.* A number of companies produce similar polyester fabric under their own trade names, for example, Dupont's Dacron and Celanese's Fortrel. Government regulations only specify the chemical content. Each company develops its own process. Do not substitute another company's product for the one supplied with your kit.

One timesaver is a covering envelope. Some aftermarket suppliers sew fabric into a sock that slips over airframe components (Fig. 11-22). The builder merely glues the sock down and shrinks it; all the fitting and cutting are eliminated. Envelopes

Fig. 11-22. *A fabric envelope would slip right over this wing panel, eliminating the need for measuring and cutting. Again, it's the usual question: Save money by doing it the hard way, or save time by spending money on envelopes?*

can save a considerable amount of time. They add about a third more to the *fabric cost*, which works out to about a 10 percent increase in total covering cost.

Fabric is also used in tape form for smoothing sharp corners and reinforcing high-stress areas. You could spend hours cutting strips from your cloth, but why bother? Order tape rolls instead. The two types are reinforcement tape, which is narrow but thick, and finishing tape, which is thin, has a very smooth weave, and is 2 or 3 inches wide.

Dopes are the other major material. Dope has two major qualities. First, it seals the weave of the cloth to make it airtight. Second, it's flexible, so the normal drumming of the fabric won't make it break away. Some processes have their own trademarked name instead of the term "dope." While the formulation might be different, the function is the same. For safety's sake, though, don't mix processes.

At least one coat of dope must contain blockers to stop ultraviolet ray deterioration owing to the sun. This usually consists of aluminum powder preadded to the dope. It's usually called "silver dope" and is the first coat applied.

Temperature is important to the doping process. If it is too cool, the dope gets thick and stiff to apply and takes forever to dry. Every 10°F drop in temperature *doubles* the drying time. The longer it stays wet, the more dust, dirt, and wayward flies end up imbedded in the surface.

Hot weather isn't good either. If the dope dries too fast, it gets a rough, dusty appearance. Retarders are available that slow the drying process.

Again, when using a particular process such as Stits or HIPEC, *always use only* the materials approved by the process designer. In addition, make sure that the surface coating—varnish, primer—of the aircraft structure is compatible with the covering process. There normally isn't any problem, but it's better to be safe.

FASTENERS

Fabric must be attached to the underlying structure. The primary concern is making sure that the fabric doesn't separate from the top surface of the wing. The traditional method is rib stitching, or making loops with a lacing cord around each rib through the top and bottom fabric.

But rib-stitching is time-consuming. The modern alternative is to glue the fabric to the ribs. Often, though, the processes require that the underlying surface be at least 1 inch wide, and the capstrips on most small wooden-winged homebuilts are narrower than that. Some FAA inspectors won't approve glue in cases where the ribs are round tubing, such as the RANS series and the CIRCA Nieuports. In these cases, alternatives to rib-stitching include small pop rivets or stainless-steel sheet metal screws.

SAFETY

The health hazards of fabric work are the usual: Use with adequate ventilation, and avoid contact with bare skin. (I swear that I've seen the same warning on a bar of soap.)

Anyway, any fabric glue or dope will make you giddy, so get some fresh air in the shop. Too much wind will blow dust around; you might have to postpone doping in these conditions. Avoid skin contact by using brushes and rubber gloves. As usual, a respirator should be worn for any type of spraying. The solvents used in the glues and dopes are toxic as well, so follow the manufacturer's recommendations.

A big hazard is fire. Dopes are flammable, and dope fumes are even worse. Eliminate all sources of ignition when applying dope, or provide extraordinary ventilation. Friction or even static electricity can ignite freshly doped surfaces. Provide air circulation, and leave the parts alone until the dope is dry.

COVERING PROCEDURE

The specific steps and sequences of covering your kitplane should be included in the plans, and specific information on using the covering systems should be available at low cost, or even free, from the process manufacturer. Let's look at the type of tasks involved.

Protection

Traditional varnishes, spar varnish, for example, are not compatible with the solvents used in the covering process. Therefore, don't use spar varnish where fabric will be attached. Check with the process manufacturer, but any two-part coating (urethane, epoxy) probably will be acceptable.

Once the airframe is protected from the fabric, protect the fabric from the airframe. Any little bump or any sharp edge might cut the surface. A classic example is the wire trailing edge used by Avid Aviation. Without protection, the cable could cut through the fabric. Watch bolt and pop rivet heads as well. Round off wooden corners prior to varnishing. The edge of most flat aluminum sheets is another problem area. These points and sharp edges must be blunted.

There are several ways to do this. The traditional one is to glue antichafe fabric tape over the offending points. Masking and duct tapes have the advantage of being self-adhesive, but they deteriorate over time. Electrical tape is more hardy, but it's expensive and awkward to use owing to its stretchiness; it's admirably suited for narrow areas where more thickness is necessary, such as along the thin aluminum edges (Fig. 11-23).

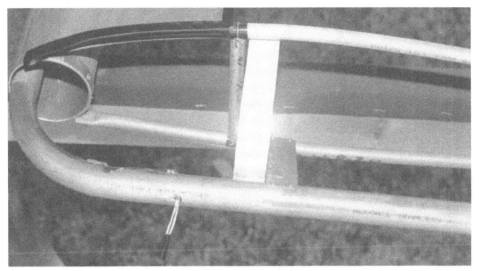

Fig. 11-23. *Electrical tape on the edges of the aluminum sheet protects the fabric. This is the tip of a RANS Sakota wing.*

Whichever method is used, apply as many layers as necessary, and taper successive layers to smooth the transition. Don't bury bolts that eventually might need removal.

Another factor in protection is the reinforcement of areas where large holes must be opened in the covering. An example is wing strut fittings. The usual practice is to pop rivet an aluminum frame (of thin sheet) around the position of the opening. When the covering is applied, it is glued to the sheet. The fabric is cut out of the center of the frame when the glue has dried (Fig. 11-24, and see Fig. 10-17 for a view of this fitting prior to applying the fabric).

On many kitplanes, rudder cables must pierce the fuselage covering. These smaller holes don't need frame-type reinforcement attached to the structure; rather, the fabric itself is reinforced locally. However, determining the exact exit point is difficult once the covering is in place.

Temporarily install the cables prior to applying the fabric. Make a cardboard template to match the structure, and cut the proper exit location on it. Note that if you make a mistake on the cardboard, just tape up the hole and cut a new one. Once the fabric is installed, place the template in the proper position and cut the hole.

Fig. 11-24. *Typical hole through fabric. Note the additional fabric added for reinforcement.*

Or a straightedge across the bare truss can establish the plane of the fabric and the intersection point of the cable. Measure the distances to several reference points on the truss. When the covering is installed, tie a pencil to a string and draw short arcs of the proper radius from the reference points. The intersection of the arcs will be the exit point.

Fabric application

Like most glues, the fabric attaching stuff doesn't work well on smooth, shiny surfaces such as some epoxy metal primers. Buff such areas with 500 grit sandpaper. Don't cut through the protection; just dull the surface a little. Wipe the dust off with a clean rag or paper towel.

The different portions of the aircraft require variations in the basic technique. Fabric pieces should be as large as practical. Most kitplane wings, for instance, require only two pieces—one top and one bottom.

Fuselages are another case. The complex shapes involved, such as the vertical stabilizer intersection, might need a smaller piece or two. Again, follow your plans.

The first step is to cut the fabric to size, with considerable extra to account for errors. Clamp it in place, and cut any slots required for fittings to pass through. These areas will be reinforced with tape subsequently. Then glue the material down in the sequence recommended by the builder's and fabric processor's manuals.

Don't sling the glue around wildly. It can drip to the other side of the structure and distort its fabric. Don't add glue to the exterior of the fabric unless called for. Otherwise, it actually might slow the curing process and require extensive cleanup before doping.

Apply the fabric as tightly as possible. The shrinking performed after installation is not intended to make up for sloppy work. The more the fabric is shrunk, the weaker it gets. Don't become fixated on a tight surface before shrinking; folds are all right, but try to eliminate large sags.

When gluing fabric to tubing, it must wrap between 180 and 270 degrees around the tube. In other words, at least halfway around the tube. Wrap no farther than three-quarters of the way, though, because it becomes tough to get the point of the iron far enough inside to seal the edge.

When fabric must overlap on a tube (such as covering a control surface or the horizontal stabilizer), there must be at least 1 inch of overlap between the two sheets. Add a 2-inch fabric tape as well.

When the joints between fabric sheets cannot wrap around the structure, the two sheets must overlap by about 2 inches if gluing is the only means used to join them. Such a seam can only be made if 2 inches of structure are underneath. The seam can't be made over a wing rib, for example, unless that rib is 2 inches wide over the entire length of the seam. Sewing is the only approved method if there isn't enough structure underneath.

However, wings are covered lengthwise. Unless you're scrimping with short stretches of fabric remnants, there's no need to join fabric anywhere but at the leading and trailing edges.

The top and bottom surfaces should overlap at least 2 inches at the leading edge. This shouldn't be any problem because every kitplane covers the leading edge with thin aluminum sheet that adds strength and impact resistance, as well as leaving a dandy surface for gluing fabric.

In all cases, seams should be covered with fabric tape that is wider than the overlapped area.

Shrinking

Polyester fabric is shrunk under heat, and an ordinary iron is the typical shrinking tool. As mentioned earlier, calibrate its dial, and make sure that it can hold a constant temperature. While a household iron is perfect for large expanses such as wings and fuselage sides, smaller RC model irons are best suited for precision work. Some processes use heat guns, but they're not really precision devices. Don't use one unless it is recommended by the process manufacturer.

One thing is sure: You will be surprised and pleased at the degree to which the fabric shrinks. Figures 11-25 and 11-26 show "before" and "after" views of ironing a repair patch. As mentioned in the last section, overshrinking isn't good. But careful attention to eliminating the biggest sags during fabric installation will reduce the

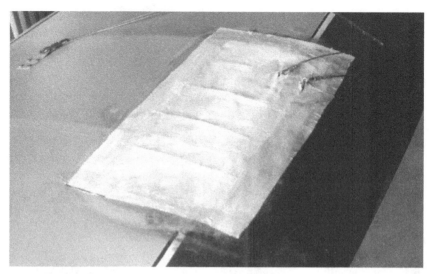

Fig. 11-25. *The Poly-Fiber fabric has been glued down but not yet heated. Note the wrinkles; these are the fold lines from the way the fabric was stored.*

danger. One type of wrinkle the iron might not take out is a small one that is set in glue. Watch these, and straighten them out before the glue dries.

Another glue-oriented problem is fabric sticking too far around a tube. For instance, the surface side of the covering might become stuck around much of the tube's circumference so that there's a long indentation where the fabric meets the tube. Slip your fingers behind the dented area and push outward while using your other hand to keep from pushing too far. Tighten the freed fabric with the iron.

Use care when shrinking the fabric because excessive tension can warp the underlying structure on some of the lighter vehicles. However, most of the processes require heat application at a certain high level, typically 300 to 500°F, to "set" the fabric. Otherwise, the fabric might become baggy in cold weather. You haven't much choice in the matter.

When possible, equalize the shrinking process by shrinking both sides sequentially. Shrink the top side of an aileron a bit and then the other side. Then shrink the top surface a bit more and so forth. Contact the kit and/or covering process manufacturer for more details.

Fortunately, any distortion is not generally permanent. The fabric can be stripped away for another try.

When the fabric is installed and shrunk, finishing tapes are glued over all seams and areas that require reinforcement, such as wingtips and sharp edges. The finishing tapes cover irregularities and make the entire job look smoother.

Rib Attachment

Certified aircraft must have positive attachment of the upper wing fabric to each rib. The traditional method is *rib stitching,* which is running a loop of cord through the fabric and around each rib.

Fig. 11-26. *The same repair area after heating.*

Typically, kitplane manufacturers substitute glue for rib stitching. According to the Poly-Fiber manual: "Over the years, the question of direct substituting rib lacing with cement bonding to the ribs has come up regularly, and the answer is no."

Basically, Poly-Fiber feels that the peel strength of the glue is overly affected by the surface preparation, solvent penetration from the finishing coats, and aging of the glue. The top of the wing is a severe low pressure area (it must be to supply lift), and fabric separation would be disastrous.

This isn't a condemnation of the Poly-Fiber process. It's one of the standards of the industry; and its developer, Ray Stits, is highly regarded.

Why, then, do the kitplane manufacturers specify cement instead of stitching? Because instances of fabric separation are rare. In other words, gluing works. Rib stitching is a tedious, thankless job. No matter what you do, the lacing shows through the fabric and detracts from the desired smooth appearance.

It's your decision. My club's Fly Baby had glued fabric and flew for more than twenty years without a problem. But tube-type ribs don't leave much surface for fabric attachment. In these cases, some sort of additional security is called for.

As mentioned earlier, the traditional method is lacing with cord. Rib stitching is one of those "ye olde" aviation arts. The lacing cord runs along the lower surface of the wing, loops over the ribs at intervals, and is tied with a *seine* knot (Fig. 11-27). I've seen many a diagram on the seine knot, none of which made it clear. It's actually fairly simple but is best learned by seeing it demonstrated. Talk to your EAA counselor.

One problem with rib stitching is the personnel required—a partner usually is needed. Other attachment methods can be done easily solo. The most popular are

Fig. 11-27. *Rib stitching demonstration at an Oshkosh forum. A good place to learn the seine knot.*

screws (either wood or sheet-metal depending on the rib material) and pop rivets (for tube-type ribs). Use a ½-inch-diameter aluminum washer to apply the force over a wide area.

The minimum attachment spacing for a given method depends on whether the area is within the propeller slipstream. For aircraft with redline speeds of less than 150 knots, spacing is 2.5 inches in the slipstream and 3.5 inches everywhere else.

Penetrating the fabric weakens it; hence a strip of reinforcement tape should be glued to the target areas before beginning. When completed, the laces or screw/rivet heads are covered by a strip of finishing tape.

Access panels and drain grommets

One nice thing about fabric covering is the ability to install large access panels without affecting structural integrity. Oversized access panels are awkward, but it's quite easy to put in nice hand-sized panels.

Of course, you can't cut big holes willy-nilly. They must be reinforced to prevent the fabric from tearing. There's a standard circular inspection plate that works quite nicely. It's convex in shape (like a contact lens) with an aluminum strap riveted across the back. When installed, the rim pinches the fabric to both ends of the strap, holding the plate in place. But apply pressure to the center, and the center snaps inward, releasing the pressure and allowing the plate to be removed.

Such inspection plates require a 3.5-inch hole in the fabric, which is reinforced by a plastic ring (Fig. 11-28). Glue the ring to the inside of the fabric, and cover it with a fabric patch. When the glue is dry, cut away the material in the center with a pencil soldering iron, and cover with the circular plate.

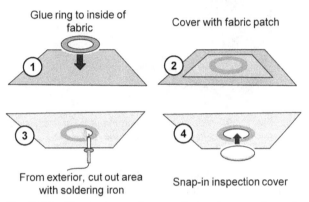

Fig. 11-28. *Circular inspection panels are cheap and easy to install. The fabric inside the ring doesn't have to be cut out until access to the area is needed.*

Note that you don't *have* to open every hole. Many folks install the rings but only cut open the hole and install a plate if they actually need to access that particular area.

While it's easiest to use these available commercial items, access panels of any size can be made (Fig. 11-29). Make them and similarly shaped reinforcement rings from aluminum. Drill a few mounting holes between the plate and the ring. Glue the ring to the fabric, and cut out the center area. Sheet metal screws then attach the plate to the ring.

Drain grommets are easier. Although the fabric covering should be nearly water-tight, moisture can leak in or simply form owing to humidity and temperature changes. The water will accelerate wood rot and/or aluminum corrosion unless grommets permit drainage.

The grommets are nothing more than plastic washers. Glue them to the inside of the fabric at points that will be low when the plane is at rest. One should be installed underneath the uphill side of the trailing edge of each rib, for example. Then punch out the center with the tip of a soldering iron or a hot nail.

The next step is dope application, which is discussed in the painting section of Chap. 12. Otherwise, the fabric covering process is complete. It's not especially difficult, but quality is important—not from the strength point of view, but because

Fig. 11-29. *Wide access panels in the tail of this Sonerai allow good access to the workings of the tail feathers.*

the fabric is the exterior surface of your aircraft. No matter how meticulous you might be on the interior, wrinkles and sags in the fabric will turn your diamond back into a lump of coal. Take your time, and be careful.

QUALITY CONTROL

Probably the biggest quality control point in fabric covering is to pick a single covering system and stick with it. The death of aviation legend Steve Wittman and his wife has been traced to mixing covering systems.

As far as working with wood is concerned, here are some points to remember:

- Where wood must be bent, bend it parallel to the grain. This is opposite to the usual practice with metal.
- Similarly, don't bother with pilot holes and reams when drilling in wood. Drill holes directly to the proper size.
- Use wood bits with your drill rather than the usual multipurpose ones.
- Maintain your wood at a moisture level of approximately 12 percent.
- Isolate all metal parts from wood—the moisture accelerates corrosion.
- Proper protection of the wood is vital. Use multiple coats of good-quality varnish.
- For fabric-covered airplanes, make sure that the material used to seal/protect the underlying surface is compatible with the covering process.
- Follow the temperature recommendations of the glue and varnish manufacturer.

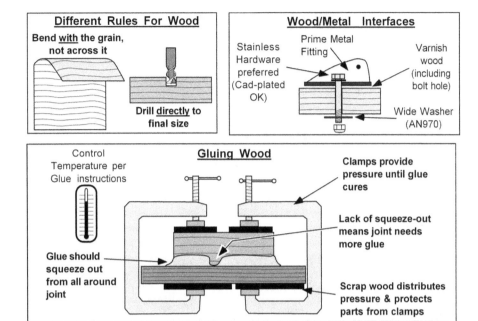

Fig. 11-30. *Wood requires proper handling to ensure integrity of the final product.*

- Don't substitute wood unless you follow the guidelines of AC 43-13.
- Clamp wood together when gluing. Certain types of glues require fairly high pressure; epoxies generally less.
- Protect the wood from clamps by using scrap wood or metal to spread the pressure.
- Sufficient glue should be applied to a joint so that it flows out everywhere around the periphery of the joint.
- Chamfer all wood edges to inhibit splintering.

Figure 11-30 summarizes the key points.

12

Completion

BUILDING A KITPLANE OF ANY TYPE TAKES A LONG TIME. The structure itself is the most daunting, but the subsystems (i.e., engine, electronics, brakes, and controls) require just as much attention and construction time. Even a simple Rotax engine installation will take 50 hours or more.

But the subsystems work has several advantages. For one thing, much of it is simple, bolt-together work. Unlike structural integrity itself, subsystem operation is tested easily after completion. Changes and adaptations then can be made as problems are found.

Probably the best reference for subsystems installation are four books by Tony Bingelis: *The Sportplane Builder, Sportplane Construction Techniques, Firewall Forward,* and *Tony Bingelis On Engines.* These books are filled with details on brake systems, electronics, interiors, occupant protection, and myriad other subjects about which you'll eventually have questions. The books are available through EAA and homebuilders' supply companies.

This chapter concentrates on the finishing details—painting, testing, certification, and first flight.

PAINTING

Painting is a major construction landmark—in most builder's minds it's when the project turns into an airplane (Fig. 12-1). Don't let the gleam in your eye result in a less-than-gleamy aircraft.

When?

Most builders are eager to get the bird looking really sharp for the first flight. We want glittering wings for the postflight photos. So once everything's together, we paint the aircraft and haul it to the airport.

There's one problem that many builders don't face: The airplane and their flying of it probably won't be perfect. The test-flight period isn't just a formality. Few airplanes make it through the 25 or 40 hours without the need for some work.

Fig. 12-1. *No matter what effort goes on inside the structure, most people will judge your aircraft by the quality of its paint job.*

It might be minor. Wheel shimmy takes its toll on wheel pants. Engine cooling often is insufficient, and cowling modifications may be necessary.

Problems occasionally are more serious. Got much taildragger time? Even 10 hours or so in a Citabria might not be enough to prepare you for the lighter, lower-powered planes. Minor cosmetic damage is a common occurrence during the test period.

I once saw a video of a builder taxi testing a Kitfox. The plane spent more time dragging wingtips than running in a straight line. By the time the builder found that his gear bungee cord tension was incorrect, he'd ground several inches off his drooped wingtips. It didn't do his nice paint job much good, either.

The only thing more difficult than a classy paint job is matching the paint after repairs. It's often a good idea to wait until the test period is completed.

This causes some problems. First, getting the plane to the airport is usually a major operation. Folding-wing airplanes aren't too bad (Fig. 12-2), but the rest need the wings removed. Therefore, if you fly the plane before painting it, you end up repeating the process two more times, one round trip from the airport to home (for painting) and back. Painting the airplane in your hangar might seem like an ideal solution, but many airports prohibit it.

Second, some materials must be protected before the first flight. Ragwings should have ultraviolet (UV)–block silver dope applied. Aluminum airplanes should be primed. The engine and plexiglass must be protected anytime paint is being applied. Waiting means that you'll have to mask them off for priming and then mask again after flight testing. Why do it twice?

Lastly, waiting to paint involves more thorough preparation when the time comes. Exhaust deposits must be cleaned off. A minor oil leak means an extensive scrubbing before paint can be applied.

It's kind of a tossup. Decide which way suits you.

Fig. 12-2. *Folding-wing airplanes make it easier to test fly the plane in primer and then bring the plane back home for final paint.*

Who?

Should you paint the plane yourself, or hand it over to a pro? It depends on your goals, confidence, and finances.

Are you just looking for a decent-looking paint job? Then you probably can do an acceptable job in your shop or yard. Thousands of homebuilders have. I repainted much of my club Fly Baby's wing in my garage with very good results.

But what if you want more than just an OK finish? Do you feel up to it? Do you have access to a good paint booth and professional equipment? Award-winners have been painted in the owner's garage or even in open carports, but if you're interested in only the best and can afford it, consider having the aircraft professionally painted.

After all, no one can be an expert at everything. And the external finish is how 99 percent of the people will evaluate your plane. You might be a magician in fiberglass, or your GP-4 might be a cabinetmaker's dream. But all your good work comes to naught without a smooth hand on the paint gun.

A professional paint job may cost at least $4,000. Or more, depending on the size of the aircraft and the complexity of the paint scheme. Some independents will do it for less, usually on a fixed-fee plus the materials' cost.

Should you go with an independent? Ask for some references; look at other planes he or she's painted. Call the owners and see if they are satisfied.

The cost of professional painting can be reduced in another way. Professional paint shops spend most of their time preparing the aircraft by sanding the surface and masking off the things that shouldn't be painted. Most places are willing to reduce the charge if the owner does most of the busy work.

There's no real craft to covering the plexiglass with aluminum foil, nor any skill required to buff with fine sandpaper. Deliver the airplane ready for painting in the truest sense—all the painter has to do is load the gun and spray—and most places will cut the fee considerably. Or apply the primer coats yourself, and turn the plane over to the pro for the fancy finish (Fig. 12-3).

Fig 12-3. *The hard part of painting an aircraft is the finish coat. By masking and priming it yourself and then hauling it to the local paint shop, you can get a first-class paint job at a discount price.*

Part of the busy work you'll leave to them is the masking off for the various trim colors, the N-number, and the like. But there are ways around it. Many local sign companies make high-tech decals that can replace trim paint. The painter can apply the base color, and you'll add the trim later, perhaps after flight testing. The cost is less than paint, especially for complex schemes.

These graphics systems go beyond simple colored tape. Complex, fairly large patterns are designed and precision-cut by computer. The backing paper is actually on the front side of the pattern; prepare the surface, press the pattern in place, and peel off the paper.

The prototype GlaStar, for instance, was mostly unpainted. The company left the wings and tail bare aluminum and left the gel-coated fuselage bar of paint. The trim stripes and all markings were custom-made vinyl appliqués. When the plane was installed on floats, the rather staid original scheme was changed to a more splashy one (Fig. 12-4). The company even considered including the vinyl graphics in the kit, harking back to the decal sheets that come with plastic models.

Safety and environment

Before going any further, I should discuss some of the safety aspects involved with painting your own aircraft. I'm not just crying wolf here. There are cases where home-builders have *died* from inhalation of paint spray. I'm not talking about massive

Fig. 12-4. *With modern vinyl appliqués, tedious masking for the trim colors is eliminated. Note that these are photos of the same aircraft. The trim colors were applied with vinyl and stripped away with a heat gun when a new scheme was desired.*

lungfuls of enamel here—some of the modern paints will produce *cyanide* in moist environments—such as a painter's lungs.

The types of protection necessary vary with the paint. Inexpensive filter masks are sufficient in some cases. These aren't the simple surgical masks; rather, they're rubber units incorporating replaceable filters. Make sure that the proper elements for your type of paint are installed.

Don't assume that a unit is designed for painting unless it is so labeled. Years ago I was handed a military-surplus gas mask while helping some friends spray enamel. I could *taste* the paint while wearing the mask—how effective do you think it was?

The best solution is a positive-pressure respirator, in which clean air from outside the painting facility is pumped to the painter's mask. Prices on these units are coming down, or check if a fellow EAA member can lend you one. They're available for rent, too.

While worrying about your lungs, don't forget the rest of you—eyes especially. It's not farfetched that you could get a blast of paint in the face while fiddling with a stubborn paint gun.

Cover your skin and hair, too. Long-sleeve shirts, long paints, and hats are called for (Fig. 12-5). Years ago I bought a box of disposable paper coveralls at a surplus store; they're a good option, but still wear old clothes underneath.

Environmental concerns have brought about changes in all kinds of painting activities. Restrictions on waste disposal have resulted in a number of paint shops closing down; no longer can the painting floor be hosed down and paint allowed to flow into ordinary drains.

Fig. 12-5. *The well-dressed painter. Note the hat, safety glasses, and long-sleeve shirt. Fresh air is pumped in to the painter's mask.*

One big change with the commercial painters is the conversion to high-volume, low-pressure (HVLP) paint systems. Traditional systems fling a narrow stream of high-pressure air and paint that has a tendency to bounce off the target and contaminate the air. HVLP guns use a slower but wider airstream to transfer the paint.

Unfortunately, the HVLP guns generally require more air than our typical compressors deliver. Fortunately, in most places, individuals are still allowed to use the traditional painting systems, instead. If you are required (or decide) to use an HVLP gun, make sure that you have access to the right kind of compressor.

Finally, painting produces waste ranging from contaminated thinner to paint-soaked rags. Make sure that you store and dispose of them properly.

Painting facilities and equipment

If you're going to do the painting yourself, the most obvious need is spray painting equipment, such as a compressor, a paint gun, an air hose, a moisture trap, and respirators. Enamel and some other paints must be applied in heavy coats to build a gloss, so find a professional-model gun rather than a hobby one. Such guns require about 8 cubic feet per minute of compressed air, which, in turn, requires at least a 2- to 3-hp compressor. Again, an HVLP system requires more.

The airplane doesn't have to be painted as one piece, but you'll need enough space to store pieces while other parts are being sprayed. Not having enough space to set up the entire plane means that you'll have to wait until the pieces are dry before you can move them out of the way. It'll take a bit of room; where will you keep the fuselage while the wings are being painted?

Keep in mind that you need more than just room for the airplane. There must be space for the painter to walk around the components safely, with the usual foot or two clearance between the gun and the surface. If the booth is too small, there's a danger of dragging the air hose across newly painted surfaces.

A common practice is to wall off a section of the garage or shed with plastic sheet. Be careful about ventilation. The respirator cleans the air going to your lungs but doesn't ensure that it contains enough oxygen. Moreover, paint fumes are flammable, and ordinary fan motors can ignite them. If you must use an ordinary fan, use it to *inflate* your painting area instead of pulling fumes out. Explosion-proof exhaust fans are available; see if anyone in your EAA chapter has a portable fan.

One common paint area is the great outdoors. Wait for windless days (or paint at dawn and dusk), and be prepared to get bugs and dust in your finish. Overspray drifting across the fence won't endear you to the neighbors either. Or combine both ideas into a plastic-covered temporary paint booth attached to the garage (Fig. 12-6).

Good lighting is a must. Years ago I helped paint some Civil Air Patrol jeeps in an old hangar. The old color was olive drab; the new color was Air Force blue. All we had for lighting was a couple of old fluorescent fixtures on the ceiling. In the darkness and spray, we couldn't tell where the old paint ended and the new began. When the jeeps were rolled outside, the effect was startling.

Fig. 12-6. *A temporary painting structure made of plastic sheeting and wood.*

The same thing could happen in your garage. You could finish the trim colors, roll it outside, and discover that the paint is too thin over the lower fuselage. Then you have to push it back in, add another couple of coats, tape the trim strips again, and so on.

Make some portable light stands from PVC tubing and fluorescent light fixtures. Stick some clear plastic wrap across the front of the lights to keep the overspray from the tubes. Use glass or heat-resistant plastic if you have incandescent lighting.

The best light source available is the sun. Roll the airplane briefly outside between coats. Get low and squint at the surface.

What kind of paint?

There are four basic kinds of paint: enamels, lacquers, polyurethane, and dope. Enamels are inexpensive and hardy but gradually oxidize and need waxing more often than other types. They take a long time to dry, so they tend to pick up a bit of dust.

Lacquers cost a bit more and can be shined to a deep gloss, but they are a bit more susceptible to chipping. They're quite thin and require multiple coats (on the order of 5 to 20) and require plenty of rubbing to build a shine. They dry fast, though, so there's little problem with dust.

Polyurethane "wet look" paints are very shiny, very durable, and very expensive. Like enamels, they take a long time to dry.

Dopes are the exclusive province of fabric aircraft—a lacquer paint with softeners added. Let's look at specific examples.

While composite, wood, and fabric airplanes always must be painted, metal airplane builders have an option: They can leave the airplane bare metal. I don't really recommend this. For one thing, few of us are able to do a 100 percent perfect job on the exterior skin. We might end up with a few scratches or a couple of dings that need a little filler. Such blemishes are hard to hide without paint.

Second, unpainted metal airplanes are a lot of work. Polyurethane-painted jobs can get by with a coat of wax every year or so. But airplanes skinned with 2024 aluminum have to be polished continuously. People who own planes that use the more corrosion-resistant 6061 alloy, such as the Murphy line and the Sonex series (Fig. 12-7), don't have to worry as much.

Dopes always have been the traditional choice for fabric airplanes for a very good reason: They are flexible enough to withstand fabric vibration without cracking. The other types dry to a hard shell, which is not very compatible with a flexible fabric covering.

However, automotive technology comes to the rescue. Now that so many cars incorporate flexible plastic bumpers and fenders, flex agents are available for these other paints. Some people even say that they aren't needed for polyurethane paints.

Dope shares one problem with lacquer paint: To get a *really* glossy finish, 20 coats or more must be applied. Each coat must be buffed before the next is added. Multiple coatings aren't a requirement—sufficient protection and coverage are achieved with just a few coats. Most ragwing owners are satisfied with the satin finish that a couple of coats brings.

Fig. 12-7. *While it's certainly feasible, few metal aircraft builders leave their plane in bare metal. This Sonex uses 6061 aluminum, which is more corrosion-resistant than 2024.*

Traditional dopes cannot be applied to metal surfaces, making the painter switch to enamel for areas such as cowlings. More modern dopes don't have this problem. Check with the manufacturer.

One big advantage of using dope is that repairs are simple and almost invisible. New dope blends almost invisibly with old. Overspray of the other coating types will get dull around the repaired area, while new dope overlaps with nary a trace.

Overspray causes problems with gloss paints, especially polyurethane. When a dried, painted surface is dusted with airborne paint, it dulls the shine. Buffing enamels and lacquers reduce or eliminate the problem, but it doesn't work very well with polyurethane.

The typical problem area is atop the fuselage. The painter sprays the left side and the left wing and then walks around to the right. If the paint on the left side is still wet, everything's all right. But if it's dried, the new paint won't blend in with the cured stuff. It curls around the top of the fuselage and dusts the left side. Later the painter notices the dull area and repaints it. You guessed it—now the *right* side looks bad. Professional shops get around this problem by using two painters at the same time.

Your kitplane's paint job is important. It's how most people will judge your plane. A well-executed simple scheme will gain more credit than a sloppy pseudotiger with Technicolor fangs.

Not only is the paint job important for appearances' sake, but there can be safety issues as well. Control surfaces of fast airplanes must be balanced; too much weight aft of the hinge line could cause flutter. It's not unknown for an overly thick coat of paint to change marginal flutter stability to an outright hazard.

It's quite possible to get acceptable results in the garage, even if you've never painted anything large before. But there are different techniques for different paints and even for painting different areas of the airplane. This is one area where good advice and training are vital. Find an experienced hand to get you started.

Don't expect to be done in a couple of days. Painting can take 100 hours or more.

RIGGING AND CENTER OF GRAVITY

A modern kitplane is a pretty foolproof item. You have to really foul up before the airplane becomes structurally unsafe.

However, a moderate inattention to detail can result in poor handling characteristics and poor performance. The two major causes of these problems are misrigging and incorrect *center of gravity* (CG).

Rigging

Ever wonder why seemingly identical airplanes can fly so differently? If you have hundreds of hours in your own Cessna 172, every other 172 feels wrong—alien. It's like wearing somebody else's shoes. Something's not right, but there's nothing you can put a finger on.

Or one plane handles crisp and stalls clean, and another of the same model flies cockeyed and stalls viciously. What's the difference anyway?

"It's out of *rig*," most pilots would guess. But what do we mean by "rig"? Like many aviation terms, it has a nautical origin. In the days of sail, a ship's speed and handling depended on the set of the standing and running rigging. The rake of the masts and the bracing of the yards made the difference between a clipper and a hooker.

Rig on aircraft has to do with the relationship of the airfoils, stabilizers, and control surfaces to the relative wind. Production aircraft and some kitplanes have the ability to adjust the lift of each wing individually to make up for slight differences in lift. These differences usually come from manufacturing variations. And as you might expect, kitplane building is rife with manufacturing variations.

A degree's worth of difference in incidence between the left and right wings can cause noticeable wing heaviness in flight that gets worse as speed increases. In contrast, a weight difference between the wings becomes less noticeable at cruise. Hence the need for precision during construction.

But it's impossible to be perfect. The ability to compensate for these problems varies. Some kitplanes offer independent incidence adjustment, like production aircraft. Wire- and strut-braced homebuilts are prime examples of adjustable airplanes. Others don't provide for it and handle problems with fixed external trim tabs.

Before the first flight, the aircraft should be rigged to the neutral position or wherever else the plans specify. When this is done depends on the aircraft. Ragwing airplanes usually are rigged before the covering is applied (Fig. 12-8).

The first step is to set the fuselage to an in-flight attitude. The plans should specify a reference point from which level is determined. On tube-type airplanes, it's generally a certain point in an upper-fuselage longeron. It might be the cockpit sill or the top of the cowling, depending on the airplane.

Once a bubble level is in place, the airplane's pitch attitude is adjusted to center the bubble. Most kitplanes (even trigear) have a nose-up attitude on the ground, so the tail must be raised. It's easy enough with conventional gear. Place the tailwheel on a stool, and add shims until level. For trigears, roll the mains onto an elevated platform, and shim the nosewheel.

Fig. 12-8. *Ragwing builders usually rig the aircraft prior to covering to minimize wrinkles.*

The instructions should indicate the correct adjustment and the method of measurement. The usual practice uses a carpenter's level and shims. For instance, the horizontal stabilizer setting might be indicated as 2 degrees. The bubble on a 24-inch level should be level when the front end of the tool rests on shims equaling about $^{13}/_{16}$ inch.

The positions of the surfaces are set by various means. Some merely add washers under bolts. (Make sure that the bolt is long enough to accommodate the number of washers that might be required.) Turnbuckles set the incidence and dihedral on wire-braced wings. Wing struts incorporate threaded ends to change their length.

Turnbuckles are also used to set the tension for control cables. A *tensiometer* (tension-measuring gauge) is nice; unless otherwise specified in the plans, set the tension to 20 to 30 pounds or so. Check the tensions at the full excursions of the controls to ensure that the cables don't go slack in either direction.

As a final note on rigging, take a look at the ailerons. Make sure that they are both in the neutral position when the stick is centered. It's quite possible for both to be up or down slightly; adjust them if necessary.

Center of gravity

As pilots, we all learned the importance of center-of-gravity (CG) location. Every airplane's paperwork includes an actual empty weight and CG, to which we add the weight and moment of the passengers, fuel, and baggage. The CG must fall within a particular range of the wing's *center of pressure* (CP).

Because we are the manufacturer of our particular homebuilt, it's up to us to determine the basic figures. The kitplane's plans may include a typical empty weight and moment. These are just samples; you *must* calculate your own.

Fig. 12-9. *Weighing to calculate the center of gravity takes several scales and usually several helpers as well.*

The procedure is simple: The aircraft is leveled, and the weight at each wheel is measured (Fig. 12-9). This, along with the position of the wheels relative to some reference point (the "datum," usually the firewall), determines the location of the CG. The manufacturer should provide the moment arm (distance fore/aft from the datum) for each wheel, and your weights are then used to determine the CG. Figure 12-10 illustrates this process, along with some guidelines for accurate results.

If the result falls within the allowable range, all is well. If it doesn't, you've got your work cut out for you. Unfortunately, most solutions require adding weight.

A forward CG location is easiest to solve. The tail cone has a long moment, so a little weight in back goes a long way. One common fix is to shift the battery box from the cowling to the baggage compartment. This adds very little weight, but rerouting the heavy cables through the cabin is a hassle.

Aft CG locations are more common and harder to solve. There isn't much moment arm forward of the CP. It's possible. I've seen a couple of homebuilts with steel or lead blocks bolted to the engine mount. Aft-mounted equipment is rare, but see if any can be shifted forward.

One traditional solution is to lengthen the engine mount. If the aircraft is close to completion, this is an act of desperation. Imagine that everything connecting the engine to the airframe has to be lengthened. Wiring. Fuel lines. Throttle, mixture, and carb heat cables. Scrap the cowling, and scratch-build a new one. The exhaust system must be rerouted. Not to mention fabrication of the new engine mount.

Often, the only practical solution is to limit the useful load so the aft CG limit isn't breached. This typically amounts to placarding a baggage area for a lower weight limit.

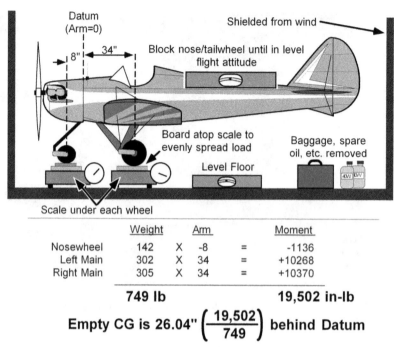

	Weight		Arm		Moment
Nosewheel	142	X	-8	=	-1136
Left Main	302	X	34	=	+10268
Right Main	305	X	34	=	+10370
	749 lb				**19,502 in-lb**

$$\text{Empty CG is } 26.04" \left(\frac{19{,}502}{749}\right) \text{ behind Datum}$$

Fig 12-10. *Guidelines for determining the CG.*

A few words about weight

So you rolled it up on the scales, and it came out a few pounds heavy, didn't it? This isn't rare. You can chalk up the advertised "empty weight" alongside those other great myths of aviation.

But how does it *really* affect you?

As far as the kit manufacturer is concerned, not at all. Because you are naturally going to limit the gross weight to the book value, right?

Let's face it. Few kitplane owners do. Most load the seats, load the tanks, and load the baggage. One small homebuilt was designed for a 450-pound gross weight. One builder won grand-champion honors with his version, which he flew at *650* pounds, almost 50 percent heavier than the prototype. Others have installed VW engines on this type of airplane and operated at 700 pounds.

Let's look at how overloads affect the operation of an aircraft. In the first place, throw all performance specifications out the window. It'll take more room to take off and land. Cruise speed and climb rate will be lower. It'll stall at a faster speed and probably more sharply as well.

The CG is a little trickier because the load point isn't even on the graph. These are all factors a pilot would expect. However, consider the effect on load factors. For instance, assume that the plane mentioned earlier is stressed to withstand 6 Gs. This means that the structure can take a load of 6 × 450, or 2,700 pounds. But with a 650-pound gross weight, this translates to a limit of a little over 4 Gs.

During normal flying, this won't make much difference. But if the builder likes aerobatics. . . . Well, let's just hope he wears a parachute.

Let's make one thing clear: I do not condone flying any airplane weighing more than the manufacturer intended. But flying beyond the limit is nothing new. The FARs even allow certain older production aircraft operated in Alaska to operate at 115 percent of their certified gross weight.

Just remember that homebuilt designs don't incorporate the generous margins of factory airplanes. Watch the CG especially; project out the CG envelope to your personal gross weight. The sides of the envelope are usually linear. And fly gently, gently.

THE LEGALITIES

At some point before completion, you should initiate the paper chase that will result in an airworthiness certificate. The forms mentioned below are available through either the FAA or EAA.

N-number and registration

You'll probably want a registration, or N-number, before painting the aircraft so that you can paint the registration number at the same time as the trim. As far as the FAA is concerned, the number must be on the aircraft at the time of inspection, so you can wait if necessary.

Then again, there's nothing illegal about reserving a number before starting construction. But some states cross-check the FAA's aircraft list to find owners who haven't paid taxes or registration fees. If you request a number early, you might get a bill from the state. The state doesn't care if the aircraft is just a pile of parts in the garage; it's registered, so you have to pay.

Reserving an N-number costs $10 per year, and it can be done online. Go to *www.faa.gov*, select "Aircraft Registration," and then select "Online N-Number Reservation Request." Or enter "N-Number Reservation" in your favorite search engine. Both the reservation and the renewals can be done online.

List your primary choice and a number of alternatives. You can request up to five characters after the *N* prefix. The last two characters can be letters. Traditionally, homebuilders tend to request the year the airplane is completed, followed by their initials: N86TD, N82EU, and the like. The FAA is loath to hand out short registrations, such as N4R, but will do so if your aircraft has limited space for full-sized numbers.

When you are ready to register your aircraft, submit a notarized affidavit of ownership to the FAA's Aircraft Registry Department along with a short cover letter either containing the previously reserved number or requesting assignment of an identification number. The affidavit is available as *AC Form 8050-88* and can be downloaded from the FAA Aircraft Registration Web page. Remember that the form must be notarized.

This form must be submitted along with AC Form 8050-1 to register your aircraft. AC 8050-1 is the Aircraft Registration Application Form; it's a multipart form not available online. It is generally easy to obtain around an airport because the same form is used to report the purchase of a plane. Fill it out, and mail it to the FAA along with a check for $5. Keep the pink copy of the multipart form—it is your temporary registration.

Send the forms and the $5 to

FAA Aircraft Registry
Aircraft Registration Branch, AFS-750
P.O. Box 25504
Oklahoma City, OK 73125

With the expanding use of the Internet, procedures are evolving. Check the FAA Aircraft Registration Department's Web page for updates to these procedures.

The N-number must be applied to both sides of the fuselage between the trailing edge of the wing and the leading edge of the horizontal stabilizer. It also can be applied to the vertical stabilizer. The minimum size is 3 inches; however, the situations under which 3-inch N-numbers are legal are rapidly diminishing.

If any of the following statements are true, the ID must be at least 12 inches high:

- You plan on flying within an Air Defense Identification Zone.

- You plan on flying to another country.

- Your kitplane cruises at 180 knots or faster.

The style or font of the lettering is restricted to simple block styles; a little slant is OK, but no Old English or italic styles, please. Color choice is left open, but it must contrast with the surrounding paint in order to be easily readable.

One interesting clause in the regulations is in 14 CFR 45.22. If the design of your homebuilt is more than 30 years old, you can display the N-number with an *X* ("NX124"), just like the experimental aircraft of the 1930s and 1940s (Fig. 12-11).

Airworthiness certificate

Three things are required to gain an airworthiness certificate: completion of *FAA Form 8130-6*, completion of *FAA Form 8130-12*, and a signoff inspection. Form 8130-12 is the eligibility statement. The aircraft is generally described, and the builder must certify that he or she built the major portion of the aircraft. This form must be notarized.

FAA Form 8130-6 is the biggie: application for airworthiness certificate. This form describes the aircraft in detail (including a three-view diagram), what it will be used for, and the restrictions placed on its operation. The kitplane manufacturer should supply the three-view; in fact, it's probably included in the instructions.

On successful completion of the signoff inspection, the inspector will sign this form and give you a temporary airworthiness certificate. More information on this process is given later in this chapter.

Radio station license

In 1996, the Federal Communications Commission (FCC) eliminated the radio station license requirement for U.S.-registered aircraft. This license may be required

Fig. 12-11. *Builders of "old fashioned" experimental aircraft can use the traditional X designator in the registration. Note the "NX" on the tail.*

if you intend to fly to other countries. FCC Form 605 is now used for all aircraft station licensing purposes. It can be downloaded from the FCC. Go to *www.fcc.gov,* or enter "aircraft radio license" in an Internet search engine.

Repairman's certificate

Part of the reason you built the aircraft was to save maintenance costs, so don't forget to apply for a repairman's certificate. Submit *FAA Form 8610-2,* which is the same form airframe and powerplant (A&P) mechanic candidates use to apply for their certificates.

AIRCRAFT PREPARATIONS

The first flight requires exhaustive preparations to ensure that the aircraft is ready. Critical elements must be checked and verified.

Fuel system

The best workmanship in the world won't help if the engine can't get any fuel. Building an airplane results in a lot of grit and garbage. Need it be said that you don't want said garbage in your fuel system? Many engine failures in homebuilts are caused by fuel system problems, often crud in the tanks and lines.

The elimination process begins during construction. Clean the tanks thoroughly after completion, and cover the inlets and ports to prevent entry of foreign material. The same applies to the lines, valves, and gascolators. Before adding gas to the tanks the first time, vacuum them out. Rig up a reducer on the end of your shop vac so that you can slip a smaller hose into the filler opening.

There'll still be some debris in the system. Your main fear should be the blockage of a fuel line or valve because it can occur unseen and takes much effort to trace down. Filters and screens make contaminants easy to find and eliminate and ensure that any particles making it through can't block the downstream lines or valves.

To start with, each tank should incorporate a finger screen inside the tank at the outlet port. These thread into a fitting on the base of the tank, and the fuel line then threads into them. The fine mesh catches the debris, and the long shape delays the point where the screen gets clogged.

Every aircraft must have a gascolator. It's the last redoubt against contamination reaching the carburetor or fuel pump. The gascolator also incorporates a quick fuel drain to draw off any water or particles in the fuel. It's always installed on the engine side of the firewall, as low as possible. Any water then will tend to collect in it.

The finger screens and the gascolator are your main lines of defense against fuel starvation due to blockage. They should be removed and cleaned at least once before the first flight and at gradually increasing periods thereafter. The finger strainers can only be checked when the tank is empty, so removal and cleaning must be planned carefully.

If air can't get into the tank, fuel can't flow out of it. Tank vents should be clear, too. Some can be cleaned with a short piece of wire.

Gravity-flow fuel systems require verification of flow rates. These systems rely on *fuel head* (pressure resulting from the fuel tank being higher than the carburetor) for flow, and if there's insufficient fuel head, the engine will be starved.

Fuel flow testing is done in the worst-case situation: simulating a climb at full power with near-empty tanks. The fuel system must be capable of supplying 150 percent of the fuel required by the engine at full power. This information should be available from the engine manufacturer. Typically, the full-power rate runs between twice and three times the cruise power consumption.

The first step is to set up the aircraft in a climb attitude. This angle varies between aircraft, but it's usually more than the typical taildragger's ground attitude. Even pulling a trigear's tail down to the ground might not produce the right climb attitude.

Roll the main gear up a ramp to sit atop platforms. Taildraggers are then ready; the tails of trigear airplanes must be pulled down and tied or a block placed under the nosewheel (Fig. 12-12). The kitplane manufacturer should have the required attitude, but the platform height probably will be around 12 inches or so.

Ensure that the tanks are empty, and disconnect the fuel line at the carburetor. The disconnected end must stay at the same general level as the carb; don't let it hang down. Tie in it place with wire or tape. Slip a short piece of neoprene fuel hose on the end, and drape it down to an empty can.

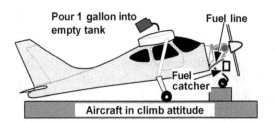

Time	Rate
2 min	30 gph
3 min	20 gph
4 min	15 gph
5 min	12 gph
6 min	10 gph
7 min	8.5 gph
8 min	7.5 gph

Fig. 12-12. *Check the rate of flow of gravity-fed fuel systems with the aircraft in a climb attitude. The rate of flow should be 150 percent of the fuel required at full power.*

Slowly add fuel to the tank under test until it comes out of the fuel line in a steady stream. Keep track of how much you've added; this is the value for unusable fuel in that tank.

Once you have a steady-state flow established, turn off the fuel valve, empty the drain can, and dump exactly 1 gallon into the tank. Position the newly emptied can under the engine, stick the neoprene drain line into it, and turn on the fuel valve and start the timer.

When the can is filled to the start value, stop the timer. Compute the actual fuel flow, convert it into gallons per hour, and compare it with the required value. Here are some sample values for 1 gallon:

2 minutes	30 gph
3 minutes	20 gph
4 minutes	15 gph
5 minutes	12 gph
6 minutes	10 gph
7 minutes	8.5 gph
8 minutes	7.5 gph

During this process, watch for leaks. The rubber used in some of the lines tends to age when exposed to air for a long period (such as during construction) and can become brittle and crack. If so, replacement will be necessary.

This process isn't really necessary for airplanes that use fuel pumps. However, you'd really like to know that the engine receives enough fuel in climbout attitude. Consider blocking the wheels/holding the tail down for a full-power test. Do this only if you can arrange a safe test setup, though.

Instruments

It's tough enough to make the first flight of a new aircraft without being faced with inoperative instruments. Altimeter failure wouldn't be too tasking. But what if the airspeed needle remains solidly on zero? How would that affect your landing? What if the oil temperature gauge is inoperative, and the engine overheats?

You'd really like to check out the operation of your instruments before the first takeoff.

Operation of most of them can be verified in common-sense ways. For the oil temperature gauge, for instance, remove the sensing bulb from the engine and dunk it into boiling water. Or use water heated to a lesser degree with a candy thermometer alongside to provide a reference. Ditto for the coolant-temperature gauge for Rotaxes.

Ordinary hand vacuum pumps can test fuel and oil pressure gauges. But don't use them on pitot-static instruments—the pump is too strong. An FAA-approved repair station can check the entire system for leaks and accuracy. It's not a bad precaution if you have the bucks. Don't spend $150,000 or more on your Lancair and begrudge the $100 or so a pitot-static system check might cost.

But you can do some checks yourself. The airspeed indicator itself is easy and a little fun to test. Stick a couple of hoses on the pitot and static ports and go for a drive. Have a passenger hold the extended pitot tube out the window and check the reading of the instrument. Remember, depending on the winds, the airspeed might not match the speedometer reading.

Similarly, an altimeter or VSI can be checked by going flying with someone else. Just hold the instruments on your lap and compare them with the aircraft's gauges. There'll be some differences—your units won't be connected to the aircraft's static system.

However, most homebuilt aircraft problems arise in the pitot-static system itself. This is the advantage of the repair station's equipment—the repair station checks the whole system from end to end. It's difficult to get that degree of accuracy in the home workshop, but you can at least ensure that the system reacts to pressure changes as it should.

The pitot system can be tested by slipping a piece of rubber hose over the end of the pitot tube. Roll up the end of the tube while a helper watches the gauge. This compresses the air in the tube and simulates the gentle pressures involved.

When the reading reaches the anticipated cruise speed, pinch the hose shut to hold the reading. The airspeed needle should hold constant for at least a minute. If it falls, there's a leak in the system. It could be due to the attachment of your rubber test hose, so check it first.

Checking the static system requires negative pressure. Block off all the static ports. Tightly roll up a couple of feet of neoprene tubing, and connect the end to a static line. Unroll the tubing and watch the VSI. It should indicate a climb. The altimeter should rise as well. When it reads 2,000 feet or so, pinch off the tube. Again, the indication should hold constant for at least a minute.

Any leaks in the pitot-static system can be chased down with a small pressure source, but *disconnect all instruments first.* Plug the ends that go into the gauges, apply pressure to the other end of the line, and wipe soapy water over all connections. The bubbles indicate the leaks.

Testing vacuum-operated instruments can wait until the flight phase because they include none of your required VFR instruments. However, if you can arrange a source of sufficient vacuum, test these instruments during construction before the panel and cockpit are closed up. Problems will be a lot easier to fix then.

The aircraft electrical system and various masses of metal affect the magnetic field of the airplane. Aircraft compasses include small magnets that are adjusted to compensate for each particular installation. Perfect correction isn't possible, so errors are tabulated and listed on a correction card mounted near the compass.

The procedure for determining these errors is called "swinging the compass." It must be performed after the engine has been broken in. The aircraft is aligned with the magnetic cardinal directions (north, east, south, and west) with engine and radios operating. Many fields have a compass rose painted onto the asphalt for exactly this purpose. If you've got a good orienteering compass, you can tape lines on the asphalt for your own compass rose.

The differences between compass heading and the actual heading are noted, and the N-S and E-W compensating magnets are adjusted as listed in the compass instructions. Once all possible error has been eliminated, the correction card is filled out.

You'd like to swing the compass with the aircraft in flight attitude. This is easy to work with trigear planes; it's almost insurmountable for taildraggers. It's not that tough to prop the tail on a sawhorse or whatever, but running the engine in this position can be a bit dicey. If you can come up with a safe way to do it, great.

Controls

Controls must operate freely over their entire ranges. Any binding or scraping must be eliminated. Try to simulate flight conditions. For instance, have someone apply pressure to the surfaces while you move them with the stick.

Have the helper lift up a wingtip while the ailerons are moved. Thirty years ago, a small problem with a certain homebuilt design was discovered: When the wing started developing lift during the takeoff run, the structure pinched down on the aileron torque tubes. The controls were free and easy on the ground but hard to move in the air. Make every effort to ensure that your controls are free and clear in any combination of attitude or loading.

Ensure that all other controls work smoothly as well. Throttle, mixture, carb heat, flaps, brakes, and the like should move freely throughout their ranges.

Make sure that there's no interference between controls. Hold full left rudder, and move the elevator throughout its range. Repeat with the right rudder, and then check the flaps and ailerons the same way.

Propeller tracking

Each propeller blade should track along the same path; otherwise, vibration and failure may result. Mistracking is usually caused by improper propeller installation or a bad prop. Tracking should be checked at some point prior to the engine's first start.

For safety's sake, remove one plug from each cylinder. Set the plane at flight attitude. Move the prop until the blades are vertical. Place a block of wood on the floor so that one edge is just barely touching the front of the bottom blade. Then rotate the prop until the other blade meets the block.

If it just barely brushes it, tracking is dead on. Otherwise, some gap exists between the second blade and the block, or the blade brushes the block away. Measure the amount of difference. If blade tracking is off by more than $\frac{1}{16}$ inch, check the installation, and contact the propeller maker.

ENGINE START

Your next actions can affect the aircraft's operation for years. Abuse the engine in the first few hours of operation, and it might use oil, run roughly, and even fail prematurely.

First start

There comes a time when the engine must be started for the first time. The FAA requires at least one hour of ground running before it'll allow the aircraft to fly. Don't "pencil whip" this requirement.

The first startup is not to be done alone. You want at least one other person around to help, preferably your EAA technical counselor or another experienced builder.

Your primary desire is safety. If you're starting it up at home, make sure that the propeller area is blocked off to prevent dogs and kids from wandering into it. Keep the propeller plane clear as well, in case the prop shatters.

The area around the engine should be spotless and afford nonslip footing. It'll be necessary to examine the engine while it's running, so you don't want a slippery floor. Keep a bag or two of kitty litter handy to soak up any spills.

Don't fill up all the tanks, and keep stored gasoline away. Have fire extinguishers handy, and make sure that everyone knows where they are. You're most likely to need a class B (burning liquids) certified extinguisher, but get the garden hose out as well in case fire gets started by something other than gas or oil.

Ensure that everyone has earplugs or hearing protectors—they are going to be working in close proximity to a very loud engine. Eye protection of some sort should be worn as well. At the very least, the propeller will be stirring up a lot of dust.

Make sure that the oil tank is full and that all engine controls are free. The engine manual should specify which type of oil to use. Two-cycle engines might require a different oil-gas mix for break-in; check the manual.

Fill the coolant tank for liquid-cooled engines. Rock the plane fore and aft and side to side to purge air from the coolant lines, tank, and radiator. Top off as necessary.

Go over the engine compartment, tightening anything you can fit a wrench to and making sure that everything that should be safety-wired is. Check the mag timing one more time, and make sure that you've set the valves properly. Nothing should be loose in the engine compartment, even if it has nothing to do with the test run. Have the tech duplicate the whole examination as a double-check.

Three instruments are vital: an oil pressure gauge, an oil temperature gauge, and a cylinder head temperature gauge. The oil gauges already should be installed on the panel and should have been checked, as mentioned earlier in this chapter. An add-on calibrated set should be used to compare their readings. Because so many break-in operations are controlled by engine temperature, accurate gauges are vital. Liquid-cooled engines need a reliable coolant temperature gauge as well.

You might or might not have a cylinder head temperature gauge installed. In any case, you'll want one for the testing phase, even if it's mounted temporarily. The temperature senders (thermocouples) must be solidly attached to the cylinder heads. Aircraft engines include a threaded hole for the thermocouple to screw into; these are called "bayonet mounts." Another kind is built into a spark-plug gasket. The gasket type is adaptable to practically any engine.

One issue that arises is how many to install. The best way is to add a thermocouple to every head and wire them into a selection switch. Many people just install one sender; in this case, it should be mounted to the aftmost cylinder on the right side. Two-cylinder Rotax engines should mount the thermocouple to the rear cylinder. In any case, the sender should be removed and installed on other cylinders on a regular basis to make sure that none are running hot.

There should be one person to monitor the gauges and at least one other watching the engine. Tie down the tail and wings, and chock the wheels. Coordinate a system by which any person can signal for a quick shutdown even if it's just frantic waving and yelling. Considering the noise and the earplugs everyone should be wearing, perhaps whistles or an aerosol-can boat horn might not be a bad idea.

Some criteria for emergency shutdown include:

- Visible flames of any sort
- Spraying liquids
- Continuous smoke from other than exhaust (Grease spots and the like might burn away, sending up momentary puffs.)
- Moderate, continuous smoke from the exhaust (Some black smoke might appear due to improper mixture, which is normal.)
- Loose components, especially those threatening to come loose and foul the prop
- Excessive vibration (Some must be expected at first, but the engine should smooth out.)
- Any other perceived safety problem (kids wandering too closely)
- Any instrument condition indicative of engine faults

Ready to start. Crank it by hand a few times to loosen it and distribute the oil. Make sure that the mags are off, of course. Follow the engine manual's instructions for first start. Watch the oil pressure gauge like a hawk, and if you do not have pressure within 30 seconds, shut the engine down pronto.

The other problem to watch for is overheating. There are many causes, but one is easy to fix: Don't run an air-cooled engine very long with the cowling off. On any design where the cylinders *don't* stick out of the cowling (à la J-3 Cub), the cowling is a necessary part of the cooling system. Running without it isn't a bad idea for the first start or so, but most aircraft require them in place for break-in.

Break-in

The engine manufacturer should have supplied a break-in plan. The actual procedure varies. Usually the engine is revved up to a particular rpm and held for a specified time, and then power is either increased to the next level or dropped to idle. The entire break-in might take a couple of hours total running time. Usually break-in can be separated into several sessions, but Rotax engines must be run to a particular rpm profile. If the sequence is interrupted for any reason, it must be rerun from the start.

Monitor the engine gauges throughout the process. Change the oil after two hours of accumulated running, and watch for the appearance of metal in the filter or strainer. Don't forget to log all times and maintenance operations in the engine log.

When the basic break-in procedure is complete, it's time to operate it under actual conditions.

GROUND TESTING AND FAA INSPECTION

With the engine apparently operating normally, the aircraft's ground handling can be investigated.

Taxi testing

Taxi testing has two objectives. The first is the pilot's familiarization with the ground handling of the aircraft and identification of any problems. The second is to predict the aircraft's trim during takeoff and landings.

During familiarization, you'll get a feel for how the runway looks with the aircraft on the ground so as to be able to judge height during landings. Learn, too, how to control the airplane on the ground. Check the steering and brakes.

Don't forget to watch the engine gauges to forestall damage from overheating or oil exhaustion. Operate the engine with full cowling; now is the time to find out if the cooling system needs modification.

There is a bit of controversy regarding high-speed taxi testing. Some consider that the hazards involved with high-speed testing overcome the benefits that can be derived from them. Typical accident scenarios include inadvertent flight or loss of control. I know local pilots who have destroyed their yet-to-fly homebuilts through either scenario.

Talk the situation over with experienced pilots of the same aircraft type or your EAA flight advisor. One or two runs to check out your brakes and steering may be all that's really necessary.

Never do high-speed taxi testing unless the airplane and pilot are fully ready to fly. Carry enough fuel, and make sure that all the cowlings and panels are in place. Don't allow yourself to become distracted during these tests—your first response to any problem should be to stop the test and protect the airplane. Think out your reactions to various problems in advance. *Maintaining control of the airplane is your primary task.*

One use of high-speed testing is to predict the aircraft's trim. Work up to a maximum of 80 percent of the predicted stall speed in 5-mph increments. Test the ailerons by rocking the wings. See how effective the rudder is. At 80 percent of stall speed, the elevators should be powerful enough to allow the aircraft to assume takeoff attitude: tail down in trigears, tail up in taildraggers. If it isn't, you may have a CG problem.

From the distance required to reach 80 percent of stall, predict where the actual takeoff point will be. Run a simulated takeoff abort, measure the distance required, and mark it off from the opposite end of the runway.

The inspector

You have two options for persons to perform the airworthiness inspection: a genuine *FAA inspector* from a local Engineering Manufacturing District Office or an independent *designated airworthiness representative* (DAR). The FAA inspector is free, but inspectors are fairly busy in some locations. The Flight Standards District Office (FSDO) located in my home town does a good job supporting local builders. But if you aren't as lucky, a DAR is the only alternative.

DARs are essentially trusted agents of the FAA; they have passed rigid requirements and have years of experience with small aircraft. They are often A&P mechanics. Having your airplane inspected by a DAR isn't getting the second team. They wouldn't be trusted if they couldn't do the job.

But they are kind of expensive. Not only must you pay $250 or so for their inspection, but you must pay for their travel and insurance as well. There are ways to reduce this cost; for instance, you could move your kitplane closer to the DAR's work location to reduce his or her travel time. Builders with folding-wing kitplanes have been known to show up at the inspector's door and set up their airplanes in the parking lot.

To help alleviate a growing shortage of DARs, the EAA worked with the FAA to start the Amateur-Built DAR (AB-DAR) program. AB-DARs must be A&P mechanics and must have built, received certification for, and flown at least 100 hours in their own amateur-built airplane. They must possess knowledge relating to the fabrication, assembly, and operating characteristics of homebuilt planes and must have performed at least three annual condition inspections on homebuilts.

The EAA recommends that you arrange for an inspector at the time the project is started. In this way, you can discuss changes and options with the inspector prior to implementing them.

The flight test area

The inspector sets the flight test area. The main concern is protection of the public; hence he or she might require that the initial flight testing be performed at some out-of-the-way airport.

In my home area, a lot of local builders are based out of a small airpark on the outskirts of town. The runway is nothing more than a 3,000-foot gash in the trees with a power substation to the south and housing developments strewn in all directions. Instead, the FAA usually specifies the first 10 hours of flight testing be performed from an ex-Army Air Force field 60 miles to the north. The runway is 5,000 feet long, and it's in the middle of farmland.

In my case, it takes 80 minutes to drive to the airport instead of five. It's awkward, and I don't like it. But if the engine fails after takeoff. . . . In any case, the FAA has been very reasonable about switching to the local airport after the aircraft has proven itself.

Unfortunately, airports themselves sometimes restrict experimental aircraft. My home field prohibits homebuilt aircraft until their test phase is completed. This policy is recent, in reaction to a particular builder's attempt to fly his poorly constructed homebuilt from the field.

As far as the flight test area is concerned, the aircraft is generally restricted to a 25-mile radius of the test airport. This area is modified by cultural and terrain features, and you might only get half the circle due to surrounding cities. Or the circle might bulge a bit to allow testing over a lake or other suitable area.

If you have legitimate reasons to change the shape or enlarge it, speak up. A fast airplane such as a Harmon Rocket (Fig. 12-13) can cross a 25-mile test area from edge to edge in less than 15 minutes; by the time one gets into stable cruise flight, it's time to turn around. Local builders of fast airplanes are routinely granted a 50-mile-radius test area.

The flight-test period is either 25 or 40 hours depending on the engine-propeller combination. If you have an approved combination—an exact-model engine as used in a particular certified aircraft combined with the exact-model propeller used in that aircraft—the test period can be set to 25 hours.

Fig. 12-13. *The size and shape of test areas can be negotiated based on local conditions and the performance expected from the aircraft. A builder of a Harmon F1 Rocket such as this one would be justified in asking for a much larger test area.*

The flight test period is referred to as the "phase I operating limitations." The inspector also will assign "phase II limitations," those which you'll have to abide by after the test flight period. In most cases, the limits are those set by FAR 91.42: no operation for hire and no operation over congested areas or in congested airways (except as specifically authorized and/or for takeoffs and landings).

Neither the FAA nor the inspector keeps track of how many hours you've put on the aircraft. When you reach the required number, make a log entry to that effect, and the aircraft is cleared to phase II limitations.

Keep in mind that the numbers just quoted are guidelines, not set by regulation. There have been cases of shorter test periods for experimental aircraft with ultralight-like characteristics. One Sorrell Hiperlight builder was assigned a test period of 10 hours.

But if you are really "pushing the outside of the envelope," your restrictions may be tighter. If the aircraft features some questionable design elements, the inspector might restrict the aircraft to the immediate proximity of the airport. Or if you are using a *real* off-the-wall powerplant, he or she might decide that 40 hours aren't sufficient and double the period of your phase I operating limitations.

It's the inspector's responsibility. Once the safety and reliability of your kit-plane are proven, there shouldn't be any problem getting your limits extended.

The inspection

How can someone examine a completed aircraft and detect all problems or errors? It's not possible. It used to be a little easier back when an inspection was required prior to closing up the interior (covering with fabric, riveting the skins, and so on). Nowadays, the FAA merely recommends periodic inspections by knowledgeable individuals, such as EAA technical counselors and A&P mechanics.

It's important to understand the purpose of the inspection. It's not to determine if the aircraft is safe. It's to ensure that it meets the regulations governing the flight testing of experimental aircraft. The safety of the aircraft will be proven in the air; it's the inspector's job to ensure adequate verification before turning the aircraft loose on the airways.

One major area is to check that the plane meets all regulations applicable to experimental aircraft and has the following:

- Basic VFR instruments, all marked with appropriate operating ranges (redlines, max/min pressure or temperature, and so on)
- All controls labeled with function and action (on/off signs for the switches, open/closed for the throttle, cold/hot for the carb heat, and so on)
- Properly registered, with the N-number applied according to regulations
- A metal identification plate listing the builder's name and address, model designation, builder's serial number, and date of manufacture
- A plate visible to both occupants explaining the experimental nature of the aircraft (the Passenger Warning placard available from EAA)
- The word "experimental" in letters at least 2 inches high visible near each entrance to the cockpit
- A plate near the tail giving the manufacturer's name, the model, and serial number of the aircraft (DEA plate; the builder's plate is adequate in many cases)
- Seat belts and shoulder harnesses installed for all occupants
- Current and correct weight and balance paperwork

If you lack any of the preceding, the inspector will not sign off on your aircraft. The inspector will examine the aircraft thoroughly. He or she is not looking for basic flaws, although there will be no hesitation to point any out. Rather, the inspection is aimed at the immediate airworthiness: All pins installed and safetied, bolts head-up and forward, cables properly tensioned, and the like.

The inspector also will want to examine the plans/instructions, your builder's log, and the photos you took during construction to satisfy the issue of who built the aircraft (Fig. 12-14). Have your receipts standing by in case the inspector asks to see them. Be prepared to explain any deviation from the plans.

The inspector probably will find a few things wrong. Most of them will be minor and can be corrected on the spot. Have your tools handy.

Once the inspector is satisfied, he or she will sign your log and airworthiness certificate. The aircraft is cleared to fly.

How to keep your FAA inspector happy

The airworthiness inspection isn't an IRS audit. You and the inspector are on the same side: You want to fly a safe airplane, and he or she wants the airplane you fly to be safe.

- Have your EAA technical counselor inspect the plane before setting up the FAA appointment. Correct all the deficiencies that he or she finds prior to the inspector's arrival.

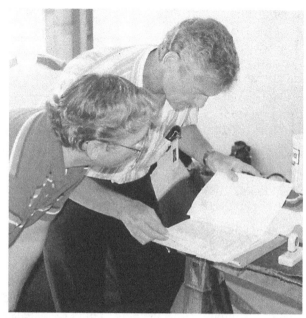

Fig. 12-14. *The inspector will go over your builder's logs and photos as well as the offical paperwork.*

- Have the aircraft ready when the inspector arrives—in a hangar with all inspection panels removed.
- Document all deviations from the plans in your builder's log.
- Use aircraft-quality materials throughout. Technically, the aircraft is experimental and doesn't have to use approved materials. But overreliance on such materials might make the inspector question the airworthiness of the aircraft as a whole. Aircraft-quality parts and hardware tell the inspector that you're more interested in safety than saving money, and that's the attitude he or she wants to see.
- Be prepared to compare any section of the aircraft with the plans. If the inspector doesn't like the design of a particular area, you'd like to prove that it was constructed per the instructions. If the change was your idea, be prepared to justify the redesign.
- Make sure that all your taxi tests and engine runups are documented in the appropriate log. The inspector will want to see log entries equaling at least one hour of engine operation.
- Have a flashlight or trouble light handy.

PILOT PREPARATIONS

The airplane's ready. Is the pilot?

All too often homebuilders stop flying during the construction process. And far, far too often the aircraft's first flight is the builder's first flight in months—if not years.

If the aircraft is an easy flier, and if everything works, both the aircraft and the pilot might come out of it in one piece. The decision is the pilot's. I suppose that a pilot has a perfect right to take those kind of chances. But I hate seeing perfectly good aircraft dinged up.

A pilot probably doesn't lose the basic skills over a long layoff. I quit flying for seven years at one point then bought a Cessna 150. An hour and a half of dual flight later, the instructor turned me loose. Flying an airplane is like riding a bicycle, except that it's harder to put playing cards in the spokes.

But any lapse in skill results in more than skinned knees. And while the primary stick-and-rudder talent might not fade very much, the reflexes for handling emergencies aren't there any more. In addition, you've probably lost the ability to do multiple things simultaneously.

Example 1. You take off carefully, mindful of every motion of the aircraft. You're totally concentrated on control during which, unnoticed, the oil pressure drops and the temperature rockets. The engine seizes on downwind. Forced landings are hard enough for those who fly regularly.

Example 2. The nylon tube to the pitot heads vibrates free of a connection soon after takeoff. The airspeed indicator drops to zero. A current pilot is less likely to dive into the ground chasing an inoperative gauge and has a better chance of executing a safe landing.

Example 3. You're intent, flying smoothly, and monitoring the engine instruments. You neglect your navigation and wander into a restricted area or Class B Airspace and the FAA pulls your license for 60 days.

Even if you manage to survive the test-flying period, your less-than-skillful handling of the aircraft can result in extra wear and tear. There are two solutions: finding someone to fly the initial tests or getting current yourself.

Who ya gonna call?

Who do you trust to make the first flight of your pride and joy? Or perhaps more to the point, who trusts *you* enough to make the first flight in your homemade airplane?

Your preferred choice would be someone who has a lot of hours in the same type of aircraft. Not only would such a person be familiar with any handling quirks the type possesses, but he'd be able to quickly tell you where your plane is deficient. If you've had a mentor through the construction of your plane, he might consent to making the first flight.

Experienced flight instructors are another good source. Flying skills are second nature, and they practice emergency procedures on a daily basis. You'd prefer one with experience in a wide variety of aircraft. Don't be insulted if she sends someone to inspect the aircraft before she agrees to take on the job.

Or does your EAA chapter have its equivalent of Chuck Yeager? Until he "retired" recently, my chapter's tech counselor specialized in first flights. He's not an ex-military pilot, nor does he have a commercial certificate or even an instrument rating. But he had 35 years of experience in building, inspecting, and flying homebuilt aircraft. He's made the first flight on Dragonflies, Fly Babies, T-18s, Jodels, and a host of other aircraft.

Pilots from any of these three categories can do an adequate job of test piloting. Or they might not be up to the challenge. You're taking a chance as well. The best bet would be someone who fits all categories: an experienced instructor who has built the same type of aircraft and has flown the test programs for variety of homebuilts. Lotsa luck.

Doing it yourself

In most cases, the builder decides to make the first flight himself. If that's the way you decide to go, see if your local EAA chapter has a flight advisor. The advisor helps the pilot evaluate his flying skills against those which the flight advisor knows are required to fly the airplane safely.

Basically, the advisor helps you to determine two things: Are you legal, and are you ready?

Legal means two things: current medical and current biannual/annual flight check. The inspector might request proof when he or she signs off on the aircraft.

There are three phases to first-flight preparation. The first is knocking the rust off your basic skills, which are probably adequate, but a refresher on basic airmanship will lay a good groundwork for the next phases. Stall recognition and recovery should be covered extensively.

The second is preparation for the special challenges presented by your kitplane. A lifetime of flying Continental and Lycoming engines won't prepare you for flying a Turbo Legend (Fig. 12-15). Try to get some stick time on the same type; ride right seat in someone else's Glasair, for example. If one isn't available, take some dual time in the closest available production equivalent. Some people recommend flying a variety of aircraft just to become accustomed to differences.

The classic example of this type of training is the conventional-gear checkout. Most of the smaller aircraft are taildraggers. Taildraggers aren't really harder to fly; it's just that they don't tolerate inattention during takeoffs and landings.

If you've never flown a taildragger before, count on needing 5 to 10 hours of dual. It can be combined with the dual needed to brush up your basic skills, so it's not an *additional* 10 hours. If you've got some Cub or Champ time in your history but nothing recent, an hour or two of refresher probably will suffice.

The last phase of preparation is training to handle the various emergencies that might occur. Even if current, any pilot can benefit from emergency drill.

Most of this should be done dual; again, it can be made part of the general refresher course. Simulated engine failures in all modes of flight are one aspect, of course. Practice aborted takeoffs on long runways with a marker or runway light simulating the real end.

Fig. 12-15. *Get the right kind of practice before flying a high-performance airplane such as the Turbo Legend.*

Rigging problems can be duplicated by running elevator and rudder trim to the limit. Have the instructor block use of rudder or ailerons to simulate control jamming, especially during maneuvers. Control problems such as these should only be practiced at altitude; don't risk an aircraft down low.

Take some aerobatic dual flight time, if available. Concentrate on spins and unusual attitude recovery. Make some flights from the test airport if it's one you aren't familiar with. Make note of the landmarks and possible emergency landing fields.

Finally, practice emergencies while sitting in the kitplane. Can you find each control and switch without looking at it? Can you get out quickly if you need to? Try it with a parachute on because they can make a difference. If it's a retractable, jack it up and practice emergency extension.

Develop emergency checklists, and commit them to memory. Repeat them aloud while sitting in the cockpit performing the actions.

You might not do everything I've outlined. And you'll probably be OK. But all this isn't just to get your reflexes tuned. It's mental preparation as well. You are soon going to be in a very scary situation—a first flight.

You've been through it once already, on the day you soloed. Preflight jitters are alleviated (but not eliminated) by *confidence.* Confidence in the airplane, and confidence in yourself. Your first solo flight occurred in an airplane in which you had a number of hours. The instructor probably rode through a couple of touch and goes before turning you loose. No question that the aircraft was in tiptop shape and that you were fully capable of flying it.

It's a similar experience for the first flight of a homebuilt. You've been working on the aircraft for months or years. The plans were followed religiously. Your EAA technical counselor visited several times and checked your work. Weight and balance are spot-on. The FAA inspector actually *smiled* as she signed off on the paperwork. The engine purrs like a lion cub. The plane should be all right.

Which leaves *you*. You've flown for 10 hours or more in the past month. The instructor failed the engine more times than you can count. You know how the plane is supposed to fly because you got a few hours practice in someone else's. You can reach all the controls blindfolded and half-asleep. You know how to handle jammed controls, engine fires, stalls, spins, and the whole panoply of aviation disasters.

With practice comes skill, and with skill comes confidence.

You'll still worry because you're a good pilot, and good pilots *know* they can't anticipate everything. But you'll be eager, too. Ready to take on whatever the plane has to give.

I'll be darned. You're ready to fly.

THE DAY

Don't tell anyone! You need your technical counselor or another experienced builder to help check the plane. Your flight advisor, too, to help coordinate the flight and stand by on the radio. Call in someone else to record the great event in pictures or on videotape if you like. But don't invite the world.

The problem is the "space shuttle *Challenger* syndrome": The feeling you *must* fly because everyone is expecting you to fly. So you ignore the crosswind, shrug off the rough-idling engine, and decide to fly a high-speed pass to show off the plane.

You must be willing to scrub the flight at any deficiency. A bunch of kitbitzers complicates matters. There's the human tendency to please the crowd. And the 10-year-old inside you doesn't want to be called chicken.

So do yourself a favor and keep the flight time secret. This can get a bit uncomfortable. You might have to round up a bunch of volunteers to unload your kitplane at the airport and assemble it. They'll all want to know when the first flight will be. But, for your own protection, decline to name the specific date. If they are EAAers, they'll understand.

The flight is often made immediately after the FAA inspection. The inspector might hang around, but it isn't required that he or she witness the event.

Everything should be perfect: weather, equipment, and pilot readiness and condition. Weather won't be perfect, of course. But you'd like a reasonably high ceiling and light winds straight down the runway.

Inspect the plane carefully. Enough gas should be carried for two hours' worth of flight. Oil and coolant should be topped off. Work the controls, and make sure that they are not binding. Set the flaps and trim to the takeoff position. Have the counselor duplicate your inspection to be on the safe side. Try to have the CG toward the forward limit.

Run the engine for 10 minutes or so, and ensure that full power is available. Make sure that all latches work properly. Phone the tower and let them know you'll be making a first flight. No, you *don't* want the equipment standing by; just let them know that you might need expedited handling if any problems occur.

You should be well rested and alert. Eat a snack if you're hungry, but no large meals. Have the flight well planned from takeoff to landing. Anticipate the worst. Mentally review your reactions to various emergencies. Wear leather shoes (preferably boots), not sneakers. Borrow a Nomex flight suit if possible; if not, wear natural fibers, not synthetics. Check with any motorcycle-riding buddies for a helmet loan.

Deciding whether to wear a parachute is often the hardest part. Is it possible to bail out of your aircraft? Is it a pusher? Can the doors be jettisoned? Can seat cushions be removed to allow you to sit normally? Can you egress quickly with a chute on?

You might not have to buy a chute—your flight advisor should know where you can borrow or rent one. Often, all you'll have to pay is just the rigger's fee for routine repacking.

Eventually, everything's ready. Fire up the engine, and taxi to the runway. Run up the engine one more time, and check the controls. Roll into position. Treat the exercise like a high-speed taxi test. Feed the power slowly. Keep to the centerline. Listen to the engine. Feel the controls. Abort at the first sign of trouble.

Rotate. Watch the attitude as the plane leaves the ground—overcontrolling is common. Move the controls deliberately to prevent *pilot-induced oscillation* (PIO). PIO is the fancy name for overcontrolling; if the nose is too high, you shove it down hard; then the nose is too low, and you pull it back up even harder. Eventually, you're a half-cycle behind the airplane. Make all inputs gentle and gradual.

Climb out at the manufacturer's best rate of climb speed or 1.3 times the predicted stall speed. Don't touch anything until the aircraft has climbed 1,000 feet. Keep the gear down. Don't throttle back unless an engine problem arises.

Once you have some altitude, fly to the test area and execute the test plan. Basically, the first flight consists of controllability, trim, and engine checks. You'll check control pressures, rigging, and engine operation. You won't actually stall the airplane, but you'll practice slow flight and approaches to landings.

Don't plan on staying up all day. A 30-minute first flight is plenty. Your adrenaline will be a'pumping, and fatigue sets in quickly. At any sign of aircraft problems, start heading back. Keep your eyes open for emergency landing fields.

Return to the airport. All you have to do is one good landing. Fly your approach at about 1.3 to 1.4 times the stall speed. Go around if necessary—more than once, if required.

The plane might handle a bit differently in ground effect, so be prepared. It's not necessary to do a perfect landing; don't get into PIO trying to grease it on.

Taxi back, shut down, climb out, and have your picture taken.

Congratulations!

You no longer have a kit—you have an airplane!

The preceding is merely an abbreviated description of your first flight activities. FAA Advisory Circular AC-90-89, "Amateur-Built Aircraft Flight Testing Handbook," developed in cooperation with the EAA, covers homebuilt testing from first flight and beyond. It can be downloaded for free, just search for "AC 90-89" using your favorite search engine.

CASE STUDY: A FIRST-FLIGHT REPORT

Back in Chapter 1 we looked at Mike Sabourin's long grind to build his Long-EZ. The moment of truth finally came on a sunny day in early September.

The inspector

The FAA inspector arrived at Mike's hangar at 11:00 a.m. sharp. The hangar door was open. Inside sat Mike's gorgeous white, turquoise, and gold Long-EZ. Also inside were two tents. Mike and his entire family had camped in the hangar for the past several weekends to eliminate the one-hour drive to and from the airport. They'd done the same the night before so as to have the maximum time to get ready for the big event.

Mike had all the required documentation lying ready on a workbench. His four photo albums documenting the building stood ready, as well. After the aircraft inspection (in which no problems were found), the inspector started going through Mike's pictures.

By that point, though, the inspector was obviously just curious. Mike's picture-perfect airplane and squared-away documentation proved that the builder knew what he was doing.

The advisor

Mike's advisor for the flight showed up a few minutes later. Tom Staggs is a former Navy E-6 pilot who flies aerobatic routines at airshows in his own Long-EZ. Tom had helped Mike prepare his aircraft; Tom also had worked on sharpening Mike's piloting skills for the first flight. He had brought his own Long-EZ to fly with Mike and monitor his aircraft from outside.

Tom took over Mike's life that afternoon—very much to the benefit of everyone involved. I looked on in wonder and admiration. While reinstalling the cowling, Mike started to fret about not finding an extra screwdriver. "Calm down, Mike," said Tom. "It doesn't make any difference. Stay focused."

The flight

Eventually, the plane was ready to go. "Mike," said Tom, "let's take a walk and talk about what you're going to do." They returned 15 minutes later. Mike pushed his EZ out of the hangar, and positioning it in front of Tom's own Long-EZ.

"Wish Daddy good luck, girls," said Maureen, Mike's wife. My camera whirred as Mike's two daughters hugged and kissed their father, the smallest standing on tiptoe to reach. Then Maureen came over. They hugged and talked quietly for a minute.

Then Mike was ready to start. While he was prepared as well as humanly possible, the tension was plain on Mike's face (Fig. 12-16). He started the engine. The two Long-EZs taxied out to the active runway.

The first run was a high-speed taxi run. The shiny EZ sped by and then slowed to exit onto the taxiway where his family waited. Mike taxied past with a big thumbs-up. The next run would be it.

Fig. 12-16. *Mike Sabourin before the first flight of his Long-EZ.*

Fig. 12-17. *Mike's joy at the conclusion of his flight is obvious. Contrast his expression here with the preceding figure.*

The EZ took the active again. The power came on. The nose bobbled slightly and then stopped with the nosewheel a foot or so above the asphalt.

Then Mike was off. The climbout was shallow and rock solid. The Sabourins listened to the air-to-air discussions on a handheld radio. "OK, Mike," said Tom's voice. "I'm joining you on your right wing. Just forget I'm here, and do what you have to do. How's it going?"

"Great, just great, the visibility's fantastic."

"Yep. What are your temperatures and pressures?"

A pause. "Hottest cylinder's 390; oil temp is 190."

"Fine, that's cooler than I'm doing. I'm going beneath you to check for any leaks." Silence. "Mike, from here, there's only one thing wrong with that airplane: It doesn't have enough scratches and bugs on it!"

"I dreamed about it—but my, oh my, am I gonna *love* this plane!"

"You sure will. Time to retract your nosewheel. OK, let's climb up to 5,000. What are your temperatures and pressures?"

So the flight went. They stayed within 10 miles of the airport. Mike practiced slow flight and ran a couple of simulated landing approaches.

Then it was back to the pattern. Two low approaches for practice, two touch and goes, and then Mike landed. Tom touched down off his right wing. Together, they taxied back to their waiting families. The tension was gone from Mike's face, replaced by a broad grin (Fig. 12-17).

May your own first flight go as smoothly!

Afterword: Get 'em flying

WHAT'S THE HARDEST PART of building a homebuilt aircraft? Participants in a worldwide computer discussion group were asked this question. The overwhelming answer:

"Starting."

But start right. Take the time to pick out the airplane that'll be right for *you*. Take it from me—there's nothing sadder than spending time building an airplane that just isn't right for you.

Don't pick a plane because it's the latest thing. Know your mission, and buy a kit that will support it. Find local help, and don't let the little setbacks get you down.

Remember, it's not called the amateur-built category for nothing. Expect to make mistakes—and plenty of them. You're getting an education. Expect a few B minuses among the A's. Airworthiness should be your top priority—don't go overboard on the accessories and the cosmetics. Airplanes are meant to fly. I'd rather see one average-quality homebuilt in the air than look at five future grand champions in the workshop.

Let's get 'em flying!

Building a kitplane is not an easy task; nor is it a quick one. There's a lot more to it than slapping some glue on some parts and wrapping them with rubber bands.

Arduous. Yes.

Difficult? At times.

Impossible? Not hardly.

Thousands of men and women built their own aircraft even before the advent of the kit.

But there will be times when it overwhelms you. The enormity of it. The apparent slow progress. The stabilizer you glued on upside down.

One cure for the homebuilder's blues is your local chapter of the Experimental Aircraft Association. This is not another plug for the EAA. I have no official connection other than occasionally holding an officer position in two very fine chapters.

But within the chapters you'll find support: somebody else building the same airplane to help you through unclear portions in the plans, a technical counselor with advice on problems, leads toward finding a low-cost engine, and willing hands and strong backs to help move projects from the garage to the airport.

Building an aircraft, even a modern kitplane, has never been a one-person job. Even if only one pair of hands touches the tools, a successful homebuilt depends on cooperation and coordination between the designer, kit manufacturer, FAA, EAA, and the builder. Your local chapter can help you tap into a mother lode of assistance and support. Contact national EAA headquarters:

Experimental Aircraft Association
EAA Aviation Center
Oshkosh, WI 54903-3086
414-426-4876 (for chapter information)
www.eaa.org

Glossary

AC 43.13—FAA Advisory Circular entitled, "Acceptable Methods, Techniques, and Practices—Aircraft Inspection and Repair."

aircraft manufacturer—According to regulation, the builder of the kitplane (you) is considered the manufacturer. See *kitplane manufacturer.*

aircraft-quality materials—Those components which meet federal quality and characteristics standards.

Alclad—A coating of pure aluminum applied to the surface of an aluminum alloy as an anticorrosion method.

alloy—Description of the makeup of a metal. Usually a numerical code, e.g., 4130 steel, 2024 aluminum

alodine—Chemical treatment for aluminum that increases its corrosion resistance. The alodine treatment gives the metal a golden sheen.

AN—Army-Navy (Standard).

anodize—Special treatment that gives aluminum extremely good corrosion resistance and leaves it with a matte black surface.

ARV—Air recreational vehicle. Generally means an aircraft built using ultralight design methods but certified in the experimental category.

back riveting—Using the rivet gun and bucking bar in reverse of the normal procedure; that is, the gun makes the shop head while the bar is placed on the manufactured head.

band saw—A floor or table-top power tool that cuts using a toothed band.

bastard—Calm down. It's really a descriptive term for a type of file.

bias—Term used in describing the direction of the weave in cloth, including fiberglass.

bid—Slang term for bidirectional fiberglass cloth. It has nearly equal strength in all directions.

bonding agent—See *resin mixture.*

brake, bending brake—Tool for making long, even bends in metal sheet.

bucking bar—Tool used to make the shop end of driven rivets.

bulkhead—Vertical sheet (metal, composite, or wood) that provides structural rigidity to the fuselage.

bungee cord—Long strips of rubber wrapped in cloth. Essentially, a big rubber band. They're used for shock absorption in small aircraft and for leaping off tall buildings with a single "spoinnngggggg."

burr—A rough protrusion of metal. Equivalent to a splinter in wood.

cad, or cadmium—Anticorrosion treatment for metal hardware.

capstrip—Flat piece of wood around the periphery of a wing rib.

chamfer—A sharp 90-degree angle converted into two 45-degree angles with a narrow, flat section between.

chip chaser—Tool for removing burrs between metal sheets.

chisel—Tool with a wedge-shaped edge used in shaping wood. A cold chisel is specially hardened for cutting metal.

chromoly—("crow-molly") Slang term for 4130 steel (chromium-molybdenum steel).

cleco—A temporary fastener inserted from one side using a special pliers.

clevis pin—Thick metal pin that closes a fork or shackle.

composite—Building material consisting of resin-soaked fiberglass layups with foam inserts to add strength.

compression load—A force that tries to collapse an object, such as a brick atop a soda straw.

compression strut—Structural element designed to withstand compression loads. Usually found between wing spars.

cotter pin—Small piece of doubled wire that acts as a safety device to prevent a bolt, clevis pin, nut, or other object from accidental disengagement.

countersink—Tool for cutting shallow cone-shaped depressions. Used to prepare metal for flush riveting and to eliminate burrs from drilled holes.

DAR—Designated airworthiness representative. Person authorized to perform FAA airworthiness inspections.

deburring tool—Device to smooth away burrs.

diagonals—Structural element on truss-type fuselages running diagonally between longerons.

dimple—Shallow depression. Can be applied by a dimpling tool or dimpling die to prepare sheet for flush riveting.

draw filing—Very long file strokes made by pulling the file along the edge of the metal. Described in Chap. 7.

drift punch—A thin rod of metal of a given diameter used to remove jammed rivets, bolts, and so on.

driven rivet—A rivet in which the shop head (see) is made by ramming the rivet shank repeatedly into a bucking bar.

dry micro—Very thick mixture of glass bubbles and epoxy/vinylester used for rough surfaces. A ratio of about 5:1 bubbles to resin.

edge margin—Minimum spacing for holes. No closer than three times the diameter between holes and two times the diameter between a hole and the edge of the metal.

emery cloth—Special kind of sandpaper. Available at any hardware store.

endless loop—see *recursion*.

epoxy—Modern two-component (resin and hardener) bonding material.

etchant—Mild acid solution used to prepare aluminum for painting.

exotherm—Heat-producing chain reaction caused by excessively large mixtures of epoxy or vinylester.

eye, cable eye—Loop in the end of an aircraft cable.

fabric—Cloth used to cover the skeleton of an aircraft.

fairlead—Device used to redirect or support aircraft cable.

fill—Orientation of the weakest axis of a woven material. Most commonly used in composite construction.

fill yarns—Threads that hold the weave together on unidirectional fiberglass cloth.

flight advisor—Experienced pilots designated by the EAA to assist homebuilders in planning their first flight and test program.

fluting pliers—Tool for applying a special "dent" on the edge of metal.

former—Nonstructural component that defines the exterior cross section of the fuselage. Stringers run between formers to define the three-dimensional shape.

4130—A steel alloy commonly used in aircraft.

fuel head—Amount of fuel pressure developed at the carburetor owing to the distance of the tank above the engine.

grain—Direction of primary strength.

green cure—A state where resin-soaked fiberglass can be trimmed easily with a razor blade or other sharp knife.

grip length—Unthreaded area on the shank of a bolt.

hard point—Area that receives special reinforcement to allow attachment of additional structure.

hardware store parts—Components that aren't certified as having met federal standards (AN, NAS, MS, and so on).

head—Portion of rivet, bolt, and so on that acts as a stop to prevent the item from passing completely into a hole. See *shop head* and *manufactured head.*

Hot-wire cutter—Device that shapes styrofoam through the use of a heated wire.

jig—A structure that holds aircraft components while they are being attached.

jig saw—A handheld tool with a reciprocating saw blade. Also called a saber saw.

joggle—Slight discontinuity in the edge of mating parts that produces a smooth, sturdy seam when joined.

kit manufacturer—The company that produces the kit for the aircraft. See *aircraft manufacturer.*

laterals—Structural element on truss-type fuselages running sideways between longerons.

layup—A single layer of fiberglass and associated resin. Single layups are rare; most require multiple layers of cloth applied sequentially.

line oil—Preservative mixture used in corrosion-proofing steel-tube fuselages.

longeron—Primary fuselage structural element for truss-type designs. Forms the corners of the fuselage "box."

manufactured head—The head the rivet comes with. See *shop head.*

micro—Shorthand for microspheres, an obsolete material. The term is also used to refer to its modern replacement, glass bubbles.

MS—Military Standard.

NAS—National Aerospace Standard.

Nicopress—System for attaching fittings to aircraft cable, developed by the National Telephone Supply Company.

ovaling—When friction or other forces change a round hole into an oblong one.

peel ply—Dacron cloth used to smooth the surface of a fiberglass layup.

pop rivet—See *pulled rivet*.

pressure cowling—A cowling that completely encloses the engine and hence must include baffling to route cooling air past the cylinders.

pulled rivet—A rivet that can be emplaced from one side using a tool to pull a mandrel to distort the shop end.

Q-cells—Tiny glass bubbles.

radius—The distance from the center of a circle to the edge. Also refers to the "rounding off" of a sharp corner.

ragwing—Slang term a for fabric-covered airplane.

recursion—See *endless loop*.

resin mixture—Shorthand used within this book for composite bonding materials in the ready-to-use condition, also called "bonding agent." For epoxy operations, it refers to the appropriate mixture of resin and hardener. In the vinylester world, it means resin with catalyst added.

rib—Wing structural element that defines the airfoil shape.

rib stitch—Wrapping cord around a wing rib to hold the fabric in place.

rivet—A piece of metal stuck through a hole between two objects and distorted to hold the parts together. See *pulled rivet* and *driven rivet*.

rivet set—Tool that transmits hammering of rivet gun to the rivet itself.

saber saw—See *jig saw*.

scarf joint—Method of joining two pieces of wood end to end.

set—See *rivet set*.

shackle—U-shaped piece of metal for attaching a cable eye to structure. The shackle uses a clevis pin to close off the open end.

shank—Portion of bolt or rivet below the head.

shear load—A sideways force, such as that imposed on a nail used to hang a picture.

sheet stock—Thin metal, aluminum or steel.

shop head—In riveting, the mushroom shape applied to the end of the rivet to hold it in place.

6061—Aluminum alloy commonly used in aircraft.

slurry—A one-to-one mixture of glass bubbles and epoxy/vinylester.

spar—Main structural element of a wing running from root to tip.

spar cap—On an I- or C-shaped spar, the top and bottom short components. The cap withstands the stresses and shouldn't be drilled unless the plans specify say to do so.

spar web—On an I- or C-shaped spar, it is the vertical component. Serves to maintain the caps at the desired distance. Web can be drilled to some extent without weakening it.

springback—Tendency of bent metal to return to some intermediate position.

stop nut—A nut that incorporates some method to prevent self-rotation and accidental disengagement.

stringer—Thin strip of wood or metal used to define an exterior surface on a fabric-covered airplane. Nonstructural.

swage—To crush down under extreme pressure.

table saw—A metal table with a circular saw blade sticking vertically out of the top.

tack weld—A spot weld intended to temporarily hold together two pieces of steel. It has no strength to speak of.

tang—A protrusion of metal intended for attachment of a cable fitting.

technical counselor—EAA member, either an A&P mechanic or an experienced builder, available for advice on the construction of a homebuilt aircraft.

tension load—A "pulling" force, such as a claw hammer pulling a nail out of a piece of wood.

thick micro, thick mix—See *wet micro*.

thimble—A teardrop-shaped device for making an eye from aircraft cable.

tube-and-fabric—Construction method where the aircraft structure is composed of steel or aluminum tubing covered by fabric.

turnbuckle—Device for tightening aircraft cable.

turtledeck—Top of the fuselage between the cockpit and the vertical stabilizer.

2024—Aluminum alloy commonly used in aircraft.

uni—Slang term for unidirectional fiberglass cloth. Most of its strength is in one direction.

uprights—Structural element on truss-type fuselages running vertically between longerons.

vinylester—A type of two-part bonding material.

Wanttaja—An aviation writer. Pronounced "Wahn-TIE-ah".

warp—Orientation of strongest axis of a woven material. Most commonly used in composite construction.

wet micro—Three or so parts glass bubbles to one part epoxy/vinylester. Also called thick micro or thick mix.

wet mix—See *slurry*.

zinc chromate—The traditional anticorrosive finish on aluminum structures.

Index

ABOUT THE AUTHOR

Ronald J. Wanttaja is an award-winning aviation writer and a systems engineer with Boeing, working in satellite orbit/constellation design and analysis, launch vehicle and onboard propulsion system trades, and operations concepts for space systems. He worked on the early design studies for the International Space Station. An Air Force veteran, he was an on-duty operator for the Defense Support Program missile early-warning satellite. As a freelance aviation journalist, he has written for *Private Pilot, Flying, Sport Aviation, Flight Line, Kitplanes*, and other publications. He is the author of the book *Airplane Ownership*, also from McGraw-Hill. His aviation writing has won several prizes, including the EAA "Bax Seat Trophy" and a journalism award from the Aviation/Space Writer's Association. Mr. Wanttaja has also written and published historical fiction.

CPSIA information can be obtained
at www.ICGtesting.com
Printed in the USA
FSHW021353170421
80457FS

9 780071 459730